T0185033

Lecture Notes of the Institute for Computer Sciences, Social Informatics and Telecommunications Engineering   291

Editorial Board Members

Ozgur Akan
  Middle East Technical University, Ankara, Turkey
Paolo Bellavista
  University of Bologna, Bologna, Italy
Jiannong Cao
  Hong Kong Polytechnic University, Hong Kong, China
Geoffrey Coulson
  Lancaster University, Lancaster, UK
Falko Dressler
  University of Erlangen, Erlangen, Germany
Domenico Ferrari
  Università Cattolica Piacenza, Piacenza, Italy
Mario Gerla
  UCLA, Los Angeles, USA
Hisashi Kobayashi
  Princeton University, Princeton, USA
Sergio Palazzo
  University of Catania, Catania, Italy
Sartaj Sahni
  University of Florida, Gainesville, USA
Xuemin (Sherman) Shen
  University of Waterloo, Waterloo, Canada
Mircea Stan
  University of Virginia, Charlottesville, USA
Jia Xiaohua
  City University of Hong Kong, Kowloon, Hong Kong
Albert Y. Zomaya
  University of Sydney, Sydney, Australia

More information about this series at http://www.springer.com/series/8197

Adrian Kliks · Paweł Kryszkiewicz ·
Faouzi Bader · Dionysia Triantafyllopoulou ·
Carlos E. Caicedo · Aydin Sezgin ·
Nikos Dimitriou · Michał Sybis (Eds.)

Cognitive Radio-Oriented Wireless Networks

14th EAI International Conference, CrownCom 2019
Poznan, Poland, June 11–12, 2019
Proceedings

Springer

Editors
Adrian Kliks (iD)
Poznan University of Technology
Poznan, Poland

Faouzi Bader
Centrale Supelec
Cesson Sevigne Cedex, France

Carlos E. Caicedo
Syracuse University
New York, USA

Nikos Dimitriou (iD)
NCSR "Demokritos"
Athens, Greece

Paweł Kryszkiewicz (iD)
Faculty of Electronics and Communication
Poznań University of Technology
Poznan, Poland

Dionysia Triantafyllopoulou
University of Surrey
Guildford, UK

Aydin Sezgin (iD)
Ruhr-University Bochum
Bochum, Germany

Michał Sybis (iD)
Poznan University of Technology
Poznan, Poland

ISSN 1867-8211 ISSN 1867-822X (electronic)
Lecture Notes of the Institute for Computer Sciences, Social Informatics
and Telecommunications Engineering
ISBN 978-3-030-25747-7 ISBN 978-3-030-25748-4 (eBook)
https://doi.org/10.1007/978-3-030-25748-4

© ICST Institute for Computer Sciences, Social Informatics and Telecommunications Engineering 2019
This work is subject to copyright. All rights are reserved by the Publisher, whether the whole or part of the material is concerned, specifically the rights of translation, reprinting, reuse of illustrations, recitation, broadcasting, reproduction on microfilms or in any other physical way, and transmission or information storage and retrieval, electronic adaptation, computer software, or by similar or dissimilar methodology now known or hereafter developed.
The use of general descriptive names, registered names, trademarks, service marks, etc. in this publication does not imply, even in the absence of a specific statement, that such names are exempt from the relevant protective laws and regulations and therefore free for general use.
The publisher, the authors and the editors are safe to assume that the advice and information in this book are believed to be true and accurate at the date of publication. Neither the publisher nor the authors or the editors give a warranty, expressed or implied, with respect to the material contained herein or for any errors or omissions that may have been made. The publisher remains neutral with regard to jurisdictional claims in published maps and institutional affiliations.

This Springer imprint is published by the registered company Springer Nature Switzerland AG
The registered company address is: Gewerbestrasse 11, 6330 Cham, Switzerland

Preface

It is our honor to introduce the proceedings of the 14th edition of the European Alliance for Innovation (EAI) International Conference on Cognitive Radio-Oriented Wireless Networks (CROWNCOM 2019), held in Poznan, Poland. This conference gathered researchers from academia, industry, and standardization bodies to present their solutions supporting cognitive-radio-based wireless networks. Adapting the scope of the event to the current investigation trends, the main topic of the conference was "Artificial Intelligence and Big Data for 5G and Beyond."

The technical program of CROWNCOM 2019 consisted of 30 papers divided into five sessions and one workshop. The technical sessions covered the following research topics: "Machine Learning for Spectrum Management," "Spectrum Sensing and REMs," "Spectrum Sharing and Management," "New Trends in Spectrum Sharing," and "Spectrum Assessment and Valuation." Complementing the technical sessions, the workshop was devoted to "Open Radio Platforms for 5G Research and Beyond" operating both in the hardware and software domains.

The technical program featured two keynotes, one tutorial, and one panel. The keynote lecture given by Marcin Dryjanski (Grandmetric), entitled "Why One-Size-Fits-All Approach Does Not Work Anymore," addressed the arising problem of heterogeneity of both services required from wireless networks and radio access technologies. The keynote given by Stefan Parkvall (Ericsson Research), entitled "5G NR – The Next Generation Wireless Access," provided an overview of the 5G New Radio technology and its evolution in the future. The tutorial, given by Navid Nikaein (Eurecom), was entitled "RAN Slicing and Control: Challenges, Technologies, and Tools." Over more than two hours, the tutorial speaker presented the challenges, solutions, and technologies enabling network slicing in 5G networks. Finally, specialists from industry, research, and standardization entities participated in a dedicated discussion panel entitled: "Timely Access to Spectrum Will Unlock Novel 5G Businesses."

This attractive program was a result of the efforts of 61 reviewers guaranteeing high-quality technical content of accepted papers. We would like to express our sincere appreciation to all program chairs (Faouzi Bader, Dionysia Triantafyllopoulou, Carlos E. Caicedo, Aydin Sezgin, Nikos Dimitriou), publicity and social media chair (Leonardo Goratti), workshops chair (Francesco Benedetto), sponsorship and exhibits chair (Vlad Popescu), publications chair (Michał Sybis), panels chair (Olivier Holland), tutorials chair (Mauro Fadda), demos chair (Pawel Sroka), posters and PhD track chair (Albena Mihovska), local arrangements chair (Krzysztof Cichoń), and Web chair (Maciej Krasicki). We are also grateful to the conference manager, Andrea Piekova, and venue manager, Katarina Srnanova, for their valuable support. Finally, we would like to express our gratitude to all the authors of submitted papers.

We believe that CROWNCOM 2019 provided a good forum for all wireless systems researchers to discuss how cognitive radio systems will help deliver the stringent requirements of future 5G networks and beyond.

June 2019   Adrian Kliks
                                                     Paweł Kryszkiewicz

Organization

General Chair

Adrian Kliks — Poznan University of Technology, Poland

Organizing Committee

General Co-chair

Paweł Kryszkiewicz — Poznan University of Technology, Poland

Technical Program Committee Chair/Co-chairs

Faouzi Bader — Centrale Supelec, France
Dionysia Triantafyllopoulou — University of Surrey, UK
Carlos E. Caicedo — Syracuse University, USA
Aydin Sezgin — Ruhr-University Bochum, Germany
Nikos Dimitriou — NCSR "Demokritos", Greece

Web Chair

Maciej Krasicki — Poznan University of Technology, Poland

Publicity and Social Media Chair/Co-chair

Leonardo Goratti — TriaGnoSys GmbH, Wessling, Germany

Workshops Chair

Francesco Benedetto — SP4TE, University of ROMA TRE, Italy

Sponsorship and Exhibits Chair

Vlad Popescu — Transilvania University of Brasov, Romania

Publications Chair

Michał Sybis — Poznan University of Technology, Poland

Panels Chair

Oliver Holland — KCL, UK

Tutorials Chair

Mauro Fadda — University of Cagliari, Italy

Demos Chair

Paweł Sroka   Poznan University of Technology, Poland

Posters and PhD Track Chair

Albena Mihovska   Aarhus University, Denmark

Local Chair

Krzysztof Cichoń   Poznan University of Technology, Poland

Contents

Workshop on Open Radio Platforms for 5G Research and Beyond

Invited Papers

PoMeS: Profit-Maximizing Sensor Selection for Crowd-Sensed Spectrum Discovery

Suzan Bayhan[1]([⊠]) [iD], Gürkan Gür[2][iD], and Anatolij Zubow[1][iD]

[1] TU Berlin, Berlin, Germany
{suzan.bayhan,anatolij.zubow}@tu-berlin.de
[2] Zurich University of Applied Sciences (ZHAW), Winterthur, Switzerland
gueu@zhaw.ch

Abstract. In a conventional network management setting, the mobile network operator (MNO) has to account for the traffic fluctuations in its service area and over-provision its network considering the peak traffic. However, this inefficient approach results in a very high cost for the MNO. Alternatively, the MNO can expand its capacity with secondary spectrum discovered opportunistically whenever, wherever needed. While outsourcing the spectrum discovery to a crowd of sensing units may be more advantageous compared to deploying sensing infrastructure itself, the MNO has to offer incentives in the form of payments to the units participating in the sensing campaign. A key challenge for this crowdsensing environment is to decide on how many sensing units to employ given a certain budget under some performance constraints. In this paper, we present a profit-maximizing sensor selection scheme for crowd-sensed spectrum discovery (PoMeS) for MNOs who want to take sensing as a service from the crowd of network elements and pay these sensors for their service. Compared to sensor selection considering the strict sensing accuracy required by the regulations, our heuristics show that an MNO can increase its profit by deciding itself the level of sensing accuracy based on its traffic in each cell site as well as the penalty it has to pay for not satisfying the required sensing accuracy.

Keywords: Spectrum discovery · Crowdsensing · Spectrum sensing

1 Introduction

Mobile network operators (MNO) crave for more radio spectrum to meet the challenging traffic requirements of their customers whose interest is moving towards video-intensive services. Rather than costly over-provisioning, an MNO can expand its capacity with secondary spectrum, which is owned by primary users (PU) but is spatio-temporally unused. However, this opportunistic utilization brings the challenge of discovering the idle spectrum and evicting the

© ICST Institute for Computer Sciences, Social Informatics and Telecommunications Engineering 2019
Published by Springer Nature Switzerland AG 2019. All Rights Reserved
A. Kliks et al. (Eds.): CrownCom 2019, LNICST 291, pp. 3–16, 2019.
https://doi.org/10.1007/978-3-030-25748-4_1

channel when the primary licensed owner appears in the band. While the regulations have moved from spectrum sensing techniques toward spectrum query from white spectrum databases (WSDB) which store all information about the PU transmitters, there is still a need for spectrum sensing as WSDBs aim primarily at protecting the PUs. Moreover, they do not coordinate spectrum sharing among opportunistic secondary users (SU) [2].

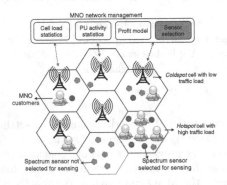

Fig. 1. An MNO can increase its capacity by using the secondary spectrum discovered by the spectrum sensors. The traffic load across the MNO cells (hotspots vs. coldspots) might differ as well as the primary users' activity.

Recent studies [3, 5, 6, 9, 12] propose crowdsensing rather than having an MNO deploy its own sensor infrastructure as the latter might lead to higher CAPEX and OPEX. In this paper, we build on previous works, e.g., [3], and address the problem of sensor selection considering the demand as well as the PU traffic activity in each cell. As depicted in Fig. 1, the MNO via its management system can collect statistics about its own traffic as well as the traffic in the PU spectrum to make more informed decision for sensor selection. While deciding on which sensors to select in each cell, the MNO has to consider the related costs of sensing, e.g., payments to the sensors, and its gain via the discovered spectrum. This gain depends on the spectrum bandwidth that will be discovered via sensors and the additional traffic that could be served with that resource. As traffic characteristics may vary among cells, particularly for small cells where the statistical multiplexing is minimal compared to macrocells, deploying sensors may not payoff if the traffic load is low in a cell. While sensing cost and efficiency are important for the MNO's profit, the regulatory bodies assert sensing accuracy requirement to protect the incumbent users, i.e., PUs, which might lead to higher sensing cost for the MNO. For example, requiring a PU detection accuracy as well as the false alarm probability below a certain threshold regardless of the PU traffic or secondary network's traffic might result in wasteful sensing by the sensors. To go beyond this inflexibility and address more realistic scenarios in this work, we explore a relaxed case where the MNO can prefer maintaining a lower sensing accuracy and then pay for the resulting PU collisions due to its lower PU detection performance. This approach provides the MNO flexibility to maintain its sensing accuracy at different levels in each cell depending on the expected PU traffic (and thereby collision-related penalties) and its traffic load (i.e., operation at different points of cost-benefit trade-off).

In this work, we make the following key contributions regarding spectrum discovery and sensor selection for crowdsensing:

– We propose spectrum discovery via crowd-sensing under a budget constraint. While there is a rich literature on crowdsourced sensing and sensor selection such as [3,6,9,12], the business aspects of this problem is largely overlooked. Our goal in this paper is to analyze how the profit model of an MNO might affect its decisions for sensor selection.
– We propose to relax the sensing accuracy constraints to save from the sensing cost, especially for cells without a high traffic demand, and yet motivate the MNO to attain a higher PU detection accuracy. We achieve this goal by introducing a penalty to the MNO if it cannot satisfy the required minimum detection accuracy. However, we believe that there should be no penalty for the divergence from the maximum false alarm probability since the MNO will inherently have incentives to minimize the false alarm probability under high traffic load. In line with that expected behavior, our simulations show that the MNO favors higher PU detection accuracy and low false alarm rate under high traffic load.
– We present a thorough analysis of our proposals for different system parameters including allocated budget, traffic load, and fraction of hot spots. Our simulations show that an MNO benefits from relaxing the strict requirements on the sensing accuracy. In our proposal, an MNO can target different accuracy levels (e.g., lower false alarm if the need for the required capacity is higher) depending on its traffic in each cell site as well as the penalty it has to pay for not satisfying the required sensing accuracy.

The rest of the paper is organized as follows. First, Sect. 2 presents the considered system model for the network and the sensors. Next, Sect. 3 introduces the cost and utility of spectrum discovery. It also formulates the profit-maximizing sensor selection (PoMeS) problem, while Sect. 4 provides several polynomial-time complexity heuristics for PoMeS. Section 5 presents a detailed assessment of the performance of the devised heuristics in comparison to the baseline which has to ensure the sensing accuracy requirements imposed by the regulatory bodies. Section 6 provides an overview of the related work on sensor selection for crowdsourced spectrum discovery while Sect. 7 concludes the paper with a brief discussion of future directions.

2   System Model

Consider an MNO with $\mathcal{A} = [A_i, \cdots, A_K]$ cell sites. Each cell site hosts users with a certain demand denoted by r_i requests/sec. Each request requires a minimum rate, e.g., $c_{\min}$ bits/sec, and for each request the user pays μ monetary units. The bandwidth of the PU's channel is denoted by B Hertz. In this channel, the PU has an activity with probability p_1^i in A_i. Hence, the MNO can use the channel with probability $p_0^i = 1 - p_1^i$ if it can discover the spectrum opportunities with perfect accuracy and without any overhead. However, depending on the length of the sensing and reporting duration as well as number of sensors, there will be an overhead as illustrated in Fig. 2. Moreover, the spectrum sensors might falsely conclude the state of the channel as occupied due to errors in sensing or fluctuations in the channel, which then decreases the amount of discovered spectrum.

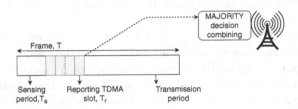

Fig. 2. A frame starts with the sensing period and continues with the reporting period. Each sensor reports its sensing outcome via a TDMA uplink during the reporting slot allocated to it. After completion of the reporting period, the BS at each cell applies majority logic to decide on the spectrum state.

The MNO has a budget of $\mathcal{B}$ (in currency units C) to pay for its crowdsensing campaign. With this budget, it needs to decide on how many sensors (and in case of heterogeneous sensors, considering their sensing capabilities and price of sensing) to employ for sensing in each cell site A_i. We denote the total number of sensors by N and the price of sensing by μ_s C per bps. After the sensors are selected (e.g., S out of N), they start sensing with the requested sensing rate (β_s) during the sensing period T_s in a frame with duration T. Each sensor reports its one bit sensing outcome, i.e., {0: idle, 1: busy}, to the base station (BS) of its cell using TDMA in a slot of duration T_r as in [8]. As a result, the time left after sensing and in-band reporting in a frame of length T equals to $T - T_s - ST_r$. Hence, the normalized sensing overhead is $\omega = \frac{T_s + ST_r}{T}$. We assume identical sensing characteristics for the sensors, i.e., they have identical local probability of false alarm (P_f) and local probability of PU detection (P_d). Given that each sensor applies energy detection, we can derive the P_f and P_d values based on SNR and noise at each sensor [1,8].

After collecting the sensing outcomes, the BS fuses this sensory data (H_i denoting sensor i's binary decision) from S sensors using majority logic, which is known to be robust against sensing errors [3]. Simply, the BS checks if $\sum_{i=1}^{S} H_i \geqslant \lceil S/2 \rceil$. If this inequality holds, it concludes that the spectrum is in use by its primary owners, hence it cannot use this spectrum. Otherwise, it can serve its users through this spectrum, e.g., via carrier aggregation with the existing licensed spectrum as currently specified in LTE. We denote the spectral efficiency of the MNO by κ bps/Hz.

Since regulatory bodies target high utilization of this scarce resource without harming the PUs, they might assert certain sensing accuracy constraints at a cell: Q_f denoting global probability of false alarm and Q_d denoting global probability of PU detection in a cell. We assume that the MNO is required to sustain Q_d^* and can target different Q_f based on the user demand in each cell site. However, for cell sites where the MNO has only low traffic activity (yet higher than the available capacity), it might consider employing a lower number of sensors which would result in lower Q_d, possibly lower than Q_d^*. In this case, we assume that the MNO will have to pay a certain penalty for not meeting the required accuracy. This approach aims at relaxing the strict requirements on the detection accuracy and giving the MNO ability to maintain different sensing accuracy levels across its cells. Yet, the MNO would essentially be driven towards higher sensing accuracy due to the penalty mechanism. In the next section, we introduce our proposal for sensor selection in a wireless network represented by this system model.

3   Profit-Maximizing Sensor Selection (PoMeS)

Let us first define the utility of crowdsensing in terms of the amount of spectrum discovered by the sensors. If there are m sensors participating in sensing, then the amount of the spectrum that will be discovered can be calculated as:

$$\mathcal{U}(m) = p_0 B \left(\frac{T - T_s - mT_r}{T} \right) (1 - Q_f(m)) \text{ Hz,} \tag{1}$$

where $Q_f(m)$ is the false alarm probability if m sensors participate in sensing. We can calculate false alarm probability for majority voting as below [3,8]:

$$Q_f(m) = \sum_{n=\lceil \frac{m}{2} \rceil}^{m} \binom{m}{n} (P_f)^n (1 - P_f)^{m-n}. \tag{2}$$

Similarly, we calculate $Q_d(m)$ as follows:

$$Q_d(m) = \sum_{n=\lceil \frac{m}{2} \rceil}^{m} \binom{m}{n} (P_d)^n (1 - P_d)^{m-n}. \tag{3}$$

We can model the profit of an MNO from each cell considering the number of requests that will be served with the discovered capacity. The discovered capacity is simply $\mathcal{U}_i \kappa$ bps. Hence, the number of requests that can be served with this capacity is:

$$R_i^{\max} = \min(r_i, \frac{\mathcal{U}_i \kappa}{c_{\min}}) \text{ requests/s.} \tag{4}$$

Consequently, we calculate the MNO's income in currency-per-second (C/s) from its customers in A_i as follows:

$$\Pi_i^+ = R_i^{\max} \mu \quad \text{C/s.} \tag{5}$$

Moreover, the MNO has to pay for the sensing service to the selected N_i sensors. If each sensor has to perform sensing with rate β_s bps and the cost of sensing is μ_s per sensing bit, then the total payment for A_i is as follows:

$$\Pi_i^- = \mu_s \beta_s N_i \ \text{C/s}. \tag{6}$$

We introduce a penalty of low PU detection accuracy for the MNO to avoid low sensing accuracy. We denote this penalty by μ_c and the gap between the required Q_d and the realized one by $\Delta Q_{d,i} = \max(0, Q_d^* - Q_{d,i})$ for cell A_i. The resulting penalty in monetary terms equals to $\mu_c \Delta Q_{d,i} R_i^{\max}$. Note that other options for the penalty function is possible, e.g., an exponential function of $\Delta Q_{d,i}$ which is more punishing compared to the linear function used here. Next, we calculate the net profit using (5) and (6), which gives us the following:

$$\Pi_i = R_i^{\max} \mu - \mu_s \beta_s N_i - \mu_c R_i^{\max} \max(0, Q_d^* - Q_{d,i}). \tag{7}$$

Now, let us present the optimization problem which will be solved by the MNO to decide on the number of sensors to be selected for each cell. Profit-maximizing sensor selection (PoMeS) problem is formally defined as follows:

$$\max_{N_i} \sum_{A_i \in \mathcal{A}} R_i^{\max} \mu - \mu_s \beta_s N_i - \mu_c R_i^{\max} \max(0, Q_d^* - Q_{d,i}) \tag{8}$$

subject to:

$$\mu_s \beta_s \left(\sum N_i \right) \leqslant \mathcal{B} \tag{9}$$

$$N_i \leqslant \lfloor \frac{T - T_s}{T_r} \rfloor \quad \forall A_i \in \mathcal{A} \tag{10}$$

$$N_i \geqslant 0 \quad \forall A_i \in \mathcal{A}. \tag{11}$$

Constraints (9) determines the maximum number of sensors to employ in the sensing campaign due to the budget constraint while Const. (10) restricts the number of sensors due to the finite size of the frame. While the constraints are linear in the decision variable N_i, the objective function is non-linear function due to non-linearity of (1), (2), and (3). Hence, our problem is a non-linear integer problem whose complexity is typically high. In addition, our problem has to account for combinations across all cells, which makes this problem computationally hard. Hence, we devise polynomial complexity heuristics in the next section.

4   Sensor Selection Heuristics

Equal Budget per Cell (EQ): This heuristic divides the budget by the number of total cells and finds the best decision for each cell under the cell's budget constraint, i.e., $\mathcal{B}/K$. Then, the maximum number of sensors that could be employed is $N_{\max} = \min(\lfloor \frac{T-T_s}{T_r} \rfloor, \lfloor \frac{\mathcal{B}}{K \mu_s \beta_s} \rfloor)$. Next, this heuristic exhaustively searches for the setting that maximizes the objective function Π_i with constraint

$N_i \leqslant N_{\max}$. If $\Pi_i^{\max}$ is nonnegative, the corresponding number of sensors is assigned to this cell. Otherwise, no sensor is deployed. Note that if the sensors have different cost and sensing accuracy, then the sensor selection problem would be more complex. Here, due to our assumption of a homogeneous setting, EQ only needs to decide on the number of sensors. Moreover, although the MNO's profit is zero for the considered time period, i.e., $\Pi_i^{\max} = 0$, the MNO may still prefer deploying sensors to this cell to increase its service availability. Because better availability might increase the reputation of the MNO, which may attract more customers in the long run. In case $N_{\max}$ is zero which is expected to happen under low budget, EQ finds the minimum required number of sensors for satisfying Q_d^*. Then, EQ selects the cells with the highest loads and employs the minimum number of sensors satisfying Q_d^* in those cells. Computational complexity of EQ is $\mathcal{O}(KN_{\max})$ as EQ calculates the profit for each cell while considering each possible number of sensors lower than or equal to $N_{\max}$.

Budget Proportional to the Serving Capacity of the cell (PROP): Rather than allocating equal budget to each cell, this heuristic allocates the budget proportional to the number of requests $R_i^{\max}$ that could be served by each cell. For the ease of computation, we set $Q_{f,i} = 0$. Then, the maximum number of requests that could be served for a cell equals to

$$R_i^{\max} = \min(r_i, \frac{Bp_0^i(1-\omega)\kappa}{c_{\min}}).$$

Then, PROP allocates to cell A_i a budget of $\mathcal{B}_i$:

$$\mathcal{B}_i = \frac{\mathcal{B}R_i^{\max}}{\sum_{A_i \in A} R_i^{\max}}. \tag{12}$$

Under $\mathcal{B}_i$, PROP exhaustively searches for the best number of sensors to be selected for each cell. Computational complexity of PROP is $\mathcal{O}(KN_{\max})$.

Incremental Gain Based Greedy Assignment (INGA): In this case, the sensor allocation starts from the cell with the highest incremental gain defined as $\Delta\Pi_i = \Pi_{i,m+1} - \Pi_{i,m}$ where m is the current number of sensors deployed in the cell. One more sensor is allocated to this cell with the highest $\Delta\Pi_i$ resulting in $m+1$ sensors in the cell. The iteration continues with the next cell which attains the highest incremental gain with one more sensor deployed. The assignment halts either when globally budget is depleted or when maximum $\Delta\Pi_i$ is negative. Complexity of INGA is $\mathcal{O}(NK\log(K))$ as INGA finds for each sensor the cell with the maximum incremental gain via a sorting algorithm.

Baseline Satisfying (Q_d^*, Q_f^*) Required by the Regulatory Body (REG): This heuristic has two variants: REG-EQ and REG-PROP where the former follows the same budget allocation approach as EQ and the latter as PROP. However, while performing exhaustive search, a solution is considered to be feasible only if the sensing constraints for both Q_d^* and Q_f^* are satisfied. If the cell budget is not sufficient to deploy the minimum required number of sensors, then

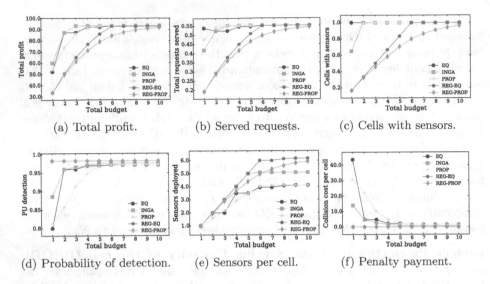

(a) Total profit.   (b) Served requests.   (c) Cells with sensors.

(d) Probability of detection.   (e) Sensors per cell.   (f) Penalty payment.

Fig. 3. Impact of increasing budget $\mathcal{B}$ allocated for the spectrum sensing service by the MNO.

sensing service is not available for this cell resulting in no additional spectrum in the cell. We consider this scheme to be the baseline which regulatory bodies have proposed in earlier proposals, e.g., (0.9, 0.1) for IEEE 802.22 [8]. Computational complexity of REG is the same as EQ and PROP, i.e., $\mathcal{O}(KN_{\max})$.

5   Performance Evaluation

Here, after describing the simulation setting and the selected parameters, we present the numerical results where we investigate the impact of various system parameters, e.g., budget, on the performance.

5.1   Simulation Setting

We simulate a cellular network with $K = 2000$ cell sites using our custom Python simulator and analyze the performance of our proposed schemes. The PU's off probability is distributed uniformly in [0.2, 0.8]. Other parameters (listed in Table 1) are set as follows: $\mu_s = 1$, $\mu = 1$, $\mu_c = 5$, $\kappa = 10$ bps/Hz, $(P_d, P_f) = (0.8, 0.1)$, and $(Q_d^*, Q_f^*) = (0.98, 0.05)$. For generating the request distribution, we first pick randomly σ of the cells as hotspots. Total requests generated from these cells will account for R_σ fraction of the requests. The rest which we call as *coldspots* will account for $(1 - R_\sigma)$ fraction of the requests. In each cell category, to have some variance in traffic, we generate the requests uniformly distributed in an interval, e.g., with 10% variance from the average load. If not stated otherwise, we set $R_\sigma = 0.6$ and $\sigma = 0.1$.

Table 1. Key simulation parameters.

Parameter	Value
Number of cell sites (K)	300
Av. num. of requests per cell (R_i)	90
Duration of frame, sensing period, and reporting slots (T, T_s, T_r)	$(10, 1/6*10^{-3}, 10*10^{-3})$ ms
PU's idle probability (p_0^i)	U(0.2, 0.8)
Sensing price (μ_s)	1
Service price (μ)	1
Collision penalty (μ_c)	5
Min. capacity requirement $(c_{\min})$	3 Mbps
Spectral efficiency (κ)	10 bps/Hz
Sensor sensing accuracy (P_d, P_f)	(0.8, 0.1)
Target sensing accuracy (Q_d^*, Q_f^*)	(0.98, 0.05)

5.2 Impact of Budget $\mathcal{B}$

In Fig. 3, we increase the total budget $\mathcal{B}$ from 1 to 10 sensors/cell with a step size of one sensor/cell. For example, when $\mathcal{B} = 1$, this means that the MNO's budget in total is $\mu_s \beta_s K$ and therefore can afford only allocating on average one sensor to each cell.

Figure 3a shows the change in the total profit of the MNO. As expected, increasing budget increases the profit. However, each scheme experiences saturation after a certain budget. This is due to the diminishing returns: deploying more sensors only increases the capacity marginally. Hence, the resulting increase in the MNO income via serving more traffic does not justify the increased payment for sensors. Another reason might be that the discovered capacity is already sufficient to serve all traffic in the cell, which invalidates more capacity addition. Observing Fig. 3b, we see that the saturation is due to the first reason as only a maximum of approximately 60% of the requests are served. We observe diminishing returns in sensing accuracy and thereby related utility with increasing number of sensors. Therefore, our schemes might prefer deploying lower number of sensors than the one required by the regulation-conforming heuristics which have to ensure $(Q_d^*$ and $Q_f^*)$. This insight is supported by Fig. 3d and e which show lower Q_d and lower number of sensors deployed for our heuristics, respectively. Figure 3f shows that under low budget our heuristics might sacrifice from sensing accuracy and pay for resulting penalty. In return, more cells can benefit from capacity expansion via opportunistic spectrum discovery. For example, as Fig. 3c shows, our heuristics employ sensors from a wide range of cells in contrast to REG-EQ and REG-PROP schemes. As an example, for $\mathcal{B} = 1$, EQ employs one sensor/cell covering all cell sites while REG-EQ employs sensors only in 17% of the cell sites as the minimum required number of sensors is six in that setting.

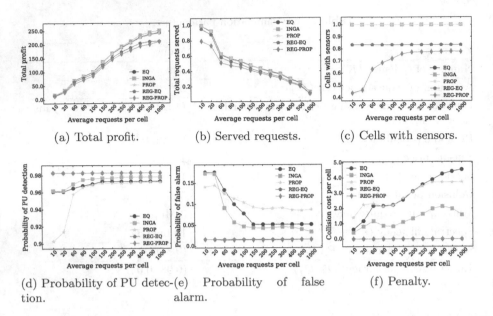

(a) Total profit.   (b) Served requests.   (c) Cells with sensors.

(d) Probability of PU detec-(e)   Probability   of   false   (f) Penalty.
tion.   alarm.

Fig. 4. Impact of increasing load (i.e., number of requests generated per cell).

Comparing all schemes in Fig. 3a, we see that when budget is low (e.g., $\mathcal{B} = 1$), INGA has the highest performance followed by PROP. Later with increasing $\mathcal{B}$, EQ gradually attains similar performance. All these schemes have a significant performance improvement (e.g., reaching 0.8x for low budget) over the regulations-conforming heuristics REG-EQ and REG-PROP. This improvement comes with a trade-off in sensing accuracy as observed in Fig. 3d where REG-EQ and REG-PROP have the highest and exactly the same performance.

5.3   Impact of Traffic Load R_i

In Fig. 4, we plot the impact of increasing load in terms of number of requests per cell. Here, we set $\mathcal{B} = 5$ sensors/cell. Under all loads, INGA succeeds the highest profit with a gap of around 5% from its closest follower PROP. All schemes have higher profit with increasing load, but there seems to be a saturation point after which the profit improves very slightly. The saturation point is reached when all the discovered bandwidth is used for the requests.

Another observation is that if the MNO prefers or has to use REG approaches, it should choose equal division of its budget across its cells as consistently REG-EQ over-performs REG-PROP for about (5–15%) depending on the setting. While increasing load improves the profit of the operator due to higher payments from the customers, some of the requests may be blocked as we see lower values of served requests in Fig. 4b. The selection of which cells to serve differs from one scheme to another. For example, schemes with proportional budget assignment prioritize hotspots, e.g., cells with higher load. Figure 4c shows that

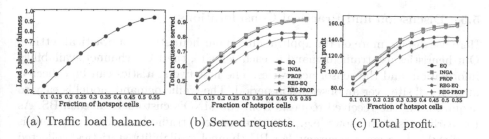

| (a) Traffic load balance. | (b) Served requests. | (c) Total profit. |

Fig. 5. Impact of increasing fraction of hotspots.

REG schemes has a confined cell span, i.e., employs sensors only a subset of cells. This is due to the limited budget which can afford 5 sensors per cell, lower than 6 sensors required for (0.98, 0.05) sensing accuracy. Hence, around 5/6($\sim$80%) of the cells are selected for sensor deployment under REG-EQ. For REG-PROP, the fraction of cells with capacity extension is lower as proportional assignment of the budget might employ more sensors in a highly-loaded cell, thus leaving other cells without any sensors.

When it comes to sensing accuracy, we observe in Fig. 4d and e that increasing load requires more spectrum, i.e., lower false alarm probability. Hence, each scheme prefers deploying more sensors, which consequently also improves probability of PU detection. Additionally, since penalty function incurs also the number of requests as its multiplier, the penalty of a lower detection accuracy becomes higher under high load. As a result, MNO is motivated to satisfy minimum Q_d^*. As Fig. 4f shows, INGA results in a lower penalty cost for possible collisions with the PU compared to other heuristics.

5.4 Impact of Fraction of Hotspots σ

In the earlier scenarios, only $\sigma = 0.1$ fraction of the cells are hotspots accounting for 60% of the network traffic. Now, we analyze the impact of traffic heterogeneity across cells. With increasing values of $\sigma = [0.10, 0.55]$, MNO's traffic becomes more homogeneously distributed across cells while the total number of requests remains the same. Figure 5a plots the observed traffic load balance which is calculated as the Jain's fairness index using the distribution of number of requests across all cell sites. As we see in the figure, MNO's traffic balance is almost 0.9 when $\sigma = 0.55$ and around 0.3 when $\sigma = 0.1$.

Figure 5b plots the change in the fraction of served requests with increasing σ. Under a more evenly distributed traffic, all schemes can serve a higher number of requests (Fig. 5b) by discovering additional spectrum in more cells. As a result, the total profit increases as shown in Fig. 5c. The relative performance of each scheme follows the same trend as observed in other scenarios: our proposals outperform REG variants under all traffic heterogeneity settings.

5.5 Discussion and Practical Considerations

Here, we discuss more on the applicability of our heuristics in a practical setting. Our heuristics use traffic demand in each cell site, the PU's channel availability information, and the sensor accuracies. The first two statistics can be collected at each cell site over an observation period. The traffic demand model is easy to acquire using the received requests from the MNO's customers at each BS. As the earlier research shows, e.g., [10], the network traffic has seasonality and is predictable to a good accuracy. For PU channel availability statistics, collected sensor data can be used to derive the PU's traffic pattern and thereby p_0. The prior work on PU traffic prediction, e.g., [7], can be used not only for estimating p_0 but also for more accurate estimation of the PU dynamics. Lastly, each sensor's accuracy can be computed at each BS by comparing it with the final sensing outcome. Obviously, it will take some time to converge to the sensor's accuracy level.

When it comes to executing payment to sensors, an MNO can prefer using smart-contract based solutions as proposed in [3]. Smart contracts are digital counterparts of traditional contracts which define the terms of an agreement as well as dispute resolution approach. However, smart contracts do not need the trust between trading parties as the contract itself is the point of trust. Hence, it can fit to our setting where the sensors and the MNO do not have to trust each other. Nevertheless, using smart contracts might entail additional monetary cost which must be paid eventually by the MNO. Hence, the MNO has to revise its profit calculation based on the additional cost of using a smart contract network.

6 Related Work

The most relevant works to ours elaborate on sensor selection for crowdsourced spectrum sensing, e.g. [3–6,9,11,12]. In [11], Ying et al. design a pricing mechanism with joint consideration of sampling value, data quality and cost of incentivized sensing. Their main contribution resides on the integration of device heterogeneity reflected in sensing data quality and costs. However, they do not consider constraints such as regulatory requirements but focus on the intricacies of REM construction under heterogeneous sensor settings. In [6], Jin et al. similarly elaborate on the participant selection in crowdsourced spectrum sensing systems and model it as a reverse auction problem. Their main focus is the privacy problem in such a system. To this end, they develop a framework for the MNO to select sensing participants in a differentially privacy-preserving manner. However, they do not address sensor selection problem under some regulatory constraints as we do in our work but rather assume that the MNO has pre-determined the sensing locations of each sensing task according to existing methods. In [12], Zhang et al. address the security of crowdsourcing-based spectrum sensing since the cooperative process is vulnerable to malicious sensing data injection attacks. Their approach considers the instantaneous trustworthiness of mobile detectors in combination with their reputation scores during data fusion for sensing decisions. However, their key concern is security rather than optimal sensor selection in compliance with regulatory requirements or profit maximization for MNOs.

There are also works which investigate the practical implications of crowd-sourced spectrum sensing. In [9], Nika et al. propose real-time spectrum monitoring with strong coverage using low-cost and commodity hardware. Although they do not work on sensor selection or sensing constraints due to regulations, their work is especially interesting as a feasibility study and employs practical hardware for large-scale low-cost sensing. In [5], Chakraborty et al. also crowd-source spectrum monitoring to low-cost and low-power commodity devices. To address the overhead drawback in the crowdsourced spectrum sensing, they propose three heuristics to select the minimum number of spectrum sensors that can best estimate the spectrum at the requested locations. In [4], they further develop a crowdsensing framework for low-cost and large-scale settings, which includes a technique for the sensor selection and fusion problem based on sensor decorrelation and clustering. However, they do not consider how the users are incentivized or the broader economical aspects.

Overall, none of these works except [3] consider the operator's business strategy and more specifically its profit. Spass [3] has a similarity with PoMeS and it also aims at maximizing MNO's profit. But, it considers a single cell. Moreover, the MNO considers the monetary overhead due to usage of a smart-contract network to pay the sensors participating in its sensing campaign in that work. Different than Spass, in this paper, we consider multi-cell setting with heterogeneous cell traffic and focus on how many sensors to select in each cell.

7   Conclusion

Opportunistic spectrum access provides ample opportunities for MNOs to increase their capacity cost-effectively whenever, wherever needed. For spectrum discovery, an MNO can buy spectrum sensing service from the sensing-capable sensors in exchange of payments for that service. However, the optimal number of sensors to be employed depends on various factors such as the MNO's own traffic in each cell, spectrum occupancy, and sensing price. In this work, we have first formulated a profit-maximizing sensor selection problem as an integer non-linear problem and then devised several heuristics with polynomial time complexity to decide on how many sensors to select in each MNO cell. Our solutions which might target a lower sensing accuracy at the expense of some monetary penalty outperform traditional approaches which have to ensure a certain level of sensing accuracy regardless of the network dynamics, e.g., needed spectrum, PU traffic, MNO traffic. Moreover, our simulations show that the MNO prefers maintaining lower sensing accuracy only when the network load is low and budget for spectrum sensing payment is limited. As future work, we plan to consider a more heterogeneous setting wherein sensors might be diverse in their sensing accuracy as well as their sensing price.

Acknowledgments. This work was partially supported by the Scientific and Technical Research Council of Turkey (TUBITAK) under grant number 116E245 and by the European Horizon 2020 Programme under grant agreement n688116 (eWINE project).

References

1. Axell, E., Leus, G., Larsson, E.G., Poor, H.V.: Spectrum sensing for cognitive radio: State-of-the-art and recent advances. IEEE Signal Process. Mag. **29**(3), 101–116 (2012)
2. Baig, G., Alistarh, D., Karagiannis, T., Radunovic, B., Balkwill, M., Qiu, L.: Towards unlicensed cellular networks in TV white spaces. In: ACM CONEXT (2017)
3. Bayhan, S., Zubow, A., Wolisz, A.: Spass: spectrum sensing as a service via smart contracts. In: IEEE DySPAN (2018)
4. Chakraborty, A., Bhattacharya, A., Kamal, S., Das, S.R., Gupta, H., Djuric, P.M.: Spectrum patrolling with crowdsourced spectrum sensors. In: IEEE Conference on Computer Communications (INFOCOM) (2018)
5. Chakraborty, A., Rahman, M.S., Gupta, H., Das, S.R.: Specsense: crowdsensing for efficient querying of spectrum occupancy. In: IEEE Conference on Computer Communications (INFOCOM) (2017)
6. Jin, X., Zhang, Y.: Privacy-preserving crowdsourced spectrum sensing. IEEE/ACM Trans. Netw. **26**(3), 1236–1249 (2018)
7. Li, X., Zekavat, S.A.R.: Traffic pattern prediction based spectrum sharing for cognitive radios. In: Cognitive Radio Systems. InTech (2009)
8. Maleki, S., Chepuri, S.P., Leus, G.: Energy and throughput efficient strategies for cooperative spectrum sensing in CRs. In: IEEE International Workshop on Signal Processing Advances in Wireless Communications (SPAWC) (2011)
9. Nika, A., Zhang, Z., Zhou, X., Zhao, B.Y., Zheng, H.: Towards commoditized real-time spectrum monitoring. In: ACM Workshop on Hot Topics in Wireless, HotWireless 2014, pp. 25–30 (2014)
10. Xu, F., Li, Y., Wang, H., Zhang, P., Jin, D.: Understanding mobile traffic patterns of large scale cellular towers in urban environment. IEEE/ACM Trans. Netw. (TON) **25**(2), 1147–1161 (2017)
11. Ying, X., Roy, S., Poovendran, R.: Pricing mechanism for quality-based radio mapping via crowdsourcing. In: IEEE Global Communications Conference (GLOBECOM) (2016)
12. Zhang, R., Zhang, J., Zhang, Y., Zhang, C.: Secure crowdsourcing-based cooperative spectrum sensing. In: IEEE INFOCOM (2013)

Tenant-Aware Slice Admission Control Using Neural Networks-Based Policy Agent

Pedro Batista[1,2]([📧]), Shah Nawaz Khan[1], Peter Öhlén[1], and Aldebaro Klautau[2]

[1] Ericsson Research, Stockholm, Sweden
{pedro.batista,shah.khan,peter.ohlen}@ericsson.com
[2] Federal University of Pará, Belém, Brazil
aldebaro@ufpa.br

Abstract. 5G networks will provide the platform for deploying large number of tenant-associated management, control and end-user applications having different resource requirements at the infrastructure level. In this context, the 5G infrastructure provider must optimize the infrastructure resource utilization and increase its revenue by intelligently admitting network slices that bring the most revenue to the system. In addition, it must ensure that resources can be scaled dynamically for the deployed slices when there is a demand for them from the deployed slices. In this paper, we present a neural networks-driven policy agent for network slice admission that learns the characteristics of the slices deployed by the network tenants from their resource requirements profile and balances the costs and benefits of slice admission against resource management and orchestration costs. The policy agent learns to admit the most profitable slices in the network while ensuring their resource demands can be scaled elastically. We present the system model, the policy agent architecture and results from simulation study showing an increased revenue for infra-structure provider compared to other relevant slice admission strategies.

Keywords: Network slicing · Reinforcement learning · Resource management

1 Introduction

Communication networks have been continuously evolving towards an ever-increasing complexity, both in integrating new technologies and supporting new verticals. The former requires a cross-domain and cross-technology network deployment and optimization while the latter imposes heterogeneous requirements on the network operators and the infrastructure that must support

This work has received funding from the H2020-MSCA-ITN-2016 SPOTLIGHT project under grant number 722788.

© ICST Institute for Computer Sciences, Social Informatics and Telecommunications Engineering 2019
Published by Springer Nature Switzerland AG 2019. All Rights Reserved
A. Kliks et al. (Eds.): CrownCom 2019, LNICST 291, pp. 17–30, 2019.
https://doi.org/10.1007/978-3-030-25748-4_2

them [5]. 5G networks, the most recent evolution of mobile communication networks, is anticipated to be platform for not only integrating new and revolutionary technologies such as Software Defined Networking (SDN) and Network Function Virtualization (NFV) but also support the requirements of new verticals through network slicing and multi-tenancy [8]. However, with such a wide gamut of technologies being integrated and support for different verticals developed, 5G networks have become extremely complex for management and control using the traditional network practices. One consequence of the 5G complexity is the large number of configurable parameters that exists in the network's cloud, radio access, control and management domains. Doing the large number of possible configurations manually is bound to trigger both suboptimal configuration setups which may lead not only to service disruption and failures but also adversely affect the revenue generation capacity of the network infrastructure. It is well recognized in this context that to handle this complexity, network automation requiring minimal human intervention will be required [5]. In existing networks, automation is generally an add-on feature that is mostly driven by pre-defined set of rules in specific context of a use case such as load balancing, mobility management, interference management etc. In 5G networks however, network automation driven by machine learning and artificial intelligence is anticipated to be a core feature that will drive most of the network management and control functions in an autonomous manner.

An important feature of 5G networks to support multi-tenancy is the network slicing concept which enables the network operator or Infrastructure Provider (InP) to facilitate different service providers in the network by providing dedicated resources [13]. The service providers in turn, offer revenue for the resources allocated to their deployed services in their dedicated slice. The network slicing concept has been considered at different scales, abstraction levels and in different network segments in the context of multi-tenant 5G networks [3,6,13,15]. However, regardless of how network slices are defined, they are eventually mapped onto a shared network infrastructure and must be managed by InP to optimize both resource utilization and revenue generated from the deployed slices.

In this work, we focus on the network slicing concept and present a reinforcement learning-based policy agent that aims to optimize the revenue generated from deploying different service slices in the network while ensuring that the deployed services can elastically scale their resource consumption footprint when needed. The rest of the paper is organized as follows. Section 2 provides the related work on network slicing and platforms supporting network slice deployment in the scope of virtualized 5G networks. Section 3 presents our proposed policy agent together with the system model and close-loop management architecture. In Sect. 4, simulation scenarios and results are presented to substantiate the increased revenue claim when compared to other relevant slice deployment strategies. The paper is concluded in Sect. 5 with a summary and discussion of future work in this scope.

2 Related Work

Virtualization is at the core of 5G network architecture where cloud platform spans across the different network segments diverging from the traditional centralized cloud architecture. A cloud-based network infrastructure inherently supports multi-tenancy and resource sharing. In such a context, there is a need for intelligent resource management to deploy network slices on the shared infrastructure. In this section, we describe some network slice management approaches that have been considered in the literature.

Samdanis et al. [13] proposes a 5G slice network management architecture. It is focused on having many players interacting over the network and the interfaces for communication among them. The architecture assumes a shared radio access network, that is divided in multiple domains, each one controlled by a domain manager, which are themselves managed by a higher-level network manager called 5G network slice broker with which the tenants communicate. To operate the network, the authors explicitly cite a set of metrics that are important for slice management. These metrics include the amount of resources allocated to a network slice such as physical resources or data rate, timing such as starting time, duration or periodicity of a request and time window, the type of resources and Quality of Service (QoS) parameters such as radio/core bearer type, prioritization, delay jitter, loss, etc. These metrics are important for understanding what the general service requirements at a high level are, such as service mobility, data offloading and disruption tolerance so it can be estimated if the current load in the system can fit the new slice.

Sciancalepore et al. [14] developed an admission control module for slice admission into a mobile network. Their model assumes that the bottleneck of the network is the physical resource (spectrum) which is to be shared among the network tenants. The information provided by the tenant to InP at the slice request include maximum resource utilization, duration of the slice and traffic class. In this context, traffic class specifies some behavior of the traffic, i.e., delay tolerance and if the bit rate should be guaranteed or not. The work considers a total of 6 traffic classes. Once deployed into the system, tenants request resources according to a Poisson process and the InP must provide them, otherwise a Service Level Agreement (SLA) violation penalty is incurred. The solution proposed to solve this problem applies a prediction of the traffic load of the requested slice. Based on this and the predictions of the previously admitted slices, the admission module can evaluate if the new slice can be placed into the system. The combination of all the possible slices in the system is modeled as a geometric knapsack problem. When the slice leaves the system, the prediction module (as part of the admission) is informed of the actual behavior of the slice so it can evaluate how accurate its prediction was and update its knowledge with new experience.

Another system for slice admission is studied by Bega et al. [2]. Their model of mobile network has physical resource (spectrum) as bottleneck and they have two types of traffic classes: elastic and inelastic. Inelastic users are characterized by having an SLA which specifies that all the requested resources must be provided when needed. Elastic users, on the other hand, do not require a specific number of resources and can cope with a variation of the number of allocated resources and their SLA is specified by an average resource availability. A slice request is composed of the slice duration, the traffic type and the slice size (in number of clients). The admission problem is modeled as a Markov Decision Process (MDP). The states are the number of elastic and inelastic users. The actions are: accept or reject the slices, while the objective is to admit as many slices as possible while guaranteeing the tenants requested SLA. Bega et al. [2] proposes the use of Q-Learning to solve the problem. They compare their solution with two heuristics and an analytical algorithm. They show that their proposed solution can adapt if the system does not behave as modeled and provide better decisions than the other proposals.

Apart from the research works targeting new approaches to network slicing and resource management, there is significant work being done on developing platforms which can be used to integrate such solutions in real networks. The seminal work on the platform side was started with the European Telecommunications Standards Institute (ETSI) NFV group that released a whitepaper outlining how network infrastructure made of physical nodes would be transformed to a software system running on general purpose servers [1]. Subsequently, the group presented the NFV Management and Orchestration Framework (MANO) that has been the reference architecture for many platforms currently being developed for virtualized network management [4]. The reference implementation of the ETSI MANO architecture is called Open Source MANO (OSM) and is actively maintained by the open source community. Similar initiatives were started by vendors and commercial entities to produce carrier grade options resulting in platforms like Open Orchestrator Project (OPEN-O); Enhanced Control, Orchestration, Management & Policy (ECOMP), and more recently their converged realization Open Network Automation Platform (ONAP) [9].

Although work on the development of MANO platforms is ongoing together with standardization on the architecture, interfaces and functionality, the need for developing intelligent solutions for core features like network slicing remains with the platforms providing an easier path towards integrating them in a realistic environment. In this work we address this issue and present a policy agent for slice admission control in virtualized 5G networks.

3   Network Slice Admission Control

This section provides an overview of a high-level network management loop that can be applied at multiple levels of the network, specially to control the admission of new slices. We present the system model considered by this paper, as well as our proposal to solve the admission problem.

3.1   The Control Loop

Network management operations in many levels of the network such as service admission or orchestration can be modeled as a closed loop operation, as depicted in Fig. 1. In general, a service arrives at some part in the network which then goes through an admission policy that decides whether it is in the interest of the network operator or infrastructure provider to accept the service or reject it. Services that are admitted into the network require some setup. For example, the service might consume some infrastructure resources, and the network may have multiple resource pools from which the resources could be provided, so the decision of which pool to use is taken in this step. Once the new service is deployed, the system enters in a general control loop where not only the deployed service instances, but also the network, is constantly monitored and optimized. The closed loop monitoring and control mechanism ensures that if a deployed service instance needs more or less resources than it did at the deployment time, then those additional resource requirements can be accommodated or unused resource be allocated to another slice in the network. Additionally, in the case of service completion or departure from the network, the closed loop operation ensures that not only the resources are taken away from the departing service but also the state of remaining services is optimized. For an intelligent admission policy, the SLA parameters observed by the departing service during its lifetime are extremely important, since a positive or negative feedback can be used by the admission policy to optimize its performance for future decisions. The management and control loop depicted in Fig. 1 can be adapted to operate in other context and various parts of the network. To demonstrate this general management concept, we will apply it to our work proposal where we use the closed loop approach to a high-level network operation in which network slices are deployed onto a shared network infrastructure.

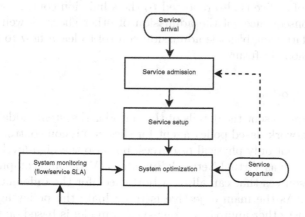

Fig. 1. Flowchart representing an overview of the general network management loop. The dashed line represents exchange of information, for example, reporting the overall satisfaction experienced by the service during its lifetime.

In this network sharing context, there are three distinct roles comprising the InP: the entity that owns the infrastructure on which the slices will be executed;

the tenant, entity that requests resources from the InP to run its services; and a user, which will consume the services from the tenant. A slice deployment request made from the tenant contains a SLA requirement and the InP must provide enough resources to fulfill the SLA. Examples of SLA include network coverage over a certain area and minimum network bandwidth. The InP has limited resources and, therefore, in some situations cannot fulfill the SLA for all the tenants requesting slices. That motivates the existence of the slice admission module. Its objective is to admit as many slices as possible into the system, with the objective of maximizing resource utilization, consequently increasing the revenue for the infrastructure provider. The constraint is that it should not allow slices that would have their SLA violated, or cause SLA violation for the other deployed services. A network slice may require resources in multiple parts of the network. For example, processing power in base stations or connectivity in the backhaul. To setup those resources, certain decisions must be made. For example, there are usually multiple paths connecting the base station to the core network, and the decision of which of those paths will provide the required connectivity for a particular slice is done at the service setup module.

Once the slice is admitted into the network, it becomes operational and the system enters in its main control loop with respect to that service. Resources in the system are constantly optimized so that their allocation to each slice matches the real-time service needs. The system is also constantly monitored, so that observed SLA by all the deployed slices is recorded and evaluated for compliance. The last significant change in the system happens when the lifetime of a slice ends, and it must leave the network. This triggers an optimization of the system, so it can optimize its resources to the slices currently deployed. The departure of a slice is also reported to the admission control, so that it can evaluate the consequences of the admission of other slices as well as the slice behavior. The latter enables the admission control to learn how to make better admission decisions in future.

3.2   System Model

In this section we present the details of the considered system model for the proposed neural network-based policy agent for slice admission control. Our system model considers not only physical resources, but also the high-level components of the network such as Virtual Network Functions (VNF) and computational and connectivity resources that can all be a bottleneck for the different slice classes in the network. As the main objective is to evaluate the policy agent for slice admission against other approaches, our system model is based on the concepts developed in Raza et al. [10].

The overall network architecture is presented in Fig. 2. Its main components, modeling a metropolitan area, are: a couple of Regional Data Centers (RDC), a few dozes of Central Offices (CO), and hundreds of Remote Radio Units (RRU). The RDC provides connectivity to external networks and has General Purpose Processors (GPP). The CO has both the Special (radio) Purpose Processor (SPP) and GPP. Those resources at the CO are more expensive than in the cloud but can deliver a lower latency. The RRU provides the radio access to the end users.

At the assumed abstraction level, there are some components that can be deployed by slices. The edge (CO) and core (RDC) both provide GPP, which can be used by slices to execute general processing functions such as a virtual Package Processor (vPP); mobile network functions, such as Package Gateway (PGW); and slice specific applications. The other resource is connectivity, which enables communication between edge and core. Each CO $c \in \mathcal{C}$ has a capacity of g_c GPP, each RDC $r \in \mathcal{R}$ has a capacity of g_r GPP and each link $l \in \mathcal{L}$ has a capacity of d_l link capacity units. In the assumed model, all the resources have integer units.

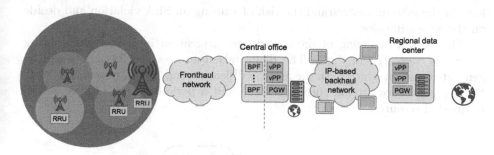

Fig. 2. Overall architecture in a flexible mobile network. Slice require functions that can be placed and consume resources in different parts of the network.

The described resources are consumed by slices that deploy the presented components in the network. The maximum number of GPP that a slice can request at each CO is k_c, and at any RDC is k_s; and the maximum number of connectivity resource between them is k_m. Deployed slices have dynamic resource requirements over time and if the requested resources cannot be provided by InP, an SLA violation occurs.

Tenant t requesting a network slice must inform InP of their immediately requested GPP at each CO j_c (which indicates the region it wants to have coverage in); the number of GPP at a RDC j_s; the connectivity between them j_m; the duration j_e; and the priority j_p. If an InP reserves enough resources for the admitted slice, it is guaranteed that no SLA penalty will be imposed and the tenant will remain satisfied. However, the tenant does not always use the maximum number of allocated resources and reserving them leads to resource underutilization. Therefore, the InP can try to understand the behavior of its tenants and sometimes oversubscribe the system by deploying additional slices so that, with managed risk of causing SLA violation, a higher revenue is achieved.

When the slice ends its service life cycle, the InP receives its revenue for hosting the service which is a fixed amount agreed at the time of admission based on slice parameters. Any SLA violation causes a decrease on this value which is proportional to the magnitude of the violation.

The described environment model allows for the study of slice admission and its consequences in the system. In the next section, it will be used to study new techniques for training Reinforcement Learning (RL) slice admission agents.

3.3 Slice Admission by Reinforcement Learning

In the network slicing scenario, the objective of the InP is to increase the revenue. In the general management loop presented in Sect. 3.1, one of the decisions that can be optimized is the slice admission. Assuming that all the other procedures (setup, scaling, etc.) are established, we study a RL agent that will learn when to accept or reject a slice into the system to maximize profit. Such an agent would have to learn how the system behaves so that it can consider the current load in the system, understand the risk of causing an SLA violation and decide on the slice admission.

The overall modeling of the experience acquisition in the network slicing system is presented in Fig. 3. The diagram emphasizes the data that will be generated and on which data the agent will learn from. In the system, there are three events: slice arrival, slice departure and a periodic check of slice health (requested resources).

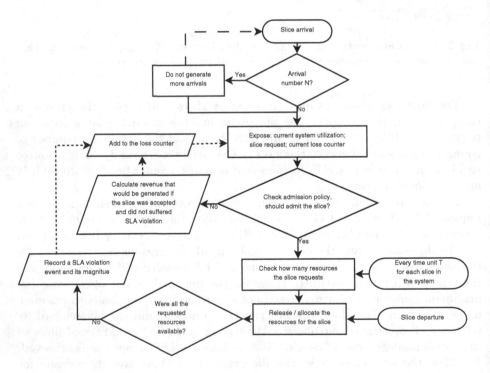

Fig. 3. Overview of the experience gathering in the network slicing system.

To make the learning suitable for RL, the system is modeled in an episodic fashion. Each episode is determined by the arrival of N slices. The objective is to accept as many of those as possible, while avoiding SLA violation.

Upon a slice arrival, the agent is consulted to make a decision. It is made aware of the current system utilization, the slice request parameters and the accumulated loss suffered by the system until the time of arrival. Based on that, it either accepts or rejects the slice. If rejected, the system interprets that some revenue could be obtained if the system had resources to admit the slice thus the rejection is interpreted as a loss. Accepted slices are allocated the number of resources requested, if those are available, if only a fraction of the resources is available, those are allocated, and an SLA violation is recorded.

The slice scaling event revisits all deployed slices in the system and checks how many resources the slices are requesting and adjusts the resource allocation. As before, if only a fraction of the requested resources is available, the fraction is allocated and an SLA violation is recorded. The scaling event is fired periodically with a period T.

When the slice finishes its execution, a slice departure event is generated, which releases resources allocated to the slice immediately. When all the N slices arrive and the accepted ones finish their execution, an episode is finished.

Aligned with previous work [7,10], our RL agent is using a policy network, which has a configurable number of inputs and hidden layers and two outputs, representing the probability of accepting and the probability of rejecting the slice. The input to the neural network is: the system utilization, the amount of each resource requested by the slice, its duration, priority and tenant. This information is encoded in a binary representation according to $\mathcal{A}_b(x)$, which creates an bit field of b bits with the first x bits equal to one and the others $b - x$ bits equal to zero. The state vector s, when slice j is requesting to enter the system, is then created by concatenating the bit fields: $\mathcal{A}_{g_c}(n_c)$, $\mathcal{A}_{k_c}(j_c)$, $\mathcal{A}_{g_r}(n_r)$, $\mathcal{A}_{k_s}(j_s)$, $\mathcal{A}_{d_l}(n_l)$, $\mathcal{A}_{k_m}(j_m)$, $\mathcal{A}_{k_e}(\min(j_e, k_e))$, $\mathcal{A}_1(j_p)$ and $\mathcal{A}_{n_t}(t)$, where n_c, n_r and n_l are the number of busy resources at c, r and l, respectively, $c \in \mathcal{C}$, $r \in \mathcal{R}$, $l \in \mathcal{L}$, and k_e represents the maximum requested duration observable by the agent.

4 Evaluation

We evaluate the proposed method using a network topology [12] shown in Fig. 4. The nodes with a high degree of connectivity were selected as RDC, some with medium degree as CO, and some nodes are connection points that provide connectivity routes.

Each RDC has a capacity $g_r = 80$, each CO $g_c = 50$ and each link $d_l = 50$. The reference behavior of slices (profiles) are the ones reported by Raza et al. [11]. The resource requirements of each slice are given by the time of the day and the type of the slice. Between 9:00 and 19:00 the high-priority slice requests 20 GPP resources at the CO, 5 connectivity resources and 5 GPP at the RDC, this is usually the busy hours at business districts where high-priority traffic is likely to occur. Other than at those hours, the high-priority slice requires only 15 GPP at the CO. The low-priority slice between 16:00 to 22:00 requires 10 GPP at the CO, 10 connectivity resources and 10 GPP at the RDC. This is likely to be the busy hour in residential districts, where low-priority slices are likely to exist. During the other hours, low-priority slices require 5 GPP at the CO, 5 connectivity resources and 5 GPP at the RDC. This configuration reproduces a network that was projected to have as main bottleneck the resources at the CO, which are the most expensive.

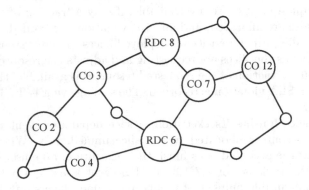

Fig. 4. Network topology used to evaluate the system.

In contrast with previous work, which considered fixed slice profiles, we assume that the resource requests of a slice is non-deterministic. The current simulator accepts integer resource requests. Consequently, we chose to use a Binomial distribution with 5 trials and a tenant-dependent success probability (u_t) to generate a *noise*. This noise is sampled every time unit and subtracted from the resource request of each resource type of the reference profile, if this subtraction leads to a negative resource request, it is understood as no resource needed (0). The slices impose penalties as discussed in Sect. 3 and the magnitude of those penalties is proportional to the tenant penalty weight (v_t).

We configured the system to run a simulation with two tenants, $t = 0$ and $t = 1$, and used $u_0 = 0.1$, $u_1 = 0.9$, $v_0 = 1$ and $v_1 = 0.1$. This setup makes tenant 0 have a higher network usage, and his penalty is higher than for tenant 1, which imposes a lower usage.

We consider as baseline an admission agent that is not aware of the tenant who is making the request [10]. The proposed strategy adds as input to the admission agent the ID of the tenant who is requesting the slice, as described in Sect. 3.3. Both agents are using a policy network with 4 hidden layers with 40 neurons in each and a ReLu activation function, following the baseline.

We trained the system for 10,000 iterations, with 25 episodes per iteration, each episode with a load of 80 Erlangs and 600 arrivals. With the trained model we ran the test for a new set of 25 episodes with the same configuration.

Alongside our proposal (Prop), we show results for four other policies. BL is the baseline, which was adapted from Raza et al. [10]. With the exception of the tenant ID, BL has the same input as Prop. RND is the random police, i.e., it chooses to accept or reject with equal probability. Fit accepts the slice if the InP has enough resources to fulfill its request at admission time. Those heuristics were defined and also used by Raza et al. [10].

The results are summarized in Fig. 5. We can observe that accepting all the slices incurs a high scaling loss which signals that some of the slices could have been rejected. Fit is too conservative and rejects too many slices causing a resource underutilization. Randomly accepting the slices essentially accepts half of them. The baseline learns a better policy and achieves a balance between rejecting some slices and handling some scaling loss. However, it does not have information on which tenant is requesting the slice. Thus, it cannot achieve the performance of the proposed solution which can fine-tune the decision to the specific tenant, and consequently find a better balance.

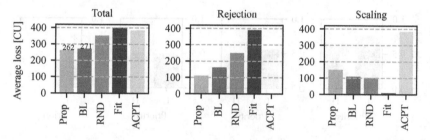

Fig. 5. Overall results. Prop is the proposed policy, BL is the baseline, RND is random, Fit accepts a slice if there are enough resources at admission time, and ACPT is accept all the slices. The left graph shows the overall loss achieved by each policy, BL has a loss of 271, while Prop has 262. In the middle, only the loss incurred by rejecting the slice is shown and, on the right, only the loss incurred by scaling.

We suppose that the baseline can still infer the tenant by the level of the usage which is indeed a function of the tenant (given u_t) and is present at its input. However, because the resource usage is noisy, it probably cannot achieve the best possible information about the tenant which is available for the proposed policy. To better understand which services the policies are choosing, we analyze the rejection probability for each class of slices.

We analyze which slices are being rejected in Fig. 6. We can see that Fit rejects more slices of tenant 0. That happens because tenant 0 usually requests more resources compared to tenant 1 (given that $u_0 < u_1$) and so it is more probable that there are enough resources for its slices. Random accepts half in any marginalization by nature. The baseline accepts more slices from tenant 0. It should have learned that rejecting tenant 0 slices incurs a higher penalty ($v_0 > v_1$) but instead it was probably inferred from the resource. However, the balance found by the baseline rejects more high-priority slices than low-priority ones, which is counter intuitive, yet this can be due to no tenant awareness, for example, the configuration of the system makes off peak hours slices from tenant 0 and peak hours slices from tenant 1 have similar resource usage and priority, but different penalties. Finally, the proposed strategy seems to reject almost all the tenant 1 slices in exchange for a higher acceptance of tenant 0. It also manages to find a point where higher priority slices are more often accepted compared to lower priority ones.

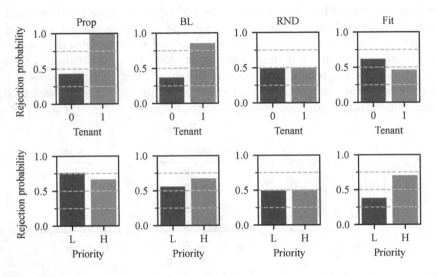

Fig. 6. Rejection probability for each policy (columns), marginalized by tenant (top line), or by priority (bottom line). ACPT policy is not shown, because it is zero for all cases.

Another way to investigate how slices are being classified is to examine the rejection probability marginalized by tenant and priority, shown in Fig. 7. We can use the same rationale for Fit: it accepts more of the low priority from tenant 1, then low priority from tenant 0, then high priority from tenant 1 and high priority from tenant 0. That is exactly the expected sequence of slices if they are ordered by the expected magnitude of its resource request when compared to each other. Random once again accepted half of each class. For the baseline, we can see that it understands that some classes are more important than others, but it seems, for example, that it cannot identify that high-priority slices from tenant 0 are more valuable than low priority ones from the same tenant. That is one of the behaviors that the proposed method could learn: it also rejects most of the slices from tenant 1, that can happen because the load is high, and there are enough slices from tenant 0, so it can be lucrative enough, at lower loads the behavior can be different.

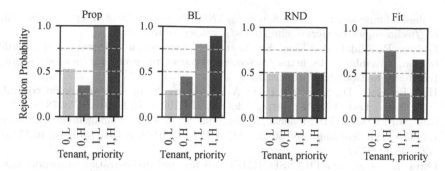

Fig. 7. Rejection probability for each policy (columns) marginalized by tenant and priority. Accept all is not shown, since it is zero for all the cases.

5 Conclusion

This work presented a reinforcement learning-based policy agent for slice admission control in virtualized 5G networks. We evaluated our proposal in a scenario where slice requests have a stochastic resource requirements footprint and admission control is non-trivial. Our network model included the concept of different tenants that give different revenues, consequently increasing the possible combinations of admission decisions. In this context, we showed that the tenant-awareness contributes to a better admission policy and brings more revenue to the InP. We evaluated our proposed admission policy in a simulated environment and compared the performance to other related strategies. As future work, we intend to evaluate the policy agent in more challenging system context where network and slices dynamics are increased. Another topic to be investigated is the proportion of tenants. The presented results only considered a uniform number of requests for each tenant but in real networks, the number of slices deployed or requested for each tenant will change over time. Finally, the performance of other machine learning algorithms in a similar dynamic system will be evaluated.

References

1. AT&T, BT, CenturyLink, Mobile, C., Colt, Telekom, D., KDDI, NTT, Orange, Italia, T., Telefonica, Telstra, Verizon: Network Functions Virtualisation. An Introduction, Benefits, Enablers, Challenges & Call for Action. Technical report, SDN and OpenFlow World Congress, October 2012. https://portal.etsi.org/nfv/nfv_white_paper.pdf
2. Bega, D., Gramaglia, M., Banchs, A., Sciancalepore, V., Samdanis, K., Costa-Perez, X.: Optimising 5G infrastructure markets: the business of network slicing. In: IEEE Conference on Computer Communications, IEEE INFOCOM 2017, pp. 1–9, May 2017. https://doi.org/10.1109/INFOCOM.2017.8057045
3. Caballero, P., Banchs, A., de Veciana, G., Costa-Perez, X., Azcorra, A.: Network slicing for guaranteed rate services: admission control and resource allocation games. IEEE Trans. Wirel. Commun. **17**(10), 6419–6432 (2018). https://doi.org/10.1109/TWC.2018.2859918

4. Dahmen-Lhuissier, S.: Open Source MANO, December 2018. https://www.etsi. org/technologies-clusters/technologies/nfv/open-source-mano
5. Gomes, P., Vidal, A., Lins, S.: Next stop: zero-touch automation standardization, November 2018. https://www.ericsson.com/research-blog/next-stop-zero-touch-automation-standardization/
6. Han, B., Feng, D., Schotten, H.D.: A Markov model of slice admission control. IEEE Netw. Lett. 1 (2018). https://doi.org/10.1109/LNET.2018.2873978
7. Mao, H., Alizadeh, M., Menache, I., Kandula, S.: Resource management with deep reinforcement learning, pp. 50–56. ACM Press (2016). https://doi.org/10.1145/3005745.3005750
8. Oliva, A., et al.: 5G-TRANSFORMER: slicing and orchestrating transport networks for industry verticals. IEEE Comm. Mag. 56(8), 78–84 (2018). https://doi.org/10.1109/MCOM.2018.1700990
9. ONAP: ONAP Architecture Overview. Technical report, December 2018. https://www.onap.org/wp-content/uploads/sites/20/2018/06/ONAP_CaseSolution_Architecture_0618FNL.pdf
10. Raza, M.R., Natalino, C., Öhlen, P., Wosinska, L., Monti, P.: A slice admission policy based on reinforcement learning for a 5G flexible RAN. In: 2018 European Conference on Optical Communication (ECOC), pp. 1–3, September 2018. https://doi.org/10.1109/ECOC.2018.8535483
11. Raza, M.R., Rostami, A., Vidal, A., Santos, M.A.S., Wosinska, L., Monti, P.: Priority-aware service orchestration using big data analytics for dynamic slicing in 5G transport networks. In: 2017 European Conference on Optical Communication (ECOC), pp. 1–3, September 2017. https://doi.org/10.1109/ECOC.2017.8346133
12. Raza, M.R., Rostami, A., Wosinska, L., Monti, P.: Resource orchestration meets big data analytics: the dynamic slicing use case. In: 2018 European Conference on Optical Communication (ECOC), pp. 1–3. IEEE, Rome, September 2018. https://doi.org/10.1109/ECOC.2018.8535581
13. Samdanis, K., Costa-Perez, X., Sciancalepore, V.: From network sharing to multi-tenancy: the 5G network slice broker. IEEE Commun. Mag. 54(7), 32–39 (2016). https://doi.org/10.1109/MCOM.2016.7514161
14. Sciancalepore, V., Samdanis, K., Costa-Perez, X., Bega, D., Gramaglia, M., Banchs, A.: Mobile traffic forecasting for maximizing 5G network slicing resource utilization. In: IEEE Conference on Computer Communications, IEEE INFOCOM 2017, pp. 1–9, May 2017. https://doi.org/10.1109/INFOCOM.2017.8057230
15. Zheng, J., Caballero, P., de Veciana, G., Baek, S.J., Banchs, A.: Statistical multiplexing and traffic shaping games for network slicing. In: IEEE/ACM Transactions on Networking, pp. 1–14 (2018). https://doi.org/10.1109/TNET.2018.2870184

Machine Learning Based RATs Selection Supporting Multi-connectivity for Reliability (Invited Paper)

Haeyoung Lee[✉], Seiamak Vahid, and Klaus Moessner

5G Innovation Centre (5GIC), Institute for Communication Systems (ICS),
University of Surrey, Guildford GU2 7XH, UK
{Haeyoung.Lee,S.Vahid,K.Moessner}@surrey.ac.uk

Abstract. While ultra-reliable and low latency communication (uRLLC) is expected to cater to emerging services requiring real-time control, such as factory automation and autonomous driving, the design of uRLLC of stringent requirements would be very challenging. Among novel solutions to satisfy uRLLC's requirements, interface diversity is widely regarded as an efficient enabler of ultra-reliable connectivity. When mobile devices are connected to multiple base stations (BSs) of different radio access technologies (RATs) and same data is transmitted via multiple links simultaneously, the transmission reliability can be improved. However, duplicate transmission of same data causes an increase in the traffic loads, leading to radio resource shortage. Considering it, efficient configuration of multi-connectivity (MC) for mobile devices is important. In this paper, the RAT selection scheme including efficient MC configuration is proposed. By adopting distributed reinforcement learning (RL), each device could learn the policy for efficient MC configuration and select appropriate RATs. Simulation results show that 20.8% reliability improvements over the single connectivity scheme is observed. Comparing to the method to configure MC for devices all the time, 37.6% improvement is achieved at high traffic loads.

Keywords: RAT selection · Multi-connectivity · Machine learning · URLLC

1 Introduction

Upcoming 5G networks are expected to support diversified services into three categories: enhanced mobile broadband (eMBB), massive machine-type communication (mMTC), and ultra-reliable and low latency communication (uRLLC) [1]. Especially, it is envisaged uRLLC could open the doors emerging various

The work presented in this paper is partly funded by the European Union's Horizon 2020 research and innovation programme under grant agreement No. 761745, and the Government of Taiwan.

© ICST Institute for Computer Sciences, Social Informatics and Telecommunications Engineering 2019
Published by Springer Nature Switzerland AG 2019. All Rights Reserved
A. Kliks et al. (Eds.): CrownCom 2019, LNICST 291, pp. 31–41, 2019.
https://doi.org/10.1007/978-3-030-25748-4_3

services, such as wireless control and automation in industrial environments [2], vehicle-to-vehicle communications, and the tactile internet, which requires to control many objects with real-time feedback [3]. For such services, the 3GPP aims at providing uRLLC for small data payloads (e.g., 32 bytes) with an outage probability of less than 10^{-5} at millisecond level latency. While the design of uRLLC of stringent requirements would be very challenging, various novel solutions have been proposed such as flexible frame structure design with shorter transmission time intervals (TTIs) [4], pre-emptive scheduling [5], and diversity for reliability improvement [6]. Especially, diversity is widely regarded as a crucial and efficient enabler of ultra-reliable connectivity [7].

As the network evolves, multiple radio access technologies (RATs) are being integrated and jointly managed, including 3GPP and IEEE families, with the vision of heterogeneity [8]. In addition, as cells are deployed closer and more heterogeneous, multiple links of different RATs would become available to user equipments (UEs) at the same time. Based on such availability of multiple links, multi-connectivity (MC) is expected to offer enough diversity and redundancy for achieving reliability [9]. Actually, the initial goal of using MC was to improve throughput performance by splitting its traffic and sending over multiple links, overcoming the capacity limitations imposed by backhaul links. By adopting packet duplication (PD) for duplicate transmission of same data [10], MC has been additionally considered as an effective solution to satisfy the stringent reliability requirement.

In heterogeneous networks (HetNets) of multiple base stations (BSs) from different RATs, for each UE, the matter to decide the suitable RATs impact on the network performance including reliability. In [11,12], the impact of the number of BSs involving multi-connectivity (called an active set) is investigated. In [11], it is shown that dynamic management of the active set of BSs (i.e., adding, removing, replacing based on thresholds of signal quality) can improve system performance in terms of radio link failure (RLF) and throughput compared to the use of fixed active set of BSs. In [12], the network load is also considered to decide the number of BSs. The impact of the network load to the effectiveness of multi-connectivity is investigated in [13]. With the assumption that UEs can access all BSs of SINR (signal to interference and noise ratio) higher than the pre-defined threshold, MC is proved more effective at the low traffic scenario in terms of throughput and RLF performance. In [14], the dual connectivity (DC) architecture is considered. The optimisation problem on RAT selection to maximize the sum throughput is formulated and the pair of serving macro and small cell for each UE is found. In [15], the utility based approach is studied considering user's satisfaction as well as service provider's satisfaction in terms of throughput. While the aforementioned works focus on improvement of mobility robustness or throughput performance, in [16], the reliability improvement through DC with data duplication is demonstrated. By the simulation results, it shows the required level of reliability and network traffic loads should be considered for each UE's DC configuration. However, how to configure DC for each UE is not investigated.

Considering the literature survey, in this paper, the approach on efficient RATs selection including MC configuration is proposed. Since the characteristics of UEs including the location and required QoS level could be different from each other in the network with dynamically changing traffic loads, the proposed algorithm employs distributed reinforcement learning. Each UE becomes an agent and learns the policy of RATs selection including MC configuration. While each UE learns the offset for effective MC configuration individually, appropriate RATs for UEs could be selected to minimizes the number of UEs in outage. It is shown that the proposed algorithm outperforms the single connectivity based RAT selection scheme by 20.8% in terms of reliability performance. Comparing to the RATs selection scheme where MC is always configured for all UEs, the proposed algorithm shows better performance by 37.6% by configuring MC only for UE at cell edge region.

The organization of the paper is as follows. Section 2 describes the system model including the architecture supporting MC. In Sect. 3, the proposed RATs selection algorithm adopting the distributed reinforcement learning mechanism is presented. Then, its performance is evaluated in Sect. 4. This paper is concluded in Sect. 5.

2   System Model

We consider packet duplication (PD) exploiting the multi-connectivity (MC) feature to improve reliability performance. With MC, a UE is able to connected to multiple BS. Since MC is an extension of dual-connectivity (DC), for simplicity, DC is considered in this paper as shown in Fig. 1. When UEs are connected to BSs, one BS acs as the master node (MN) to establish the control interface to the core network and another BS becomes the secondary node (SN). The MN

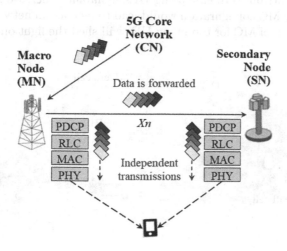

Fig. 1. Packet duplication via multi-connectivity to enhance reliability.

and SN are assumed to be interconnected by means of Xn interface. While DC can be applicable only for UEs in Radio Resource Control (RRC) connected mode, MN can initiate DC setup. The data transferred from the core network to MN is duplicated at the packet data convergence protocol (PDCP) layer of MN and the entire data is forwarded to SN via Xn interface. Thus, the same data becomes to be transmitted via both MN and SN to the UE. For resource scheduling, both MN and SN have flexibility and no restrictions are imposed at the RLC and MAC layer. The lower layers at MN and SN work independent of each other without coordination [10]. Such data duplication process can be carried out as long as the UE remains into the coverage area of both nodes.

As depicted in Fig. 2, the downlink transmission of the OFDM-based heterogeneous cellular network with multiple UEs is considered. The system consists of a set of BSs including macro BSs (MBSs) and small cell BSs (SBSs) and a set of users (UEs) capable to connect to multiple networks simultaneously. While macro BSs and small cell BSs could support different RATs, they can communicate based on Xn interface for DC [9]. Based on reference signal received power (RSRP) based cell selection, UEs measure the reference signal power from each BS, and could be connected to either the largest one or two BSs. When UEs are located close to the serving BS (e.g., UE 1, UE 2, UE 4 and UE 5), a strict reliability requirement could be met with a single highly reliable link of the strong signal power. For cell-edge UEs (e.g., UE 3), the received strongest signal level would not be strong enough to fulfil their QoS requirements. In this case, multi-connectivity employing PD could be set up to improve the reliability so that UEs can be connected to two BSs, one from macro BS and another from small cell BS. Then, the macro BS becomes master node (MN) while small cell BS becomes secondary node (SN).

While the channel quality is an important factor to decide MC configuration of UEs, the network traffic load could also impact on the reliability performance. Although MC contributes to enhancing UEs' reliability, increase in the number of UEs exploiting MC configuration would lead to increase in network traffic load [16]. Configuration of MC for too many UEs will shed the light on the benefit of

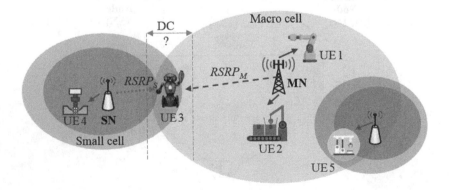

Fig. 2. Downlink transmission with multiple connectivity in heterogeneous networks.

MC. Thus, in order to decide MC configuration for UEs, the network traffic load needs to be considered as well as UEs' QoS requirements and channel quality.

In this paper, RATs selection based on MC configuration is investigated and the scenario of dual-connectivity (DC) is focused. Based on RSRP values from macro and small cell BSs, UEs's connection can be determined as follows.

- Connect to macro BS (MBS)   for $rsrp_M \geq rsrp_S + \beta$
- Connect to small cell BS (SBS)   for $rsrp_M \leq rsrp_S - \beta$
- Connect to MBS and SBS by DC   for $rsrp_S + \beta > rsrp_M > rsrp_S - \beta$

Here, $rsrp_M$ and $rsrp_M$ denote RSRPs from MBS and SBS, respectively. In the case the cell range extension (CRE) bias is given, $rsrp'_M = rsrp_M - CRE$ can be considered instead of $rsrp_M$. In order to provide DC configuration to UEs, a DC offset value β is considered with RSRP values. Optimal DC offsets β could be changed by various factors, such as the available radio resource among BSs and by the location of UEs and BSs. Since the optimal offset values vary from one UE to another, offset values could be defined by each UE [20].

3   RATs Selection with Reinforcement Learning

While the Reinforcement Learning (RL) mechanism uses experiences of agents and could§learn automatically from the environment without any training data on field, it allows an online learning. In our work, Q-learning is chosen since it enables learning the best policy without any priori knowledge of its environment.

In RL, at time epoch t, the agent in the state s_t selects and performs an action a_t. After the action a_t, it observes the environment and receives a reward cost C for this specific action. While RL accumulates costs obtained by action, it considers instant cost as well as cumulative costs in the future. With learning, it is aimed to find the optimal policy for selecting an action in a given state that minimizes the value of total cost. In Q-learning, in order to learn this policy, an agent utilises a value-function, Q-function, $Q(s_t, a_t)$. It is defined as follows [21]:

$$Q(s,a) = E\left\{ \sum_{t=0}^{\infty} \gamma^t C(s_t, a_t) \mid s_0 = s, a_0 = a \right\}, \tag{1}$$

where γ, $C(s_t, a_t)$, s_0 and a_0 denote a discount factor $(0 \leq \gamma \leq 1)$, the cost of the set of state s_t and action a_t, intial state, and initial action, respectively.

While it is really difficult to obtain the optimal policy by solving (1), RL could be exploited to find the optimal policy by using Q-table updates. In Q-table, each table entry, $Q(s_t, a_t)$, is associated with a state-action pair and the Q-learning algorithm maintains Q-table of values that represent the goodness of taking a particular action when in a given state. It is enough to converge this learning if all Q-values of the sets of states and actions are continued to be updated. Q-learning realizes (1) by updating Q-table as follows.

$$Q(s_t, a_t) \leftarrow (1 - \rho)Q(s_t, a_t) + \rho[C_{t+1} + \gamma \min_{a \in A} Q(s_{t+1}, a)], \tag{2}$$

where ρ is the learning rate of the range $0 \leq \rho \leq 1$ indicating what extent the learned Q-value will override the old one. When $\rho = 0$, the agent never learns. When $\rho = 1$, the new knowledge of the most recent Q-value is only considered. C_{t+1} represents the delayed cost, which is obtained for an action a_t taken. As the value of the discount factor γ in $[0,1]$ is higher, the future cost $\min_{a \in A} Q(s_{t+1}, a)$ is weighted more than the delayed cost C_{t+1}. By updating Q-table in (2), the agent learns the optimal policy for selection an action.

In this paper, the state, the action, and cost are defined as follows.

- **State:** The state of time epoch t is defined with the received power from BSs as:

$$s_t = \{rsrp_M, \ rsrp_S\} \text{ where } s_t \in S, \tag{3}$$

where $rsrp_M$ and $rsrp_S$ denote the reference signal received power (RSRP) from MBS and SBS, respectively. When there are multiple MBSs and SBSs, one MBS and SBS of the strongest RSRP can be selected. To make Q-table small and to convergence faster, two power values are quantized. S denotes the set of all states.

- **Action:** The action of time epoch t is defined as:

$$a_t = b_i \text{ where } b_i \in A \tag{4}$$

where b_i denotes the DC configuration offset value β and A is the set of all possible offset values (i.e., all possible actions).

- **Cost:** The cost of time epoch t is defined as:

$$c_t = n, \tag{5}$$

where n denotes the number of UEs in outage.

Each UE monitors the level of RSRP from BSs and selects one MBS and SBS of the strongest signal power. In other words, each UE observes its state. The received power value is quantized to manage Q-table size small and to convergence faster and each UE compares these quantized signal powers with its Q-table's states. If the UE cannot find the received powers from its Q-table, the new state of received powers is added to Q-table. Among those sets whose received powers are equal to the received powers, UEs can choose an action a_t based on ε-greedy exploration and exploitation policy [21]. In ε-greedy policy, at every decision epoch, a UE in state s_t explores with probability $\varepsilon(s_t)$, and stored Q-values is exploited with probability $1 - \varepsilon(s_t)$ as follows.

$$a_t = \begin{cases} \min_{a \in A} Q_t(s_{t+1}, a) , & \text{probability } 1 - \varepsilon(s_t) \\ rand(a) & , \text{probability } \varepsilon(s_t). \end{cases} \tag{6}$$

The exploration rate $\varepsilon(s_t)$ is defined by using $\lambda(s_t, a_t)$ which is the number of visits of state-action pair (s_t, a_t), as follows.

$$\varepsilon(s_t) = \frac{1}{\log\left(\sum_{a_t \in A} \lambda(s_t, a_t) + 3\right)}. \tag{7}$$

In (7), $\varepsilon(s_t)$ in $(0, 1)$ has a logarithmic decay. This approach aims to control the frequency of exploration so that the best-known action is taken at most of the times. Exploring is not stopped to enhance the long-term learning performance, but rather decreased gradually over time. For convergence, the learning rate $\rho(s_t, a_t)$ is set by using $\lambda(s_t, a_t)$ as follows.

$$\rho(s_t, a_t) = \frac{1}{\sqrt{\lambda(s_t, a_t) + 3}}. \tag{8}$$

According to above definition, each UE decides the appropriate offset value for MC configuration that minimize the number of UEs in outage. Then, each UE is connected to selected RATs by comparing RSRPs from BSs with the MC configuration offset. After BSs allocate resource to UEs, BSs calculate the number of UEs in outage UEs and send information to UEs. For resource allocation, resource block (RB), the block of subcarriers, is considered as the basic radio resource unit [9] and it is assumed that one RB is allocated to each UE.

After resource allocation, BSs send the information on the number of UEs in outage to UEs, and each UE updates Q-value based on (2).

The procedures of the proposed algorithm are explained in Algorithm 1. While Step 1 to 10 and 13 are conducted by the UE side, Step 11 and 12 are carried out by the BS side. Repeating the above steps makes Q-table values of all sets of states and actions converge, and then agents can make right actions.

Result: β, the offset value for MC configuration for each UE

Initialisation: *Q-table* with a very high number;

Learning procedure: while *Q-table converges* **do**

    1. UE selects the MBS and SBS of the strongest signal power;

    2. UE compares the (quantized) received powers with Q-table's states;

    if *no equal received powers on Q-table* **then**

        | 3. UE adds new received powers to its own Q-table;

    end

    4. Calculate $\varepsilon(s_t)$ and generate a random number δ in [0,1];

    if $\delta \leq 1 - \varepsilon(s_t)$ **then**

        | 5. UE chooses one value that has the lowest Q-value;

    else

        | 6. UE chooses one value randomly;

    end

    7. UE uses chosen offset value β as an action;

    8. UE compares $rsrp_M$ and $rsrp_S$ added by β;

    10. UE decides RATs to be connected;

    11. BSs allocate RBs to each UE;

    12. BSs calculate the number of outage UEs and pass it to UEs;

    13. Each UE updates the chosen set's Q-value based on (2).

end

Algorithm 1: The proposed RAT selection algorithm

4 Performance Evaluation

We evaluate the performance of the proposed RAT selection algorithm via simulation. Table 1 shows the initial configuration parameters. In order to compare the performance of the proposed algorithm, two reference schemes are considered: (1) single connectivity based RAT selection (labeled 'SC') where one BS of the strongest RSRP is selected, and (2) dual connectivity based RATs selection (labeled 'DC-Always') where DC is always configured to all UEs. The average number of UEs in outage is considered as the performance indicator. Considering the practical scenario, traffiic of short message size [18] and path loss model for open production space is chosen [19]. Furthermore, as interval of DC configuration offset, we use 2 dB for Q-learning to make Q-table small. The maximum value of the offset is set to 32 dB, thus the actions have 17 levels.

Firstly, we investigate the impact of the DC offset value β on the reliability performance as depicted in Fig. 3. With two reference schemes, DC based algorithms are studied using different DC offset values of 4,10,20, 30 dB. While all DC based algorithms produce better performance than the algorithm 'SC', it is observed that increase in a DC offset value contributes to enhancement of the reliability performance. However, the algorithm 'DC-Always' is shown to be superior. In this case, while 20 UEs uniformly distributed are assumed, BSs do not have difficulty in allocating RBs to all UEs. Since increase of traffic loads from DC configuration does not lead to overload BSs, configurating DC for all UEs could enhance their reliability.

Table 1. Simulation parameters

Parameters	Values
No. of MBS/SBS	1/1
Transmit power	MBS: 10 dBm, SBS: 0 dBm
Carrier frequency	3.5 GHz
Channel bandwidth	5 MHz
No. RBs	25
Noise density	$-174\,\text{dBm/Hz}$
No. UEs	20–30 (uniform dist.)
UE traffic [18]	40 bytes, 1 ms of message inter-arrival time
Path loss model [19]	LoS: $32.45 + 20\log_{10}(d_{3D}) + 20\log_{10}(f) + X_\sigma, \sigma = 3$ nLoS: $32.45 + 24.7\log_{10}(d_{3D}) + 20\log_{10}(f) + X_\sigma, \sigma = 5.17$ (where $dref = 10\text{m}, K = dref^{1.5}, d_{3D} \geq 1\text{m}$) For $d_{2D} \leq dref$, $pr_{LoS} = 1$ For $d_{2D} > dref$, $pr_{LoS} = \left(-\frac{d_{2D} - dref}{K}\right)$
Resource allocation	Round robin
DC offset value β	$[0, 2, ..., 32]$ dB

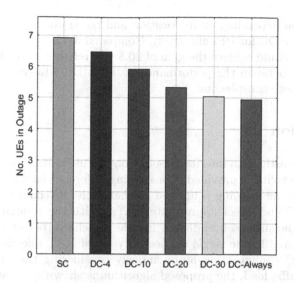

Fig. 3. Comparison of the average number of outage UEs from algorithms, SC, DC with various offset (4 dB, 10 dB, 20 dB, 30 dB), and DC-Always at low traffic loads

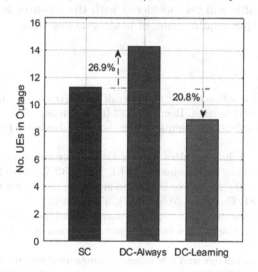

Fig. 4. Comparison of the average number of outage UEs from algorithms based on SC, DC-Always, and the proposed algorithm adopting ML at high traffic loads

Figure 4 shows the performance of the proposed approach adopting distributed reinforcement learning with two reference schemes. While 30 UEs are considered in this simulation, the increase in data traffic from all UE's DC configuration can cause resource shortage which spoils the benefit in reliability. Thus, the 'DC-Always' scheme becomes insuperior to the 'SC' scheme by 26.9%. In the proposed algorithm, 'DC-Learning', while each UE learns the optimal policy in

DC configuration depending on its location and the traffic loads, the algorithm tends to configure DC for UEs effectively. Compared to the 'SC' scheme, the proposed algorithm could achieve the gain of 20.8% in reliability. For larger number of UEs, the gap between the performance of 'SC', 'DC-Always', 'DC-Learning' could become more conspicuous.

5   Conclusion

In this paper, investigation into RATs selection in a multi-RAT network supporting multi-connectivity is provided. Considering different characteristics of UEs, distributed machine learning is applied so that each UE could learn the policy to configure MC and select appropriate RATs. With the simulation results, it is shown that the proposed approach is able to achieve better reliability performance compared to the single connectivity based RAT selection. While UEs could select MC configuration autonomously considering their characteristics and network traffic load, the proposed algorithm is shown to be superior to the mechanism to configure MC for all UEs all the time. In this paper, all UEs are assumed to have the same QoS requirements. In future research, multiple UEs of heterogeneous traffic will be considered with the resource allocation method to satisfy different QoS requirements of heterogeneous UEs.

References

1. ITU-R: IMT Vision - Framework and overall objectives of the future development of IMT for 2020 and beyond. Rec. M. 2083-0, September 2015
2. Clear5G: Deliverable 1.1 system specifications and business perspectives (2018). http://Clear5G.eu/
3. Pocovi, G., et al.: Achieving ultra-reliable low-latency communications: challenges and envisioned system enhancements. IEEE Netw. **32**(2), 8–15 (2018)
4. Pedersen, K.I., et al.: A flexible 5G frame structure design for frequency-division duplex cases. IEEE Commun. Mag. **54**(3), 53–59 (2016)
5. Esswie, A.A., Pedersen, K.I.: Opportunistic spatial preemptive scheduling for URLLC and eMBB coexistence in multi-user 5G networks. IEEE Access **6**, 38451–38463 (2018)
6. Ji, H., et al.: Ultra-reliable and low-latency communications in 5G downlink: physical layer aspects. IEEE Wirel. Commun. **25**(3), 124–130 (2018)
7. Sutton, G.J., et al.: Enabling technologies for ultra-reliable and low latency communications: from PHY and MAC layer perspectives. IEEE Commun. Surv. Tut. (2019)
8. Andrews, J.G., et al.: What will 5G be? IEEE JSAC **32**(6), 1065–1082 (2014)
9. 3GPP: NR; overall description; stage-2. Technical specification 38.300, v15.3 (2018)
10. Rao, J., Vrzic, S.: Packet duplication for URLLC in 5G: architectural enhancements and performance analysis. IEEE Netw. **32**(2), 32–40 (2018)
11. Tesema, F.B., et al.: Evaluation of adaptive active set management for multi-connectivity in intra-frequency 5G networks. In: Proceedings of IEEE Wireless Communication and Networking Conference (2016)

12. Ba, X., et al.: Load-aware cell select scheme for multi-connectivity in intra-frequency 5G ultra dense network. IEEE Commun. Lett. **23**(2), 354–357 (2019)
13. Alexandris, K., et al.: Utility-based resource allocation under multi-connectivity in evolved LTE. In: Proceedings of 86th IEEE Vehicular Technology Conference (2017)
14. Shi, Y., et al.: Dual connectivity enbled user assocation appraoch for max-throughput in the downlink heterogeneous network. Wirel. Pers. Commun. **96**(1), 529–542 (2017)
15. Escudero-Garzás, J.J., et al.: An Analysis of the network selection problem for heterogeneous environments with user-operator joint satisfaction and multi-RAT transmission. Wirel. Commun. Mobile Comput. (2017)
16. Mahmood, N.H., et al.: Reliability oriented dual connectivity for URLLC services in 5G new radio. In: 15th International Symposium on Wireless Communication Systems (2018)
17. 3GPP: Evolved universal terrestrial radio access (E-UTRA) and evolved universal terrestrial radio access network (E-UTRAN); Overall description; Stage 2. TS 36.300, V14.4.0 (2017)
18. 3GPP: Study on communication for automatic in vertical domains. Technical report 22.804 v16.2.0 (Rel 16) (2018)
19. 3GPP: Scenarios, frequencies and new field measurement results from two operational factory halls at 3.5 GHz for various antenna configurations. Nokia, 3GPP TSG RAN WG1 Meeting, R1–1813177 (2018)
20. Kudo, T., Ohtsuki, T.: Cell range expansion using distributed Q-learning in heterogeneous networks. EURASIP J. Wirel. Commun. Netw. **1**(61), 1499–1687 (2013)
21. Sutton, R.S., Barto, A.G.: Reinforcement Learning - An Introduction, 2nd edn. The MIT Press, Cambridge (2017)

Main Track

Novel Filter Bank Based Cooperative Spectrum Sensing Under RF Impairments and Channel Fading Beyond 5G Cognitive Radios

Sener Dikmese[✉], Kishor Lamichhane, and Markku Renfors

Electrical Engineering, Faculty of Information and Communication Sciences,
Tampere University, Tampere, Finland
{sener.dikmese,markku.renfors}@tuni.fi, klamichhane@outlook.com

Abstract. Cognitive radio (CR) technology with dynamic spectrum management capabilities is widely advocated for utilizing effectively the unused spectrum resources. The main idea behind CR technology is to trigger secondary communications to utilize the unused spectral resources. However, CR technology heavily relies on spectrum sensing techniques which are applied to estimate the presence of primary user (PU) signals. This paper mostly focuses on novel analysis filter bank (AFB) based cooperative spectrum sensing (CSS) algorithms to detect the spectral holes in the interesting part of the radio spectrum. To counteract the practical wireless channel effects, collaborative subband based approaches of PU signal sensing are studied. CSS has the capability to eliminate the problems of both hidden nodes and fading multipath channels. FFT and AFB based receiver side sensing methods are applied for OFDM waveform and filter bank based multicarrier (FBMC) waveform, respectively. Subband energies are then applied for enhanced energy detection (ED) based CSS methods. Our special focus is on sensing potential spectral gaps close to relatively strong primary users, considering also the effects of spectral regrowth due to power amplifier nonlinearities. The study shows that AFB based CSS with FBMC waveform improves the performance significantly.

Keywords: Cognitive radio ·
FFT/AFB based cooperative spectrum sensing · Energy detector ·
Frequency selective channel · OFDM and FBMC · PA non-linearity ·
Frequency selectivity

1 Introduction

With the growing attention on wireless communications, radio spectrum scarcity is becoming modern days' challenge. Higher demand of spectral bandwidth is pushing spectrum usage to utmost limits. However, the limitations of traditional

© ICST Institute for Computer Sciences, Social Informatics and Telecommunications Engineering 2019
Published by Springer Nature Switzerland AG 2019. All Rights Reserved
A. Kliks et al. (Eds.): CrownCom 2019, LNICST 291, pp. 45–58, 2019.
https://doi.org/10.1007/978-3-030-25748-4_4

wireless technology lead to spectrum wastage, inviting opportunistic usages of those valuable unused resources [1]. These studies have mainly focused on technologies that solve the problem of spectral scarcity by using opportunistically the frequency band to establish secondary communication. Such technology is commonly known as cognitive radio (CR) technology, which defines new dimension to the modern communication systems advocating environment-adaptive radio transmission [2]. CR keeps track of the radio transmission environment continuously while it dynamically varies its transmission parameters so as to adjust its operation to the surroundings.

Spectrum sensing based CR technology is considered as highly interesting topic in wireless communications. Spectrum sensing, in other words, involves tracking of the PU activity so as to estimate the spectral holes. Different sensing algorithms find the availability of spectral holes as an opportunity to enable the secondary communication [3,4]. Recent studies have suggested a wide variety of spectrum sensing techniques, but none of them is fully satisfying in terms of all relevant metrics like implementation complexity, reliability, and loss in secondary system throughput. Especially, spectrum sensing under low signal-to-noise ratio (SNR) is widely covered in the literature, under conditions where the noise dominates the weak PU signal [3–5]. Under these conditions, the spectrum sensing becomes critically sensitive to imperfect knowledge of the power and characteristics of noise and interferences [6–8].

Spectrum that is originally assigned to the PU can be used by a secondary user (SU) if and only if PU becomes idle. Since SUs can only use spectrum as an opportunity in terms of the spectrum sharing, spectrum sensing has a great role to play in CR technology [3,4,9]. Regarding the importance of the radio scene analysis function, basic spectrum sensing methods show numerous limitations. Shadowing, hidden node problems, etc., always make spectrum sensing challenging. A PU transmission may be unobservable for a CR sensing station while its signal is fully usable by a nearby PU receiver. In order to make the spectrum sensing function reliable, efficient, and to counteract both multipath and hidden node problems, cooperative spectrum sensing (CSS) is considered as a vital solution. CSS involves two or more cooperative radio receivers in decision making during spectrum sensing. Recent research [10] suggests possibility of collaboration among number of CR users to enhance the detection performance. Our studies further exploit the collaborative approach of spectrum sensing commonly termed as CSS. The studies of this paper mainly focus on subband based spectrum sensing methods that add the collaboration among a number of CR receivers to enhance the sensing performance and to counteract practical wireless channel effects. CSS exploits the diversity among a number of CR receivers having different multipath channel profiles and experience different large-scale fading (shadowing) characteristics towards the PU transmissions [11].

More specifically, the contributions of this paper are listed below:

- Conceptually and computationally simplified CSS methods based on subband energies are developed. Subband energies are evaluated either using fast Fourier transform (FFT) or analysis filter bank (AFB).

- Subband ED based CSS methods are applied on the traditional OFDM and on filterbank multicarrier (FBMC) waveforms, the latter one as a candidate beyond-OFDM/beyond-5G scheme.
- The effects of both power amplifier (PA) non-linearities and practical wireless channels on subband ED based CSS are investigated, comparing the performance of FFT and AFB based CSS schemes.

The remainder of the paper is organized as follows: Sect. 2 gives a general idea about traditional CSS methods. Novel FFT and AFB based CSS methods are presented in Sect. 3. Signal models including PA non-linearities, frequency-selective fading, and shadowing are described in Sect. 4. Numerical results for sensing performance are shown in Sect. 5. Finally, closing remarks are given in Sect. 6.

2   Traditional Cooperative Spectrum Sensing

Practical wireless channels show characteristics like noise uncertainty, multipath fading, and shadowing. In order to mitigate the effects of practical wireless channels, cooperation among many CR users, i.e., CSS is considered. CSS is regarded as a potential solution to mitigate effects of both multipath and shadowing which causes the hidden node problem. It also enhances the detection performance and reliability [5].

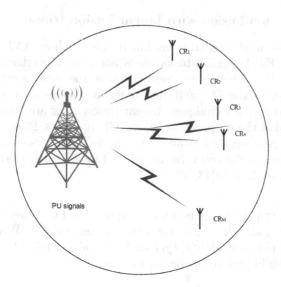

Fig. 1. CR topology including a PU transmitter and M CR stations.

As suggested in the literature [10], the cooperative scheme can help to mitigate effects of the hidden node problem. Spatial diversity among multiple receivers is achieved, as illustrated in Fig. 1. The concept of exploiting SUs' spatial diversity to counteract the hidden node effects and enabling cooperation among SUs is coined as CSS and has reached growing attention in recent years. Our studies consider the ED based CSS approach with hard decision combining. Hard decision combining refers to the combination of binary decisions from spatially separated CR receivers at the fusion center (FC) [12,13].

CSS uses two or more CRs to combine their sensing results so as to increase the reliability of the sensing decision. Combining the results from different CR receivers is performed at the FC. Different rules may be considered for combining the individual sensing results in order to achieve the highest accuracy at the FC, depending on the radio environment. With an increase in the number of CR users, the CSS procedure becomes more complicated. On the other hand, sufficient number of sensing CR stations is needed to reach sufficient sensing performance. So there is a trade-off between sensing performance and complexity of the CSS system. With increased number of CR receivers, the sensing performance enhances significantly, considering performance parameters like sensing time and reliability.

Cooperative schemes can be classified as hard and soft fusion schemes. When CRs provide binary information about the presence of PUs, the FC applies hard fusion rules. If the CRs may provide reliability information about their sensing results in the form of non-binary soft decisions, the FC applies a soft fusion scheme [13].

2.1   Hard Decision Fusion with Linear Fusion Rules

Hard fusion can be implemented using linear rules such as *AND rule, OR rule,* or *Majority rule.* Hard decision fusion does not need to exchange data among secondary nodes. Soft schemes generally improve the sensing result by sending richer information to the FC. Soft fusion techniques increase the complexity compared to hard fusion techniques. Linear fusion rules are commonly applied by the FC to exploit the cooperation among CR receivers. Binary decision from independent CR receivers is forwarded to the FC. FC processes the decisions from all CRs to make the collective decision. Linear fusion rules are based on the general k-out-of-M *rule* [13,14].

Or Rule is one of the fusion rules which is applied at FC. When at least one SU detects the PU signal, *OR rule* declares the presence of PU. With M SUs, the cooperative detection probability $P_{D,t}$ and false alarm probability $P_{FA,t}$ after the decision at the FC are computed as follows:

$$\text{OR-Rule}: \begin{cases} P_{D,t} = 1 - (1 - P_D)^M \\ P_{FA,t} = 1 - (1 - P_{FA})^M \end{cases}. \tag{1}$$

Here P_D and P_{FA} are the detection and false alarm probabilities of individual SUs reported to the FC, which are assumed to be equal for all CR stations.

And Rule declares the presence of a PU signal if and only if all SUs detect the PU signal individually. With M SUs, the cooperative detection probability $P_{D,t}$ and false alarm probability $P_{FA,t}$ at a FC are computed as follows:

$$\text{AND Rule}: \begin{cases} P_{D,t} = P_D{}^M, \\ P_{FA,t} = (P_{FA})^M. \end{cases} \tag{2}$$

Majority Rule: According to various studies, both *AND rule* and *OR rule* are limited in terms of detection and false alarm probabilities. *Majority rule* is another case of the generalized k-out-of-M rule. If at least half of the SUs report the presence of PU, FC declares the presence of PU, otherwise it declares that the spectrum is free to use for CR transmission [11,12]. Considering an even number M of sensing stations in the CSS, the cooperative detection and false alarm probabilities of the *Majority rule* can be written as follows:

$$M/2+1 \text{ -out-of- } M: \begin{cases} P_{D,t} = \sum_{j=M/2+1}^{M} \binom{M}{j} P_D^j \cdot (1 - P_D)^{M-j} \\ P_{FA,t} = \sum_{j=M/2+1}^{M} \binom{M}{j} (1 - P_{FA})^{M-j} \cdot P_{FA}^j. \end{cases} \tag{3}$$

Generalized k-out-of-M Rule: The generalized form of the linear rule can be defined by requiring k SUs out of M to report the presence of a PU signal. Here the number k can take any value between 1 to M, based on the requirements. As discussed earlier, the special cases $k = 1$, $k = M/2+1$, and $k = M$ are equivalent to *OR rule*, *Majority rule*, and *AND rule*, respectively. Nevertheless, the number k can be optimized according to the targeted detection and false alarm performance [12,14]. The cooperative detection and false alarm probabilities of this rule are as follows:

$$K_{of}N: \begin{cases} P_{D,t} = \sum_{j=k}^{M} \binom{M}{j} P_D^j \cdot (1 - P_D)^{M-j} \\ P_{FA,t} = \sum_{j=k}^{M} \binom{M}{j} (1 - P_{FA})^{M-j} \cdot P_{FA}^j. \end{cases} \tag{4}$$

3 Novel FFT and AFB Based Cooperative Spectrum Sensing

FFT and AFB based techniques are applied to a wideband signal to generate equally spaced subband signals. Subband energies are then calculated and the subband energy detector (SED) decides on the presence of PU signal(s) within the processed frequency band based on the subband energies [4,6,8]. The entire procedure is represented in Fig. 2. The receiver front-end collects the PU signals

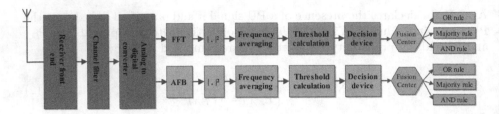

Fig. 2. Block diagram of alternative filter bank (AFB) and fast Fourier transform (FFT) based spectrum analysis methods for subband energy based cooperative sensing schemes.

which is followed by a channel filter and analog-to-digital converter (ADC). The subband signals can be obtained either via FFT or AFB and then processed accordingly [6].

A subband signal can be represented as follows,

$$Y_k[m] = \begin{cases} \mathcal{W}_k[m] & \mathcal{H}_0, \\ S_k[m]H_k + \mathcal{W}_k[m] & \mathcal{H}_1. \end{cases} \tag{5}$$

Here $S_k[m]$ is the transmitted signal by PU as it appears at the mth FFT or AFB output sample in subband k and $\mathcal{W}_k[m]$ is the corresponding channel noise sample. $\mathcal{H}_1$ denotes the *present hypothesis* of a PU signal whereas $\mathcal{H}_0$ denotes the *absent hypothesis* of a PU signal. When the AWGN only is present, the white noise is modeled as a zero-mean Gaussian random variable with variance σ_w^2, i.e., $\mathcal{W}_k[m] = \mathcal{N}\left(0, \sigma_w^2\right)$. The OFDM and FBMC signals can also be modeled in terms of zero-mean Gaussian variables, $S_k[m] = \mathcal{N}(0, \sigma_k^2)$, where σ_k^2 is the variance (power) at subband k. The subband energy is calculated from the subband signals of Eq. (5). The integrated test statistics to be used in the SED process is calculated as

$$T(y_{m0}, k_0) = \frac{1}{N_t N_f} \sum_{k=k_0-[N_f/2]}^{k_0+[N_f/2]-1} \sum_{m=m_0+N_t+1}^{m_0} |y_k[m]|^2. \tag{6}$$

Here N_f and N_t are the averaging window lengths in frequency- and time-domains, respectively. Assuming flat PU spectrum over the sensing band, the probability distribution of the test statistics can be expressed as

$$T(y_{m0}, k_0)|\mathcal{H}_0 \sim \mathcal{N}\left(\sigma_{w,k}^2, \frac{\sigma_{w,k}^4}{N_t N_f}\right) \tag{7}$$

and

$$T(y_{m0}, k_0)|\mathcal{H}_1 \sim \mathcal{N}\left(\sigma_{x,k}^2 + \sigma_{w,k}^2, \frac{\left(\sigma_{x,k}^2 + \sigma_{w,k}^2\right)^2}{N_t N_f}\right). \tag{8}$$

This yields

$$P_{FA} = P_r(T(y) > \lambda | \mathcal{H}_0) = Q\left(\frac{\lambda - \sigma_{w,k}^2}{\sigma_{w,k}^2 / \sqrt{N_f N_t}}\right) \tag{9}$$

and

$$P_D = P_r(T(y) > \lambda | \mathcal{H}_1) = Q\left(\frac{\lambda - \sigma_{w,k}^2(1 + \gamma_k)}{\sigma_{w,k}^2(1 + \gamma_k)/\sqrt{N_f N_t}}\right). \tag{10}$$

Here, $\gamma_k = \sigma_{x,k}^2 / \sigma_{w,k}^2$ is the SNR of subband k and $\sigma_{w,k}^2 = \sigma_w^2 / N_{FFT}$ and $\sigma_{x,k}^2$ denote noise variance and the PU signal variance in subband k, respectively. With FFT/AFB based processing, it is possible to tune the sensing frequency band to the expected bandwidth of the PU signal and sense multiple PU channels simultaneously. For given P_{FA}, the threshold λ can be calculated as

$$\lambda = \sigma_{w,k}^2 \left(1 + \frac{Q^{-1}(P_{FA})}{\sqrt{N_f N_t}}\right). \tag{11}$$

As described earlier in this paper, three linear fusion rules have been proposed for combining the binary decisions received by the FC. Here, three different fusion rules, *OR rule*, *Majority rule* and *AND rule* are considered. The false alarm and detection probabilities with three different linear fusion rules are calculated as follows;

$$P_{FA,t} : \begin{cases} = 1 - (1 - P_{FA})^M & OR \\ = P_{FA}^M & AND \\ = \sum_{j=M/2+1}^{M} \binom{M}{j} P_{FA}^j \cdot (1 - P_{FA})^{M-j} & MAJ. \end{cases} \tag{12}$$

and

$$P_{D,t} : \begin{cases} = 1 - (1 - P_D)^M & OR \\ = P_D^M & AND \\ = \sum_{j=M/2+1}^{M} \binom{M}{j} P_D^j \cdot (1 - P_D)^{M-j} & MAJ. \end{cases} \tag{13}$$

Here $P_{FA,t}$ is cooperative false alarm probability and P_{FA} is non-cooperative false alarm probability. Similarly, $P_{D,t}$ and P_D are cooperative and non-cooperative detection probabilities, respectively.

4   System Model

4.1   Waveforms and Spectrum Sensing Schemes

OFDM with cyclic prefix, i.e., CP-OFDM is the dominating multicarrier technology in the field of wireless communications. Additionally, discrete wavelet multitone (DWMT), cosine modulated multitone (CMT), filtered multitone (FMT), and OFDM with offset-QAM (OFDM/OQAM, also known as FBMC/OQAM) are commonly considered alternative forms of multicarrier techniques [15]. FBMC waveforms, especially FBMC/OQAM have been widely considered as candidates for beyond-OFDM multicarrier systems. It is particularly suitable for dynamic opportunistic spectrum use and CR [16,17]. FBMC/OQAM shows better spectral efficiency compared to CP-OFDM. Such FBMC/OQAM systems utilize a signal model with real valued symbol sequence at twice the QAM symbol rate, instead of complex QAM symbols. Polyphase filter banks in trans-multiplexer configuration constitute the core elements of the transmission link. Specifically, synthesis filter bank (SFB) and AFB are used at the transmitter and receiver sides, respectively [17].

On the receiver side, the FFT of an OFDM receiver or AFB of an FBMC receiver can be used also for spectrum sensing purposes, providing SED capability without additional processing elements. Subband based ED can be used in wideband spectrum sensing which covers multiple PU frequency channels or even the whole service band. For FBMC, the PHYDYAS prototype filter with overlap factor $K = 4$ is used [17]. Such filter bank reaches about 50 dB stopband attenuation, providing efficient detection of narrow spectral gaps between PU channels [15,16].

In the following, we consider two scenarios: (i) Sensing CP-OFDM signal using FFT-based SED and (ii) Sensing FBMC/OQAM signal using AFB-based SED. The two waveforms have the same number of active subcarriers with common subcarrier spacing.

4.2   Power Amplifier Model for PUs

Various interference leakage effects due to RF imperfections affect critically the spectrum sensing performance in practice. The most significant issue in this context is the spectral regrowth due to the nonlinear PA of the PU transmitter. For a practical PA model, we consider the linear time-invariant (LTI) portion of the Wiener PA model, which has a pole/zero form of the system function given by [4]

$$H(z) = \frac{1 + 0.3z^{-2}}{1 - 0.2z^{-1}}. \tag{14}$$

This is extracted from an actual Class AB PA with fifth order nonlinearity. In this study, 5 dB backoff is assumed with this PA model.

The potential spectral hole between two relatively strong PUs is shown in Fig. 3 for both scenarios, as determined by the corresponding sensing process, i.e., FFT for CP-OFDM and AFB for FBMC. While FBMC/OQAM has superior spectral containment, AFB on the receiver side enhances the resolution of spectrum sensing, making it possible to detect potential narrowband PUs within the sensing band.

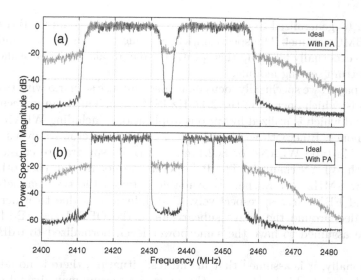

Fig. 3. Effects of the PA model on (a) OFDM and (b) FBMC based PU spectra. A Wiener behavioral model with fifth order nonlinearity and 5 dB backoff is used for the PA.

4.3   Channel Model

This study applies frequency selective multipath channel models together with log-normal shadowing model. All PU and CR channels use *Indoor* and *SUI-1* frequency selective channel models having 90 ns RMS delay spread with 16 taps, and 0.9 μs delay spread with 3 Ricean fading taps, respectively [17,18].

The log-normal path loss can model is as follows:

$$PL = PL_0 * \left(\frac{d_j}{d_0}\right)^a * \varphi, \tag{15}$$

or in dB scale as,

$$PL_{dB} = PL_{0dB} + 10 * a * log(d_j/d_0) + \varphi_{dB}. \tag{16}$$

Here, PL_0 is the path-loss at the reference distance d_0, a represents the path-loss exponent, d_j represents the distance of jth CR receiver, and φ represents the shadow fading with Gaussian distribution, zero mean, and standard deviation σ.

We consider the two scenarios mentioned above, while the CR waveform is always FBMC. For log-normal fading $\sigma = 9$ dB and the path-loss exponent $a = 2$.

5   Numerical Results

In this study, the potential spectral hole between two relatively strong OFDM or FBMC channels is illustrated in Fig. 3. We focus on two cases, one with a gap between two OFDM channels and another one between two FBMC channels. The results can be generalized to cases where the gap is between an OFDM channel and an FBMC channel. The spectrum leakage due to transmitter nonidealities can lead to eventually filling up this spectral gap and raising the false alarm rate of the spectrum sensing module.

In this paper, we specifically focus on a spectrum use scenario with two active PU channels, which operate in the 2.4 GHz ISM band. This is an unlicensed frequency band which is utilized by various applications, including WLAN signals, cordless phones, Bluetooth wireless devices, and even microwave ovens. OFDM based 802.11g-type WLANs, or 802.11g-like FBMC spectra are considered at 3rd and 8th WLAN channels. The PU spectra do not overlap each other, and a 5 MHz or 8 MHz spectral hole is available between the two channels in the OFDM and FBMC cases, respectively. The difference is due to wider guardbands needed around the active subcarriers in the OFDM case. Both active signals are assumed to have the same power level, normalized to 0 dB in our scenario.

Additionally, it is assumed that in the test situation, there is no additional signal in the spectral hole. However, the spectrum sensing may give false alarms due to interference leakage from the PUs due to their non-linearity. Such an effect is found to be dependent on the SNR values of the PUs, as observed at the sensing stations. A smaller subband spacing of 81.5 kHz is used for spectrum sensing and CR transmissions, instead of the 312.5 kHz sub-carrier spacing of WLAN. This improves the spectral resolution of spectrum sensing and CR operation. The frequency window is chosen as $N_f = 5$ to increase the detection performance. Then the effective sensing subband width is 407.5 kHz. With FBMC/OQAM, the 8 MHz spectral gap contains 19 sensing subbands and with CP-OFDM, the 5 MHz gap contains 12 sensing subbands. A SU reports the presence of a PU if the sensing threshold is exceeded in any of the subbands.

The required sample complexity for $P_D = 0.99$ and $P_{FA} = 0.1$ at the target SNR $= -3$ dB is determined with the aid of Eq. (8) in [6] as $N_t = 91$ for non-cooperative spectrum sensing as a reference case. The same sample complexity is considered in our CSS study and desired cooperative $P_{FA,t} = 0.1$ is assumed with all fusion rules. The PA non-linearity introduces interference leakage to the spectrum gap between the PUs, as illustrated in Fig. 3, and the width of the spectral hole is reduced.

Results from ideal and practical PA cases using the *Indoor* frequency selective channel with log normal path loss model are shown in Figs. 4 and 5, respectively. The figures show the cooperative false alarm probability as a function of the average SNR of the PUs as observed at the sensing stations. The average SNR is assumed to be the same for all sensing stations and for both adjacent PUs. FBMC based WLAN-like signal model shows much-improved spectral containment compared to OFDM based WLAN under both the ideal and practical PA cases. Furthermore, AFB-based SED shows significant enhancement over the FFT-based one. Looking at the cooperative false alarm probability close to the target level of 0.1, the *OR rule* makes the CSS process least sensitive to interference leakage from the adjacent relatively strong PUs, while the *Majority rule*

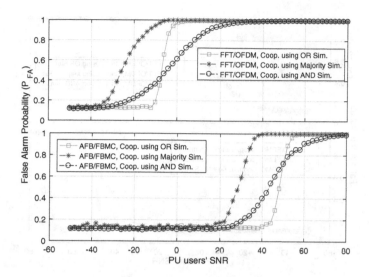

Fig. 4. Comparison of cooperative false alarm probabilities between FFT-based SED for CP-OFDM PU (upper) and AFB-based SED for FBMC/OQAM PU (lower). Three different linear fusion rules are applied under *Indoor* channel & log-normal fading, linear PU transmitter, time record length of $N_t = 91$, and $M = 8$ sensing stations.

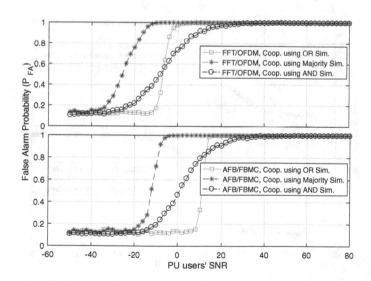

Fig. 5. Comparison of cooperative false alarm probabilities between FFT-based SED for CP-OFDM PU (upper) and AFB-based SED for FBMC/OQAM PU (lower). Three different linear fusion rules are applied under *Indoor* channel & log-normal fading, PU non-linearity effects, time record length of $N_t = 91$, and $M = 8$ sensing stations.

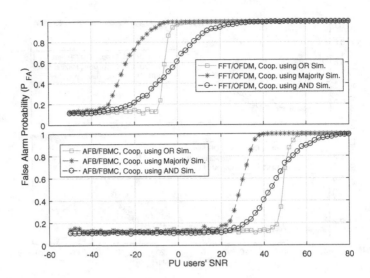

Fig. 6. Comparison of cooperative false alarm probabilities between FFT-based SED for CP-OFDM PU (upper) and AFB-based SED for FBMC/OQAM PU (lower). Three different linear fusion rules are applied under *SUI-1* channel & log-normal fading, linear PU transmitter, time record length of $N_t = 91$, and $M = 8$ sensing stations.

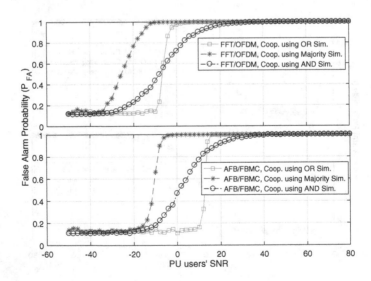

Fig. 7. Comparison of cooperative false alarm probabilities between FFT-based SED for CP-OFDM PU (upper) and AFB-based SED for FBMC/OQAM PU (lower). Three different linear fusion rules are applied under *SUI-1* channel & log-normal fading, PU non-linearity effects, time record length of $N_t = 91$, and $M = 8$ sensing stations.

shows highest sensitivity. This is true for both scenarios and both PA models. FBMC/OQAM with AFB based SED is quite robust towards the interference leakage with linear (or well linerised) PA. Also with the used practical PA model, the benefit over CP-OFDM with FFT-based SED is quite significant, allowing about 20 dB stronger PUs at the adjacent frequencies.

Corresponding results for the *SUI-I* channel are shown in Figs. 6 and 7. The results and conclusions are quite similar with the *Indoor* channel case.

6 Conclusion

In this paper, enhanced energy detection based cooperative spectrum sensing was studied. The proposed subband-based scheme allows to effectively detect potential PU signals with widely varying bandwidths in possible spectral gaps close to strong PU channels. In such scenarios, the sensing task becomes difficult due to interference leakage from the relatively strong adjacent channels. The results demonstrated highly reduced sensitivity for filter bank based waveforms and sensing schemes towards such interference leakage, compared to basic CP-OFDM and FFT-based sensing. As for the CSS schemes, it was found that the *OR rule* is clearly least sensitive to the interference leakage, compared to the *AND rule* and *Majority rule*.

In addition to FBMC, these result are generally applicable to all advanced PU waveforms with improved spectrum containment, including filtered OFDM schemes which have received wide interest in the 5G-NR context.

This study does not cover soft decision based CSS, which remains as an important topic for future work. Moreover, analytical derivations for different subband based CSS schemes, including the interference leakage effects, is a rather complicated issue, and is left as a topic for future studies.

Acknowledgment. This work was supported in part by the Finnish Cultural Foundation.

References

1. Federal Communications Commission: Spectrum policy task force. Report ET Docket no. 02–135, November 2002
2. McHenry, M.: Frequency agile spectrum access technologies. In: FCC Workshop on Cognitive Radio, Washington US, May 2003
3. Dikmese, S., Renfors, M., Dincer, H.: FFT and filter bank based spectrum sensing for WLAN signals. In: Proceedings ECCTD 2011 Conference, Linkoping, Sweden, August 2011
4. Dikmese, S., Srinivasan, S., Shaat, M., Bader, F., Renfors, M.: Spectrum sensing and spectrum allocation for multicarrier cognitive radios under interference and power constraints. EURASIP J. Adv. Signal Proc. **2014**, 68 (2014)
5. Kandeepan, S., Giorgetti, A.: Spectrum sensing in cognitive radio. In: Cognitive radios and enabling techniques. Artech House Publishers, Boston (2012)

6. Dikmese, S., Sofotasios, P.C., Ihalainen, T., Renfors, M., Valkama, M.: Efficient energy detection methods for spectrum sensing under non-flat spectral character-istics. IEEE J. Selec. Areas Commun. **33**(5), 755–770 (2015)

7. Dikmese, S., Sofotasios, P.C., Renfors, M., Valkama, M.: Maximum-minimum energy based spectrum sensing under frequency selectivity for cognitive radios. In: Proceedings of CROWNCOM 2014 Conference, Oulu, Finland, June 2014

8. Dikmese, S., Sofotasios, P.C., Renfors, M., Valkama, M.: Subband energy based reduced complexity spectrum sensing under noise uncertainty and frequency-selective spectral characteristics. IEEE Trans. Signal Process. **64**(1), 131–145 (2016)

9. Dikmese, S., Ilyas, Z., Sofotasios, P.C., Renfors, M., Valkama, M.: Sparse frequency domain spectrum sensing and sharing based on cyclic prefix autocorrelation. IEEE J. Sel. Areas Commun. **35**(1), 159–172 (2017)

10. Ganesan, G., Li, Y.: Cooperative spectrum sensing in cognitive radio, part I: two user networks. IEEE Trans. Wirel. Commun. **6**(6), 2204–2212 (2007)

11. You, C., Kwon, H., Heo, J.: Cooperative TV spectrum sensing in cognitive radio for Wi-Fi networks. IEEE Trans. Consum. Electron. **57**(1), 62–67 (2011)

12. Atapattu, S., Tellambura, C., Jiang, H.: Energy detection based cooperative spectrum sensing in cognitive radio networks. IEEE Tran. Wirel. Commun. **10**(4), 1232–1241 (2011)

13. Hossain, M.S., Abdullah, M.I.: Hard decision based cooperative spectrum sensing over different fading channel in cognitive radio. Int. J. Econ. Manag. Sci. **1**(1), 84–93 (2012)

14. Althunibat, S., Di Renzo, M., Granelli, F.: Optimizing the K-out-of-N rule for cooperative spectrum sensing in cognitive radio networks. In: IEEE GLOBECOM 2013, pp. 1607–1611, December 2013

15. Bansal, G., Hossain, J., Bhargava, V.K.: Adaptive power loading for OFDM-Based cognitive radio systems with statistical interference constraint. IEEE Trans. Wirel. Commun. **10**(9), 2786–2791 (2011)

16. Cui, Y., Zhao, Z., Zhang, H.: An efficient filter banks based multicarrier system in cognitive radio networks. Radioengineering **19**(4), 479–487 (2010)

17. Dikmese, S.: Enhanced spectrum sensing techniques for cognitive radio. Ph.D. Dissertation, Tampere University of Technology, Tampere, Finland, March 2015

18. Jain, R.: Channel models: a tutorial. In: WiMAX Forum AATG, pp. 1–6 (2007)

Performance Evaluation of Windowing Based Energy Detector in Multipath and Multi-signal Scenarios

Johanna Vartiainen[1]([message]) , Heikki Karvonen[1] , Marja Matinmikko-Blue[1] , and Luciano Mendes[2]

[1] Centre for Wireless Communications, University of Oulu, Oulu, Finland
{johanna.vartiainen,heikki.karvonen,Marja.Matinmikko}@oulu.fi
[2] Radiocommunications Research Center, Inatel, Santa Rita do Sapucaí, Brazil
luciano@inatel.br

Abstract. Connectivity in remote areas continues to be a major challenge despite of the evolution of cellular technology. 5th Generation (5G) technology can address remote connectivity if lower carrier frequencies are available, which calls for shared use of spectrum to enable cost-efficient license-free solution. Therefore, spectrum sensing has its own role in future wireless systems such as mobile 5G networks and Internet of Things (IoT) to complement database approach in dynamic spectrum utilization. In this paper, a windowing based (WIBA) blind spectrum sensing method is studied. Its performance is compared to the localization algorithm based on double-thresholding (LAD) detection method. Both the methods are based on energy detection and can be used in any frequency range as well as for detecting all kind of relatively narrowband signals. Probability of detection, relative mean square error for the bandwidth estimation, and the number of detected signals were evaluated, including multipath and multi-signal scenarios. The simulation results show that the WIBA method is very suitable for future 5G applications especially for remote area connectivity, due to its good detection performance in low signal-to-noise ratio (SNR) areas with low complexity and reasonable costs. The simulation results also show importance of the used detection window selection since too wide detection window degrades the detection performance of the WIBA method.

Keywords: Signal detection · Spectrum utilization · 5G system · Overlapping · Sampling

1 Introduction

5th Generation (5G) technology can be considered to be an extensive revolution of the mobile communication systems that brings a whole new era to connectivity. Near-future 5G system brings spectrum efficiency, scalability, intelligence,

© ICST Institute for Computer Sciences, Social Informatics and Telecommunications Engineering 2019
Published by Springer Nature Switzerland AG 2019. All Rights Reserved
A. Kliks et al. (Eds.): CrownCom 2019, LNICST 291, pp. 59–72, 2019.
https://doi.org/10.1007/978-3-030-25748-4_5

low latency and advanced security features. It enables Internet of Things (IoT) [1] which connects massive number of objects like computers, services and devices like sensors and mobile phones together. At the same time, it has potential to serve currently underserved remote areas. As the amount of disposable communication radio channels is limited, the effectiveness of the use of radio spectrum must be optimized. 5G system includes the use of higher frequencies (e.g. 3.5 and 28 GHz) as well as aggregated use of licensed and unlicensed bands. Traditionally, in cellular systems wireless spectrum is made available through an inflexible spectrum allocation, where frequency bands are permanently allocated to some licensed (primary) users (PU). 5G-RANGE project [13] proposes that a cost-efficient remote and rural area connection can be enabled by using 5G cognitive radio (CR) networks where unlicensed (secondary) users (SU) are able to use temporarily unused frequency bands (i.e. frequency holes) in licensed frequency channels leading to increased spectrum efficiency. The main requirement for shared frequency use is that the licensed user must not be interfered by SU(s). Two high-level approaches are used for that purpose: database and spectrum sensing. As databases collect and store information about licensed users such as TV and program making and special events (PMSE) signals (e.g. wireless microphone signals) in some geographical area, spectrum sensing can be used to find out (detect) which frequency bands are available for a transmission by observing the radio environment.

In 5G scenarios, spectrum sensing can be used to enhance the traditional database approach by bringing more accurate information about the actual spectrum usage and thus increase the potential and reliability of shared spectrum access. 5G communication application areas for spectrum sensing include, e.g., mobile cellular systems [2], device-to-device (D2D) communication [3], and IoT [4]. Sensing can be used when the information in database or from geolocation method (like GPS) is inaccurate, or there is no connection to the database at all, like in disaster-related events or in remote areas. In addition, secondary users (SU) may use sensing when defining are there other SUs present [5]. 5G can be tailored to be used for remote area connectivity where the use of TV white spaces (TVWS), i.e., Very High Frequency (VHF) and Ultra High Frequency (UHF) bands, with database can be enhanced with spectrum sensing. In rural and remote areas the challenge is that distances are long and, thus, signal-to-noise ratio (SNR) levels are low.

As 5G systems will include a huge number of devices, especially in IoT scenarios, design complexity and costs should stay in a reasonable level. Energy detection (ED) is a cost-efficient sensing technique that is recommended to be used especially in cooperative sensing, where users collaborate and exchange their sensing information [6]. 5G cooperative sensing based on ED methods has been studied, for example, in [7]. The problem is that conventional ED does not perform well at low SNR values.

In this paper, a novel ED-based spectrum sensing method is studied, namely the windowing based (WIBA) signal detection method [8]. This recently proposed method is an efficient blind spectrum sensing method that is able to iteratively estimate the noise level by using adaptive thresholds. The WIBA method uses overlapping blocks in spectrum sampling to increase its detection performance. In [8], where the WIBA method was proposed, only the probability of detection and the number of detected signals in one-signal case were studied. In this paper, the method is studied more comprehensively. The effect of the detection window length M to the detection performance in different channel situations is studied. Relative mean squared error (RMSE) for the bandwidth estimation is considered, as well as detection probability over multipath channels. In addition, multi-signal situation is considered. The results are compared with the ones achieved by the widely studied localization algorithm based on double-thresholding (LAD) method [9], which has been found to outperform conventional ED methods [10].

2   System Model

Connectivity in rural and remote areas is a true challenge because most of today's technologies aim for coverage below 10 km radius. In a sparsely populated area, a 10 km cell will only cover a small number of subscribers, resulting in very high fees per user. Another problem for realizing remote connectivity is the high cost of the spectrum licenses, which increases the investments to deploy a mobile network and hinders its economic feasibility. 5G in remote areas requires the use of lower frequency bands to reach wider area coverage, e.g., 50–100 km. Therefore, the upcoming 5G millimeterwave bands are not the first target for remote area connectivity. Instead, the use of so called TV white spaces has the potential to be used for 5G networks, to provide cost-efficient solution in remote areas, after its main boom over a decade ago. Administrations have developed rules for the use of TVWS and selected geolocation database approach as the means to protect the incumbent TV broadcasting usage, see e.g. [11]. In these approaches, devices wishing to access the TVWS need to inquire a database and report their location to be allowed to use a channel such that the incumbents are protected. The Federal Communications Commission (FCC) defined Citizen Broadband Radio Service (CBRS) for shared commercial use of the 3.5 GHz band with the incumbent military radars and fixed satellite stations [12]. CBRS system includes the use of spectrum sensing in conjunction with database to avoid unlicensed users interference to military radar systems. The use of spectrum sensing to complement database approach in TVWS has been studied to some extent but has not been adopted by other system so far.

The opportunistic use of the TVWS demands protection of the incumbents. While several standards that employ cognitive radio approaches rely on geolocation database to inform the base station (BS) about the spectrum opportunities in a given region, the spectrum sensing can be used in conjunction with the database approach to enhance the reliability and increase shared spectrum access opportunities. Database information may be inaccurate due to software based propagation estimation which can lead to erroneous results in varying terrain shapes that are present in remote area scenarios. Spectrum sensing will be used also to detect other SUs at the same region. In addition, there are situations where the use of spectrum sensing can provide benefits such as in the presence of unauthorized transmissions (e.g., pirate TV transmissions). Figure 1 summarizes the high-level system model for combined spectrum sensing and database approach. In the 5G-RANGE project [13], this approach is proposed to be a feasible solutions for remote area system which targets to dynamically exploit free spectrum holes available at TV bands. Next, we introduce the developed spectrum sensing method, which is considered as a one feasible option for 5G-RANGE system, in detail.

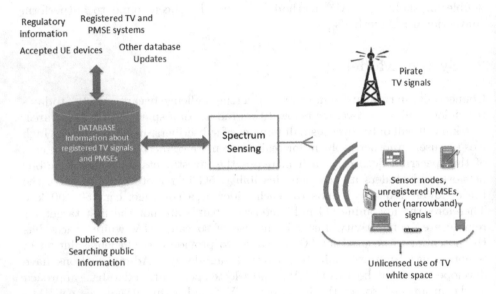

Fig. 1. System model for spectrum sensing to complement database approach.

3   Spectrum Sensing

Here will be described the WIBA method which is considered to be used for spectrum sensing in 5G-RANGE system. In the performance evaluation, a well-known LAD method is used as a point of comparison and will be introduced shortly in this section.

Both the methods are blind spectrum sensing algorithms that are able to iteratively estimate the noise level by using adaptive thresholds. They can be applied to a wide set of situations. The signals to be detected must be narrowband with respect to the analyzed bandwidth (BW). The narrower the signal, the better the method perform, hence it is reasonable to make an assumption that the BW has to be at most 50% of the analyzed BW [8,10]. According to [10], as the signal's BW gets wider, SNR must be higher in order to achieve an acceptable sensing performance. Note, that the methods can be used in any frequency band (kHz–GHz).

The signal detection is based on the estimated noise level, therefore information about the noise level or present signal(s) are not needed. The noise is assumed to be a white Gaussian process. Even though the assumption is that the noise is Gaussian, it has been shown that the signal can be found even if the noise is not purely Gaussian [10]. A detection threshold is used to divide received samples into two sets: one set contains estimated noise-only samples, and another set contains estimated signal samples and noise. Threshold selection is addressed by the constant false alarm rate (CFAR) principle, which means that the used detection threshold parameter is calculated *a priori* using a pre-selected desired false alarm rate P_{FA} and the statistical properties of the noise [14,15].

In this paper, it is assumed that the samples x_i, taken in the frequency-domain are zero mean, independent Gaussian distributed (i.i.d.) complex random variables. The energy of sample x_i is $y_i = |x_i|^2$, which follows a chi-squared distribution. By assuming a chi-squared distributed variables with $2M$ degrees of freedom, the threshold parameter T can be found by solving [16–18]

$$P_{FA} = e^{-TM} \sum_{k=0}^{M-1} \frac{1}{k!} (TM)^k, \tag{1}$$

where P_{FA} is the pre-selected false alarm rate. Note that (1) does not depend on the noise variance. When $M = 1$, variables follow chi-squared distribution with two degrees of freedom, and 1 leads to a threshold parameter

$$T = -\ln(P_{FA}). \tag{2}$$

Example threshold parameter values T for different values of P_{FA} and M are presented in Table 1. For example, when $M = 1$ and $P_{FA} = 0.01$, then $T = 4.605$. Note that the threshold parameter is constant for specific M and P_{FA}, and can be calculated beforehand.

Table 1. Threshold parameter values T for different P_{FA} and M values.

P_{FA}	$M = 1$	$M = 4$	$M = 10$	$M = 100$
0.1	2.303	1.670	1.512	1.130
0.01	4.605	2.511	1.878	1.247
0.001	6.908	3.266	2.266	1.338

3.1 WIBA Method

In the WIBA method, overlapping is used in spectrum sampling. Assume that N energy samples y are obtained during the channel sensing. The observed samples are divided into L overlapping blocks (i.e. detection windows) with length M.

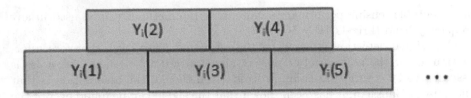

Fig. 2. Illustration of 50% overlapping when there are L overlapping blocks and the length of one block is M.

An example case, where the degree of overlapping between two blocks is 50%, is illustrated in Fig. 2. Samples in each block are summed up among themselves, so each block $Y_i(l)$, $l = 1, \cdots, M$ consists of samples $\frac{kM}{2} + 1, \cdots, \frac{kM}{2} + M$, $k = 0, \cdots, L - 1$. The signal detection threshold is [8]

$$T_h = T \frac{1}{L} \sum_{i=1}^{L} Z_i, \qquad (3)$$

where T comes from (1) and Z_i is the total energy in ith block, i.e., $Z_i = \sum_{l=1}^{M} Y_i(l)$ when $i = 1, 2, ..., L$.

3.2   The LAD Method

The LAD method [9,10] utilizes iterative forward consecutive mean excision (FCME) threshold setting process. Therein, the threshold is $T_h = T\bar{y}$, where threshold parameter T comes from (2) and $\bar{y}$ is the mean of energy samples. Threshold setting procedure is described more detailed, e.g., in [17]. After calculating two FCME thresholds, the upper and lower ones, using two different threshold parameters, the LAD method uses clustering to group adjacent samples assumed to be from the same signal. The LAD method clusters together adjacent samples above the lower threshold. The cluster is accepted to be caused by a signal if at least one of the samples is also above the upper threshold. The performance of the LAD method can be improved using an ACC parameter that allows p (usually $p = 3$) samples to be below the lower threshold between two accepted clusters [10].

4   Simulation Results

In the computer simulations, the WIBA method was studied and compared to the well-studied LAD method which has been found to outperform general ED methods [9,10,17]. In this work the effect of the detection window length M to the detection performance in different channel situations was studied. RMSE for the bandwidth estimation was evaluated, as well as detection probability over multipath channels in multi-signal situations. It was assumed an AWGN

channel and the measured signal, occupying 5–30% of the channel BW, was based
on BPSK modulation. The BPSK signal was band-limited by a RC filter with
a roll-of factor of 0.22. The number of frequency domain samples $N = 1024$.
SNR was defined as a total signal power per total noise power, i.e., over N
samples. The probability of detection P_d was defined so that the signal is defined
to be detected if threshold is crossed at its center frequency. The amount of
Monte Carlo iterations were 1000. The WIBA method used $P_{FA} = 0.01$, 50%
overlapping, M varied, and $L \approx 2\frac{N}{M}$. The used threshold parameter T depend on
M as shown in Table 1. Detection window length M was defined to be optimal
when it equals to the signal bandwidth. Table 2 shows optimal detection window
lengths M for signals with different bandwidths. For example, window length
$M = 52$ samples is optimal for signal with 5% BW (= 52 samples). The LAD
threshold parameters were 13.81 ($P_{FA} = 10^{-6}$) and 2.66 ($P_{FA} = 0.07$) [10], and
$M = 1$ (= no windowing). An adjacent version of the LAD method with ACC
parameter $p = 3$ was used.

Table 2. Optimal detection window lengths M for signals with different bandwidths
(samples/%).

Detection window length M	Signal BW samples/%
10 samples	BW 10 samples/1%
40 samples	BW 40 samples/4%
52 samples	BW 52 samples/5%
102 samples	BW 102 samples/10%
204 samples	BW 204 samples/20%
306 samples	BW 306 samples/30%

4.1   One Signal Scenario

In [8], an initial performance evaluation of WIBA was done by studying the
probability of detection and the number of detected signals in one-signal case.
Based on those results it was concluded that a very long window is preferred
instead of the very short one when considering performance in terms of P_d.

In this paper, BW estimation accuracy is studied. Relative mean square error
(or root mean squared relative error, RMSRE) of BW estimation is defined to be

$$RMSE_\gamma = \sqrt{\frac{1}{N} \sum_{i=1}^{N} \left(\frac{\gamma_i - \hat{\gamma}_i}{\gamma_i} \right)^2}, \tag{4}$$

where γ_i is the BW and $\hat{\gamma}_i$ is the estimated BW.

Table 3 shows the results when there is one signal with 10, 20 or 30% BW,
and $M = 52, 102, 204$ and 306. Results for optimal window lengths are in bold.
For example, when the signal BW is 10% and $M = 102$, RMSE is 100% for

Table 3. Relative Mean Square Error (RMSE) [%] in the one signal scenario for 10, 20, and 30% bandwidth when $M = 52, 102, 204$ and 306.

BW % (samples)	WIBA, $M =$				LAD ACC
	52	102	204	306	
BW 10% (102)	58	**100**	300	500	8
BW 20% (204)	15	50	**100**	198	6
BW 30% (306)	7	1.5	33	**100**	13

WIBA method. On the other hand, RMSE for LAD ACC method is only 8%. It can be noticed that using WIBA method, too long window degrades the BW estimation accuracy because in that case, the detected signal does not cover the whole window.

In Fig. 3, RMSE vs. SNR is presented for a signal occupying 10% of the overall BW (corresponding to the first line in Table 3). Figure 3 also shows at which SNR values each method achieve $P_d = 0.9$. Note that the WIBA method has $P_d = 0.9$ when -13 dB $\leq$ SNR ≤ -11 dB, depending on the M, while the LAD ACC method achieves $P_d = 0.9$ when SNR $= 5$ dB. That is, the performance difference is 16–18 dB. Because the WIBA method is able to operate in low SNR region (SNR < -10 dB), it is feasible for remote area scenarios, where long distance propagation makes received signal's strength weak. However, the LAD method has better BW estimation accuracy. It can be seen that, for the WIBA method, RMSE rises with the SNR when M is large. This is because the fact that as the detection performance of the LAD method depends on the bandwidth of the detected signal, the detection performance of the WIBA method depends also on the length of the used detection window.

4.2   Multi-signal Situation

In this scenario, it is assumed that two RC-BPSK signals are present in the channel. The results are presented in Table 4, considering that there are one or two signals occupying 10% and 5% of the channel BW, respectively. For example, when $M = 102$ and there are two signals with BWs corresponding to 10% and 5%, the performance of the WIBA method is at most 1 dB worse when compared to the one signal scenario. Optimal values for M are 102 for 10% BW signal and 52 for 5% BW signal. Note that M does not effect the LAD ACC performance because there is no windowing. Based on Table 4, multi-signal situation has only slight effect to the performance of the methods.

In Fig. 4, the number of detected signals vs. SNR is presented. There are two signals with 5% and 10% BWs, and $M = 10, 40, 52, 102$ and 204. This figure also shows at which SNR each approach achieve $P_d = 0.9$. For example, when $M = 52$, $P_d = 0.9$ when SNR $= -12$ dB. The window is very short when $M = 10$ and $M = 40$. Optimal window lengths are $M = 52$ for 5% BW signal and $M = 102$ for 10% BW signal. When $M = 40, 52$ and 102, the WIBA method

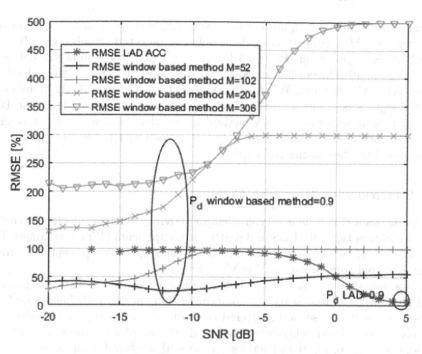

Fig. 3. RMSE vs. SNR results for the case when bandwidth of the signal is 10%.

Table 4. Required SNR [dB] for $P_d = 0.9$ when there is one or two signals present.

Window length M	# of signals	Signal BW	WIBA method $P_d = 0.9$	LAD ACC method $P_d = 0.9$
$M = 102$	Two	10%	-13 dB	3 dB
		5%	-13 dB	-1 dB
$M = 102$	One	10%	-13 dB	1 dB
		5%	-14 dB	-2 dB
$M = 52$	Two	10%	-12 dB	3 dB
		5%	-14 dB	-1 dB
$M = 40$	Two	10%	-11 dB	3 dB
		5%	-14 dB	-1 dB
$M = 10$	Two	10%	-5 dB	3 dB
		5%	-10 dB	-1 dB
$M = 10$	One	10%	-5 dB	1 dB
		5%	-11 dB	-2 dB

estimated the number of signals correctly when $P_d = 0.9$. It can be seen that too short window ($M = 10$) estimates the number of detected signals correctly only when SNR is larger: when $P_d = 0.9$ (SNR = −5 dB), the number of detected signals is 2.7, and achieves 2 when SNR = 1 dB. This corresponds the behaviour of the LAD ACC method. When using the LAD ACC method, the number of detected signals is about 2.2 at its best. As can be seen from Fig. 5, the BW estimation accuracy of the WIBA method may suffer if the window is too wide ($M = 204$, for instance). Large M means that closely spaced signals can be seen as one signal by the sensing technique.

4.3   Scenario with Multipath Channel

Multipath channel can be a very challenging environment for spectrum sensing since it includes typically LOS and scattered components (Rician channel). Let a_i, $i = 1, \cdots, K$ be the average amplitude of each signal component. The total energy of signal components is $E = \sum_{i=1}^{K} a_i^2$.

In the simulations, there were LOS component and two scattered components ($K = 3$). The first scattered component had energy 3 dB below the LOS component, while the second scattered component had energy 6 dB below the LOS component. Used delays were 2, 20 and 100 samples for the first scattered component, and 10, 40, 70 and 100 for the second scattered component.

In Fig. 6, detection probability vs. SNR in multipath channel case is considered. Signal BW is 10%, $M = 102$ (optimal), and there are two multipath components with different delays in samples. It can be seen that the multipath enhances the detection performance by 1–2 dB, regardless of the sample delays. This is because constructive summation increases the energy of the signal, and this affects the detection when using ED based methods. Here, SNR is defined to include only LOS energy. If SNR includes energy of LOS and scattered components, the performance is 1–2 dB worse, and the performance equals to the non-multipath performance.

Next, the bandwidth estimation accuracy is studied. In Fig. 7, RMSE vs. SNR is presented in the presence of multipath. Here, signal BW is 10% of the channel bandwidth and $M = 102$ (optimal). This figure also shows the minimum SNR values when the $P_d \geq 0.9$ is achieved. For example, when there is no multipath and the WIBA method is used, a SNR = −13 dB is required to achieve $P_d = 0.9$. As a comparison, the LAD ACC method requires SNR = 1 dB to achieve $P_d = 0.9$. The difference between the WIBA and the LAD ACC methods is 14 dB. However, it can be noticed that the LAD method has better BW estimation accuracy. The multipath has about 1–3 dB effect to the RMSE performance.

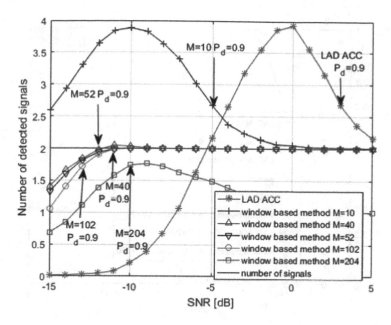

Fig. 4. Number of detected signals vs. SNR results. There are two signals with 10% and 5% bandwidths to be detected.

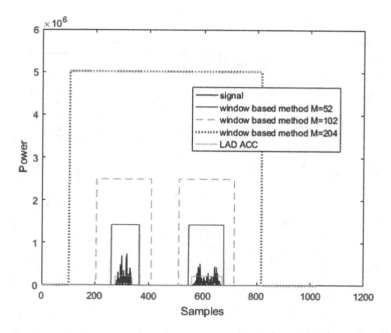

Fig. 5. One snapshot of two simulated signals with 5% and 10% bandwidth. $M = 52$ (optimal for 5% BW signal), 102 (optimal for 10% BW signal), and 204.

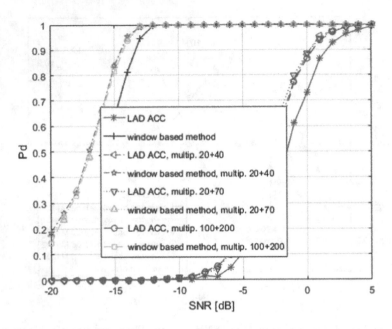

Fig. 6. Probability of detection vs. SNR in multipath channel case. Signal bandwidth is 10% and $M = 102$. SNR is calculated for LOS component.

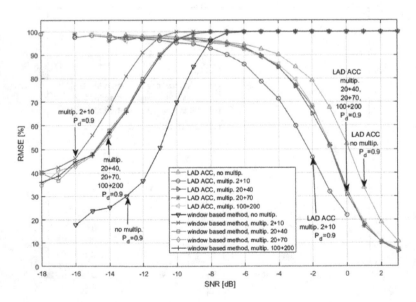

Fig. 7. RMSE vs. SNR results in the presence of multipath components. BW of the signal is 10% and $M = 102$.

5 Conclusions

Remote area connectivity problem can be solved by using lower frequencies and making shared license-free spectrum access possible to enable cost-efficient solution for low user density areas. Traditional database approach can be enhanced by including spectrum sensing to more accurately characterize the current spectrum usage in order to identify more opportunities for shared spectrum access. In this work, the performance of a spectrum windowing based energy detection method WIBA was studied, and comparison was made with the well-studied LAD ACC method. Probability of detection, relative mean square error for the bandwidth estimation, and the number of detected signals were evaluated. From the simulations results, one can conclude that the WIBA method has better detection probability than the LAD ACC method. The WIBA method is able to operate with SNR below −10 dB, depending on the signal and window lengths. The WIBA method is suitable for 5G applications especially for rural and remote areas due to its good detection performance in low SNR areas. The effect of the detection window length to the detection performance in different channel situations was also studied. Too long detection window degrades the performance of the WIBA method. The LAD ACC method outperforms the WIBA method in terms of bandwidth estimation accuracy. Therefore it can be concluded that if signal detection at a given frequency band is enough for the system, WIBA method is preferred. If BW estimation accuracy is important, LAD ACC could be used after WIBA method to improve the BW estimation.

Acknowledgment. This research has received funding from the European Union Horizon 2020 Programme (H2020/2017–2019) under grant agreement N0. 777137 and from the Ministry of Science, Technology and Innovation of Brazil through Rede Nacional de Ensino e Pesquisa (RNP) under the 4th EU-BR Coordinated Call Information and Communication Technologies through 5G-RANGE project. In addition, this research has been financially supported in part by Academy of Finland 6Genesis Flagship (grant 318927) and CNPq-Brasil.

References

1. Bockelmann C., et al.: Towards massive connectivity support for scalable mMTC communications in 5G networks. In: IEEE Access, vol. 6 (2018)
2. Li, B., Li, S., Nallanathan, A., Zhao, C.: Deep sensing for future spectrum and location awareness 5G communications. IEEE J. Sel. Areas Commun. **33**(7), 1331–1344 (2015)
3. Atat, R., Liu, L., Chen, H., Wu, J., Li, H., Yi, Y.: Enabling cyber-physical communication in 5G cellular networks: challenges, spatial spectrum sensing, and cybersecurity. IET Cyber-Phys. Syst. Theory Appl. **2**(1), 49–54 (2017)
4. Ejaz, W., Ibnkahla, M.: Multiband spectrum sensing and resource allocation for IoT in cognitive 5G networks. IEEE Internet Things J. **5**(1), 150–163 (2018)
5. Vartiainen, J., Hoyhtya, M., Vuohtoniemi, R., Ramani, V.V.: The future of spectrum sensing. In: International Conference on Ubiquitous and Future Networks, ICUFN 2016, pp. 247–252 (2016)

6. Akyildiz, I.F., Lo, B.F., Balakrishnan, R.: Cooperative spectrum sensing in cognitive radio networks: a survey. Phys. Commun. 4(1), 40–62 (2011)
7. Liu, X., He, D., Jia, M.: 5G-based wideband cognitive radio system design with cooperative spectrum sensing. Phys. Commun. 25(2), 539–545 (2017)
8. Saarnisaari, H., Vartiainen, J.: Spectrum window based signal detection at low SNR. In: International Conference on Military Communications and Information Systems (ICMCIS), Poland (2018)
9. Vartiainen, J., Lehtomaki, J.J., Saarnisaari, H.: Double-threshold based narrowband signal extraction. In: IEEE Vehicular Technology Conference (VTC), pp. 1288–1292 (2005)
10. Vartiainen, J.: Concentrated signal extraction using consecutive mean excision algorithms. Ph.D. Dissertation, Acta Univ Oul Technica C 368. Faculty of Technology, University of Oulu, Finland (2010)
11. ECC: Technical and operational requirements for the possible operation of cognitive radio systems in the white spaces of the frequency band 470–790 MHz. In: ECC Report 159 (2011)
12. FCC: Amendment of the Commission's Rules with Regard to Commercial Operations in the 3550–3650 MHz Band. In: Report and Order, FNPRM, FCC-15-47 (2015)
13. 5G-RANGE Project. http://5g-range.eu/. Accessed 1 Mar 2019
14. Miller, K.S.: Complex Gaussian process. SIAM Rev. 11, 544–567 (1969)
15. Neeser, F.D., Massey, J.-L.: Proper complex random processes with applications to information theory. IEEE Trans. Inf. Theory 39, 1293–1302 (1993)
16. Lehtomaki, J.J., Vartiainen, J., Juntti, M., Saarnisaari, H.: CFAR outlier detection with forward methods. IEEE Trans. Sig. Process. 55, 4702–4706 (2003)
17. Saarnisaari H., Henttu P.: Impulse detection and rejection methods for radio systems. In: Military Communications Conference (MILCOM), pp. 1126–1131 (2003)
18. Proakis, J.G.: Digital Communications, 3rd edn. McGraw-Hill Inc., New York (1995)

Transmitter Classification with Supervised Deep Learning

Cyrille Morin[1]([⊠]) (iD), Leonardo S. Cardoso[2,3] (iD), Jakob Hoydis[2,3] (iD),
Jean-Marie Gorce[2,3] (iD), and Thibaud Vial[3]

[1] Univ Lyon, Inria, INSA Lyon, CITI, Villeurbanne, France
cyrille.morin@inria.fr
[2] Nokia Bell Labs, Nozay, France
jakob.hoydis@nokia-bell-labs.com
[3] RTone, Lyon, France
thibaud.vial@rtone.fr

Abstract. Hardware imperfections in RF transmitters introduce features that can be used to identify a specific transmitter amongst others. Supervised deep learning has shown good performance in this task but using datasets not applicable to real world situations where topologies evolve over time. To remedy this, the work rests on a series of datasets gathered in the Future Internet of Things/Cognitive Radio Testbed [4] to train a convolutional neural network (CNN), where focus has been given to reduce channel bias that has plagued previous works and constrained them to a constant environment or to simulations. The most challenging scenarios provide the trained neural network with resilience and show insight on the best signal type to use for identification, namely packet preamble. The generated datasets are published on the Machine Learning For Communications Emerging Technologies Initiatives web site (Datasets and usage and generation scripts can also be found there: https://wiki.cortexlab.fr/doku.php?id=tx-id.) in the hope that they serve as stepping stones for future progress in the area. The community is also invited to reproduce the studied scenarios and results by generating new datasets in FIT/CorteXlab.

Keywords: Transmitter identification · RF fingerprinting · Deep learning

1 Introduction

Communication systems' constant evolution requires a constant search for new techniques that allow to squeeze out every bit of performance out of the system, a constant need to improve spectral efficiency in order to achieve the theoretical maximum capacity given by Shannon's law. This requirement is all the more important in systems that transmit many small packets, like Internet of Things

© ICST Institute for Computer Sciences, Social Informatics and Telecommunications Engineering 2019
Published by Springer Nature Switzerland AG 2019. All Rights Reserved
A. Kliks et al. (Eds.): CrownCom 2019, LNICST 291, pp. 73–86, 2019.
https://doi.org/10.1007/978-3-030-25748-4_6

(IoT), where currently headers may outweigh the number of payload bits transmitted. Furthermore, headers are currently the only barrier against transmitter identification errors and transmitter impersonation on edge devices that don't have the resources to use cryptographic protocols. Indeed, security issues are of utmost importance in this new era where hackers are able to easily attack transmissions even at the physical layer. Hence, a new means to provide secure identification of transmissions is needed, both to improve security as well as to render headers a thing of the past.

In the last five years, supervised deep learning (SDL) has imposed itself as the tool to achieve state-of-the-art performance in many fields, starting with image processing to voice recognition, product suggestion, and more generally data analysis and signal processing in physics, medicine and consumer products. SDL really shines in cases where labelled data is plentiful and mathematical models are not known. In the radio communication world, data is generated by the Terabyte per second all over the world, but for traditional applications, precise models already exist. The existence of those good models explains why SDL is not yet widely used in radio communication. Yet it's starting to gain traction in the last couple of years, especially with channel decoding [6] and spectrum monitoring [7]. Indeed, the tool promises increased performance in areas where models are yet to be derived and algorithms are not yet practical for real time implementations. The task of identifying transmitters has attracted some attention in the last years, with two different approaches:

First, the confirmation of identity, by comparing a received signal with a previously authenticated one to verify if those characteristics match. This approach facilitates handling of changing environment by constantly updating the stored authenticated signal, as long as transmissions are more frequent than channel variations. The fingerprints for each device is not stored inside of the identification system but in the recorded signals. It can allow the identification of emitters unseen at training time, but increases processing times: comparisons need to be made with every known device. In [12], the authors study channel responses in a simulated building floor to emulate spatial variations. In [9], channel response is also used, but this time using a Gaussian Mixture Model to study similarities between samples gathered on real radio devices.

In second come the classifiers, and these are the more recent and numerous works, where a system is tasked not to give a similarity score between current and previous samples but to directly output the identity of the transmitter amongst a pool of previously seen radios. A convolutional neural network (CNN) is used in [10] to estimate IQ imbalance parameters of incoming signal from a simulation environment and these parameters are used in a classical Bayesian decision process. Then [2] leverages parameters estimated routinely by decoding systems: DC and frequency offsets, IQ imbalance and channel information as input to a neural network. Matlab simulations show 99% accuracy with up to 10 000 devices. In [5] the authors use 7 consumer Zigbee transmitters to gather data. They remove the decoded data from the received signal to isolate channel and transmitter information effects before feeding it to a CNN and achieve up to 90% classification accuracy. However, one could argue that some hardware effects depend on the transmitted signal, for example amplifier non linearities, so removing it could be detrimental to the identification process. Different machine learning techniques are evaluated in [13] over a dataset of signals gathered with real USRP devices accounting for 6 radio interfaces. They develop an architecture called a multi layer perceptron made of a network of small neural networks and introduce the use of wavelet transforms with the goal of learning to classify with as few training samples as possible. The work in [3] focuses on amplifier characteristics and measures non linearity variations between 7 USRP cards. The data is used to train a network on hundreds of simulated devices. It also shows the effects of local oscillator leakage on classification accuracy. Finally, in [8] attention is switched to IQ imbalance and DC offset to perform classification. A CNN is trained on dataset made with 16 different USRP devices but it is not able to cope with changes in environment between experiments. For this reason, artificial impairments are added to the signal and increase classification accuracy. Still in the same context, the work in [1] deals with wired communications to increase security and prevent intrusion of malicious systems in the network inside a car.

As previously stated, some works base the identification of transmitters on characteristics outside of the scope of the transmitter radios themselves using, for example, the channels [9,12]. In realistic applications, however, radios are rarely at fixed locations and channel characteristics evolve with the surrounding environment. A better identification method would be to base the identification on the radio frequency (RF) *signature* of the radios themselves, as addressed in [2,3,8,10]. However, in those works, either simulations or simple datasets were used which may not provide a biased free classifier that actually focuses on RF signatures. Furthermore, in those works, no attention is given on the quality of the dataset itself for the identification task at hand (absence of bias, reproducibility, interference with outside sources), and in [8], artificial impairments remove possible security claims: if identification is done on software added elements, any malicious software can do the same and impersonate the user.

To counter the problems encountered by previous works, herein, an extensive dataset campaign was generated aiming to train a more robust classifier. The datasets were carefully crafted to avoid biases like channel, transmitted power, packet structure, and receiver position through different measurement campaigns in a controlled environment. We use the Future Internet of Things/Cognitive Radio Testbed [4] (FIT/CorteXlab) to gather experimental datasets to train neural network classifiers able to identify emitters based on their hardware characteristics. The datasets are available online, as well as the scripts used to generate them, so that anyone can reproduce them on FIT/CorteXlab.

The remainder of this paper is organised as follows. Section 2 deals with the identification problem itself and the characteristics that make it particular. It also describes the FIT/CorteXlab testbed, used for the measurement campaign. Section 3 describes the dataset generation process required to train a signature based CNN. Then, in Sect. 4 the CNN structure is described as well as the training process. Results are presented and discussed in Sect. 5. Finally, conclusions and perspectives are drawn in Sect. 6.

2 The Identification Problem

Current authentication schemes for packet transmission are based on transmitting an identification number inside of a frame header. This poses two problems:

- The number needs spectral and energy resources to be transmitted. In IoT protocols, these resources are already very limited.
- The identification is trivial to spoof. Meaning that the transmission needs to be authenticated again at higher levels with cryptographic means, or in cases where computing power or energy is limited it cannot be done.

2.1 Characteristic Elements of a Point-to-Point Transmission

Channel Effects. First and most obvious is the impact of the channel between emitter and receiver. In the standard case of a transmission with a fixed average power, the reception power is a direct indicator of the path loss and thus of the distance between emission and reception. This power can then serve as a coarse indicator of the emitter. In this case, multi-path parameters can also be easily measured. These are highly dependent on the position of the emitter.

These two elements have a high impact on the signal and are simple to measure, but they are dependent on the system topology which is expected to change over time, not on intrinsic properties of the radio cards. This could still be used as a way to detect impersonation by looking at sudden changes in these parameters, as studied in [11,12].

Power Amplifier Imperfections. Homodyne radios suffer from IQ imbalance. The In-phase and Quadrature components of the signal do not go through the same path with the same components. This means that they may not be amplified by the same gain and the resulting constellation gets skewed. The second part of the imbalance comes from the fact that the two parts are not mixed with sine waves at exactly 90° and the resulting constellation gets rotated. This effect is used greatly in [8] and [10] but it's mostly present in devices with a homodyne architecture as those used in software refined radio (SDR) and not in superheterodyne radios that make up the majority of consumer devices.

Amplifiers of RF signal are non linear components. They are setup so they function mostly in their linear region but, even there, their response curve is not perfectly linear. The parameters corresponding to this are slightly different from one radio chip to another, even amongst the same product line [3].

Local Oscillator Imperfections. The local oscillator is tasked with translating the baseband signal to the carrier frequency. It is however not able to set itself to the exact same frequency as the receiver. This creates a frequency offset that may be characteristic of an emitter The offset is not only dependent on intrinsic elements of the radio card, but also on temperature so it can change over time at a rate that can also be indicative of the emitter, if it's high enough to be measured.

The frequency translation operation that uses the oscillator to bring baseband signals to carrier frequency can present a local oscillator leak in homodyne devices. In this case, a peak is sent at the oscillator frequency, whose power does not depend on the actual signal sent, but only on the amplifier settings. The phenomenon is used in [3] to identify emitters even when, considering data sent, $SNR = -\infty$.

2.2   Avoiding Dataset Bias

The aforementioned elements are not all desirable: The channel effects are only distinctive of the position of the emitter and may not be reliable when the environment becomes realistically dynamic. Some of the other imperfections are specific to homodyne radio cards. This means that they can appear in the USRPs that will be used, but not on many mainstream consumer grade superheterodyne radios. So reducing the reliance on these imperfections allows better generalisation of the results to other radio hardware architectures.

The USRPs are calibrated to remove IQ imbalance and DC offsets and the emission gain is set to reduce local oscillator leakage to a minimum. But channel effects cannot be removed as simply while still maintaining realistic over-the-air radio propagation (no cable). Instead of removing them, they can be randomised: When a parameter is random and varies over a range that overlaps these of the other emitters, one specific realisation of that random parameter cannot give insight on who is the transmitter. In this case, the neural network (NN) will not learn to depend on that parameter to perform identification.

2.3   The FIT/CorteXlab Platform for Learning

The FIT/CorteXlab testbed is a SDR testing facility in France that enables testing of physical (and higher) layer techniques for future wireless systems. It counts with 42-node high-performance SDR nodes, whose frequency range is roughly from 400 MHz to 4 GHz, at 20 MHz of maximum bandwidth. It is fully accessible to the wireless research community and uses GNU Radio as its programming environment. One of its main characteristics, the one that makes FIT/CorteXlab particularly well suited for machine learning (ML), is its shielded room. It allows for reproducible experiments since the shielding provides a fully a controlled environment. Also important for ML is the presence of a server equipped with GPUs connected via high speed data links to the SDR nodes, which allow on-the-fly training and exploitation of SDL techniques.

3   Dataset Generation Process

The data generation process is modular to allow for testing of a wide range of scenarios with various amount of bias. There is a standard core process that can be parameterised to create different scenarios. The signal processing chain is written as GNU Radio flowgraphs for emission and reception that provide a light API for configuration. All the high-level management of the experiments is done in small python and bash scripts for easy configuration and reproduction of scenarios.

3.1   Core Process

There are 21 emitters, one receiver, and one scheduler. Each of these, except the scheduler, uses a National Instruments USRP N2932 SDR working at 5 Msample/s and 433 MHz. All the emitters have to transmit packets to the receiver and use the same frequency band. The scheduler's role is to ensure that there is no interference between packets from different emitters without needing to implement carrier sensing algorithm. It also causes packets from different emitters to be sent in close temporal proximity. This is useful if the environment is not static: one specific environment configuration could be indicative of a specific emitter if it was not the case. Every millisecond the scheduler selects randomly one among the possible emitters, then sends a packet to it via UDP.

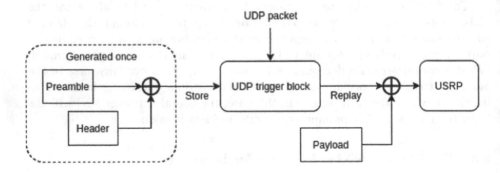

Fig. 1. Simplified emitter flowgraph

The emitters have their USRP set to burst mode. This means that when they are not actively transmitting, their amplifiers are off so they do not emit anything. Even local oscillator leakage is prevented by this. Upon reception of a scheduler packet, an emitter wakes up its USRP, waits for the amplifier to stabilise and sends a frame (Fig. 1). The frame is composed of three parts:

- A known preamble for detection and time synchronisation with the frame.
- A header: an OFDM frame containing the identification of the emitter.

– The payload that we are interested in, containing either noise, a random or a static QPSK modulated sequence of 560 samples long. In all these cases, the payload does not contain any emitter specific information.

There is a time gap between the header and the payload to give the amplifier time to stabilise after the header and avoid interference between the two parts if there is a significant amplitude difference between the two parts (Fig. 2).

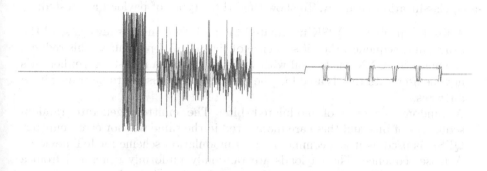

Fig. 2. Frame samples sent to USRP for emission, with zeroes for amplifier wake up, preamble, header, and payload with guard intervals

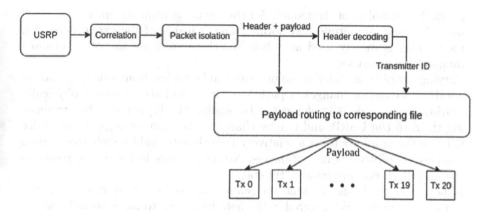

Fig. 3. Reception flowgraph

The receiver's USRP stays on for the duration of the experiment. It uses a correlator to detect the presence and the timing of the known preamble, then isolates the number of samples corresponding to the length of the header and the preamble. An OFDM packet receiver is used to decode the header and the identification number is used to send the payload samples to be recorded in the corresponding file (Fig. 3).

The grouping of the recorded files for one experiment in one scenario forms a dataset. Experiment run times are set to gather about 50000 packets per emitter.

3.2   Scenarios

An experiment scenario has two main parameters to allow testing and selection of different elements:

Type of Signal. The goal here is to identify the emitter based on how it sends data, not what it sends. But every type of data does not necessarily yield the same classification accuracy. To show this, three types of payload are tested:

- A fixed sequence of QPSK modulated bits: All the emitters always send the exact same sequence: the bit sequence of the 802.15.4 preamble. This reduces the noise the CNN has to deal with while still being a realistic case: here it's not the user data that is used but the preamble and this has to be transmitted anyways.
- A random sequence of modulated bits: The emitters generate random sequences of bits and these are modulated in the same way for every emitter. QPSK is used as it is a commonly used modulation scheme for IoT devices.
- A noise sequence: The payloads are randomly uniformly generated from a noise source. This allows to test if the modulation choice has an impact on the performance or if any modulation would work.

Transmission Complexity

- Plain: The simplest of the modes. All the payloads from all emitters are sent with the same amplitude and nothing moves inside of the experimentation room. This is mostly used as a benchmark to compare the other channel randomising scenarios.
- Varying amplitude: The emission amplitude varies from one payload to another to emulate changes in path loss for each emitter without physically moving them. This is implemented by scaling the IQ samples before sending them to the USRP, and not by changing the gain settings of the device because the amplifier takes a relatively long time to stabilise whereas scaling samples in software is instantaneous. Nothing moves inside of the room, so the multipath parameters are still static.
- Robot: The payloads are sent the same way as in the previous scenario. A robot is introduced, covered with metallic sheets to increase radio waves reflections and set to move randomly inside the room. This introduces new and constantly changing reflections to randomise the multipath parameters (Fig. 4).

4   Learn to Classify

4.1   System Architecture

We use a CNN type network with five layers of convolution and six dense layers. It takes 600 complex samples organised in a matrix of 600×2 float numbers

Fig. 4. The Turtlebot robot with the metallic sheets, inside of FIT/CorteXlab

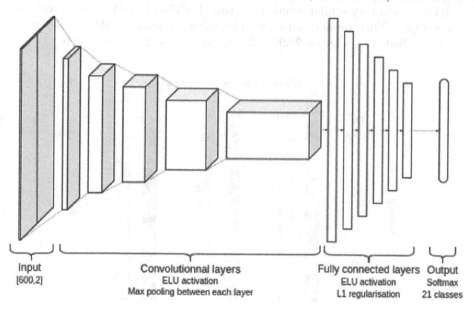

Fig. 5. Neural network architecture

for the Cartesian coordinates of the complex numbers. And it outputs a vector of 21 numbers corresponding to the likelihood that the input was from one of the 21 transmitters. Each layer has an exponential linear unit (ELU) activation function except for the output layer which uses a softmax activation (Fig. 5).

4.2   Training Phase

Before training, datasets are randomly shuffled and then split into 3 parts: 70% used for training, 10% for validation and hyperparameter tuning and the last 20% for testing. Networks are trained over all the gathered datasets with the same architecture. Training is done on mini-batches of 128 examples over more than 30 epochs, with an Adam optimiser, 0.001 of learning rate, l1 regularisation on the dense layers while minimising the categorical crossentropy loss function.

Hyperparameter tuning was done by training networks with increasing amount of dense and convolutional layers, with varying batch sizes, learning rates and regularisation on a dataset that was found to be hard to train on: Varying amplitude and a payload of random bits. They were trained for ten epochs and the best performing was selected.

5   Results

The first step is to establish a benchmark with the simplest scenario. This actually shows a comparison with the network in [8] without artificial impairments. In [8], they use a very similar setup of 16 static USRPs placed in a room with no moving object. Their network achieves a classification accuracy of 98.6% whereas the one studied here achieves 99.9% with more classes (21 instead of 16).

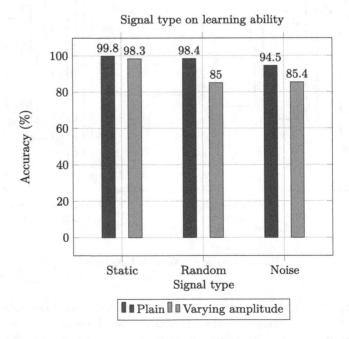

Fig. 6. Accuracy reached by networks trained on plain or varying amplitude scenarios and with various signal types. The accuracy is measured on the test set from the training dataset.

Figure 6 presents the classification accuracy achieved for the three signal types, while comparing the Plain and Varying amplitude scenarios. Having randomness in the transmitted signal degrades classification accuracy and this degradation increases with the scenario complexity. However, we can also see that having a completely random noise does not cause a significant performance loss over a QPSK modulated signal, even though the latter has a limited constellation size compared to noise. Thus a higher order modulation should not cause further accuracy losses, but the ideal case is to use a static signal, as would be a frame preamble.

Figure 7 studies the impact of environment modification on classification accuracy for the three scenarios. In it, one can observe that, the more complex a scenario is, the harder it is for a network to train on it, by a slight margin but the better it is at resisting changes in the environment. The training datasets were generated one after another, with no changes inside the FIT/CorteXlab room, then a metallic stool was introduced inside the room and the other datasets were generated. This ensures that the change in signal propagation is exactly the same for the 3 scenarios. The Plain scenario suffers from a big loss of accuracy, as was noticed in [8], but the Robot one is very resilient to this kind of change.

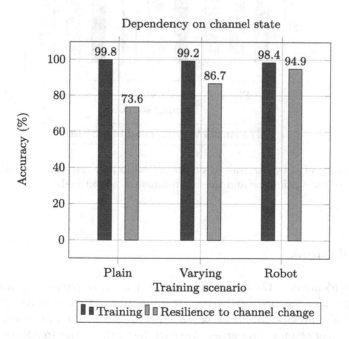

Fig. 7. Accuracy of networks trained one scenario with static signal and tested, either on test data from the training dataset or on a dataset with the same scenario but with a chair added to the room.

Finally, Fig. 8 focuses on the ability to generalise to other scenario types. A network is trained on each scenario and tested on all of them. We can observe a decrease in accuracy when a simple scenario is tested on a more complex one. From this, one can infer that the more random the channel is, the more the trained network is able to cope with a change of scenario and also with a change in the environment.

The key takeaway from these results is as follows: if one were to implement a transmitter identification system in a production setting, its goal should be to train the neural network with the maximum amount of channel variability.

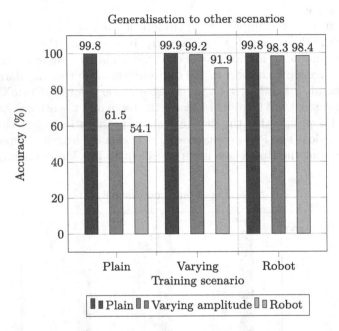

Fig. 8. Accuracy of networks trained on one scenario and tested on the others. Here the signal is of the static type and the environment is untouched.

6   Conclusions

The task of identifying transmitters based on hardware physical characteristics and imperfections has been gathering attention in the literature recently. These identification strategies have been based aspects such as IQ imbalance, amplifier non linearities or channel properties. Most of the works in the area involve the use of machine learning, and more specifically deep learning with good identification accuracies. However, the data used to train the neural networks from prior works can not guarantee bias avoidance against unwanted elements, such as channel effects.

In this work, instead of focusing on a specific radio imperfection, neural networks were trained on raw IQ samples so as not to overlook any effect. On the other hand, data was generated with the goal of minimising the role of the unwanted channel on the identification task by randomising the various channel parameters.

The considered neural network architecture shows state-of-the-art performance when tested on similar scenarios as previous works. An exploration of various parameters was done with results that shows that the identification task is simpler when transmitted signals do not change and that unknown signals only incur a small performance decrease, independently of the modulation type. They also show that training data with increased complexity do not impede learning ability but provides increased robustness against environment modifications. Finally, from the experiments and the results, we see that FIT/CorteXlab provides the stability and reproducibility necessary for machine learning approaches.

We plan to extend this setup to use packets recorded from not just one but from many receivers to increase the randomness of the perceived channel with the transmitters and evaluate the robustness of this approach to environment modification. Another direction of exploration is to verify that higher order modulation schemes do not impair classification accuracy with respect to what observed with QPSK.

Acknowledgement. This work was supported by Inria Nokia Bell Labs ADR "Analytics and machine learning for mobile networks". Experiments presented in this paper were carried out using the FIT/CorteXlab testbed (see http://www.cortexlab.fr).

References

1. Avatefipour, O., Hafeez, A., Tayyab, M., Malik, H.: Linking received packet to the transmitter through physical-fingerprinting of controller area network, January 2018. http://arxiv.org/abs/1801.09011
2. Chatterjee, B., Das, D., Sen, S.: RF-PUF: IoT security enhancement through authentication of wireless nodes using in-situ machine learning, May 2018. http://arxiv.org/abs/1805.01048
3. Hanna, S.S., Cabric, D.: Deep learning based transmitter identification using power amplifier nonlinearity, November 2018. http://arxiv.org/abs/1811.04521
4. Massouri, A., et al.: CorteXlab: an open FPGA-based facility for testing SDR & cognitive radio networks in a reproducible environment. In: 2014 IEEE Conference on Computer Communications Workshops (INFOCOM WKSHPS), pp. 103–104. IEEE (2014). https://doi.org/10.1109/INFOCOMW.2014.6849176
5. Merchant, K., Revay, S., Stantchev, G., Nousain, B.: Deep learning for RF device fingerprinting in cognitive communication networks. IEEE J. Sel. Top. Signal Process. **12**(1), 160–167 (2018). https://doi.org/10.1109/JSTSP.2018.2796446
6. Nachmani, E., Beery, Y., Burshtein, D.: Learning to decode linear codes using deep learning, July 2016. http://arxiv.org/abs/1607.04793
7. O'Shea, T.J., Clancy, T.C., McGwier, R.W.: Recurrent neural radio anomaly detection, November 2016. http://arxiv.org/abs/1611.00301
8. Sankhe, K., Belgiovine, M., Zhou, F., Riyaz, S., Ioannidis, S., Chowdhury, K.: ORACLE: Optimized Radio clAssification through Convolutional neuraL nEtworks, December 2018. http://arxiv.org/abs/1812.01124

9. Weinand, A., Karrenbauer, M., Sattiraju, R., Schotten, H.D.: Application of machine learning for channel based message authentication in mission critical machine type communication, November 2017. http://arxiv.org/abs/1711.05088
10. Wong, L.J., Headley, W.C., Michaels, A.J.: Emitter identification using CNN IQ imbalance estimators, August 2018. http://arxiv.org/abs/1808.02369
11. Xiao, L., Greenstein, L., Mandayam, N., Trappe, W.: Fingerprints in the ether: using the physical layer for wireless authentication, July 2009. http://arxiv.org/abs/0907.4877
12. Xiao, L., Greenstein, L., Mandayam, N., Trappe, W.: Using the physical layer for wireless authentication in time-variant channels, July 2009. https://doi.org/10.1109/TWC.2008.070194
13. Youssef, K., Bouchard, L.S., Haigh, K.Z., Krovi, H., Silovsky, J., Valk, C.P.V.: Machine learning approach to RF transmitter identification, November 2017. http://arxiv.org/abs/1711.01559

Preferential Radar Method for Dynamic Assignment of Wi-Fi Channels

Jesús David Sandoval Posso[(⊠)]
and Carlos Andres Giraldo Castañeda[(⊠)]

College of Engineering, Pontificia Universidad Javeriana Cali,
Valle del Cauca, Colombia
{jdposso, cgiraldoc}@javerianacali.edu.co

Abstract. A variety of methods has been developed for the optimization of channels in the bandwidth of Wi-Fi networks. Some methods will consider the existence of a single network and allow dynamic allocation of Wi-Fi channels to access points based on different criteria [1–4]. However, in many cases there is more than one Wi-Fi network and some methods were designed in this context and try to reduce the interference between access points and coexisting networks [5, 6]. This article presents a new method identified as a preferred radar algorithm (RP). The RP is based on a search algorithm that decreases the global interference of the environment for each access point. The modified ICORAL and RP methods were implemented allowing to improve the global interference of the environment. The RP and ICORAL methods were fully characterized with radar system embedded in a university environment. The RP shows a significant improvement of at least 30% more interference decrease than the ICORAL.

Keywords: Dynamic channel assignment · Cognitive radio · Wi-Fi

1 Introduction

In recent years, many wireless devices use the Internet to transmit and receive information from the network, better known as Internet of Things (IoT). It is generally based on the communication principles of the IEEE 802.11 standards, or Wi-Fi network [6]. The number of wireless devices has increased significantly and has generated channel allocation problems and optimization of the use of the 2.4 GHz band. The Bluetooth, Wi-Fi, NFC and ZigBee protocols have extended spectrum techniques in the 2.4 GHz band [7]. To minimize this problem, several approaches have been proposed, seeking dynamic solutions that optimize the ISM band. Among these techniques is the Dynamic Allocation of Cisco DCA Channels, which is taking decisions of channel change according to the historical study of the behavior of the network and some improvement parameters [2]. The dynamic allocation of multi-factor MFDCA channels [8], propose a focus on energy efficiency and the channel width adaptation algorithm seeks to reduce interference and improve transfer speed [9]. However, these techniques assume the existence of a single network where all the access points belong to it and it is not considered that more than one network can coexist in the environment.

© ICST Institute for Computer Sciences, Social Informatics and Telecommunications Engineering 2019
Published by Springer Nature Switzerland AG 2019. All Rights Reserved
A. Kliks et al. (Eds.): CrownCom 2019, LNICST 291, pp. 87–99, 2019.
https://doi.org/10.1007/978-3-030-25748-4_7

Some techniques approach the problem where several networks interact in the same medium. As it is the case of the coral algorithm [3] the latter monitors the network with a mobile unit and reassigns the non-overlapping frequency channels of the 2.4 GHz band generated coral shaped structures in populations of access points in channels 1, 6 and 11. In this article, we present a method where the allocation of channels is independent of the networks that coexist in the medium. The method is based on cognitive radio and makes use of radar external to the networks that monitors the medium 360° and locates the access points referring to it for the reassignment of channels, decreasing the global interference. This new method is called the preferred radar algorithm.

2 Methods of Channel Dynamic Assignment

2.1 Improved Coral Algorithm (ICORAL)

The methodology of the improved coral algorithm is based on the principle of assigning each access point a channel without spectral overlap. This is achieved by maintaining a 5-channel difference between assigned access points. The rules that govern the method are described below. The flow chart of the preferred radar method is shown in Fig. 1.

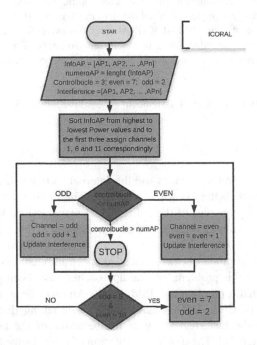

Fig. 1. Flow diagram of the ICORAL dynamic assignment method.

Rules of the ICORAL methodology:

1. Sort from highest to lowest power the APs monitored by the radar.
2. Non-overlapping frequency channels are assigned in 1, 6 and 11 to the APs of greater power with respect to the radar.
3. For the following APs, pairs of non-interfering channels (2,7), (3,8), (4, 9) and (5,10) are assigned.
4. A new evaluation of the next AP to be studied is started and the process from step 2 onwards is repeated until analyzing all the APs registered by the radar.

2.2   Preferential Radar Algorithm RP

The methodology of the preferential radar algorithm is based on the principle of assigning to each access point a channel that minimizes the impact of global interference of the operating environment and independent of the network to which it belongs. The rules that govern the method are described below. The flow chart of the preferred radar method is shown in Fig. 2.

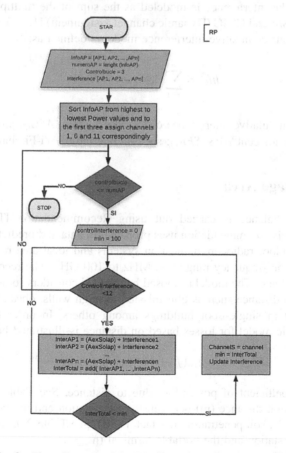

Fig. 2. Flow diagram of the RP dynamic assignment method.

Rules of the RP methodology:

1. Sort from highest to lowest power the APs monitored by the radar.
2. Non-overlapping frequency channels are assigned in 1, 6 and 11 to the APs of greater power with respect to the radar.
3. For the following APs, the cumulative interference metric of each of the Wi-Fi band channels with the ancestors already assigned to the environment is studied.
4. The value of the accumulated interference generated by each channel is stored and the channel with the least interference weight in the environment is identified so that it is assigned to the AP studied.
5. A new evaluation of the next AP to be studied is started and the process from step 3 onwards is repeated until analyzing all the APs registered by the radar.

3 Accumulated Interference Model

Taking as reference the interference model of several methods of dynamic channel allocation where the interference is modeled as the sum of the multiplication of two factors, such as RSSI and COIC (Dynamic channel assignment) [1, 3, 5, 6, 8] and based on this principle our cumulative interference model is defined as:

$$In_{i,j} = \sum_{j=1}^{n} ASC_{i,j} \times SFR_{i,j} \qquad (1)$$

Where: $In_{i,j}$: cumulative interference of the access point, $ASC_{i,j}$: area of overlap in coverage between adjacent APS, $SFR_{i,j}$: spectral overlap of Wi-Fi channels.

4 APs Coverage Area

The radioelectric balance is carried out using Recommendation ITU-R P.1238-9 (06/2017) [10]. This recommendation uses propagation data and prediction methods in the planning of indoor radiocommunication systems and local area radiocommunication networks in the frequency range 900 MHz to 100 GHz. The recommendation is used in the Wi-Fi band. The model proposed by the ITU considers some coefficients of power loss due to distance such as transmission through walls, obstacles, loss factors that can manifest in single-floor buildings among others. In order to simplify the problem. The basic model for losses based on distance is illustrated below:

$$L_{total} = 20log_{10}(f) + Nlog_{10}(d) + L_f(n) - 28 \qquad (2)$$

Where: N: Coefficient of power loss due to distance. See Table 1, f: frequency (MHz), d: separation distance (m) between the base station and the portable terminal (where d > 1 m), Lf: soil penetration loss factor (dB). See Table 2, n: number of floors between the base station and the portable terminal (n ≥ 1).

Table 1. Coefficient of loss of power, N, for the calculation of the loss of transmission in interiors. [11].

Frequency (GHz)	Residential building (dB)	Office building (dB)	Commercial building (dB)
0,9	-	33	20
2,4	28	30	-
5,2	-	31	-
5,8	-	24	-

Table 2. Ground penetration loss factors, Lf (dB), being the number of floors penetrated, for the calculation of the transmission loss in interiors ($n \geq 1$) [11].

Frequency (GHz)	Residential building (dB)	Office building (dB)	Commercial building (dB)
0,9	-	9 (1 floor) 19 (2 floor) 24 (3 floor)	-
2,4	10 (apartment) 5 (home)	14	-
5,2	13 (apartment) 7 (home)	16 (1 floor)	-
5,8	-	22 (1 floor) 28 (2 floor)	-

For the calculation of the coverage distance, omnidirectional propagation antennas are assumed with dipole antennas G = 1.68, transmission powers of 20 mW and reception sensitivity −80 dBm for all the APs monitored by the radar. The ASC factor is calculated as a measure of the interception between adjacent APs coverage areas as shown in Fig. 3.

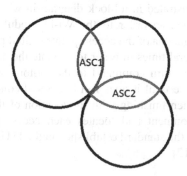

Fig. 3. Overlapping coverage area.

5   Spectral Overlap of Wi-Fi Channels

The SFR factor is obtained by observing the Wi-Fi channel distribution monitored by the radar. Where it is observed that a separation of 5 channels is necessary so that there is no spectral overlap. As the separation between channels decreases the percentage of overlap increases as observed in Table 3 and Fig. 4.

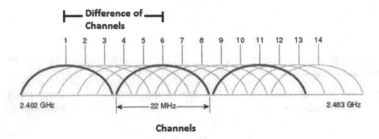

Fig. 4. Channels Wi-Fi.

Table 3. Overlap frequency in channel Wi-Fi.

Difference between channels	Overlap %
0	100
1	77,27
2	55,55
3	31,82
4	9,09
>5	0

6   System Design

In Fig. 5, the system is illustrated in a block diagram in which the computer controls the whole system and sends an order to the Wi-Fi module so that the signals are captured by means of the antenna of the access points (initial position of the radar). The scanning is performed several times in order to validate the acquired signals. After the first scan, the PC sends the turn command to the motor controller so that it moves 10 degrees counterclockwise until it goes full circle and concludes the monitoring of the radar environment. When finishing with the location of the access points the user interface draws the environment and locates each access point and its respective coverage area according to the standard established in the ITU-R P.1238-9 (06/2017) of the Wi-Fi radio link [10, 12].

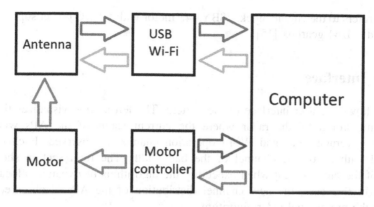

Fig. 5. System block diagram.

Figure 6 shows the global system with each of its parts. The pieces that make up the comet 17 radar are: 1. Antenna Ettus LP0965, 2. Base for the antenna, 3. Base for switch, 4. Coupling of the motor shaft, 5. Switch, 6. Stepper Motor, 7. Arduino Uno, 8. USB - Wi-Fi, 9. Power driver for Step-Step motor, 10. First level of the frame, 11. Frame cover, 12. Second level of the frame

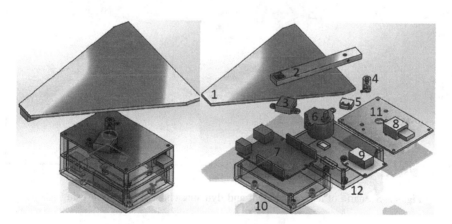

Fig. 6. Radar Comet and radar break down.

In the block diagram, there is the block of the antenna that is represented in the exploded view of the radar as item number 1. Directional logarithmic PCB antenna with gain of 6 dBi and a bandwidth from 850 MHz to 6.5 GHz [13]. The Wireless-N 150Mbps reference adapter is located in the USB-Wi-Fi block. It has a Realtek 8188CU reference chipset that accepts the IEEE802.11 n, g and b, part number 8. The pieces number 5, 7 and 9 refer to the motor controller block in the block diagram, which consists of an Arduino Uno, a switch to determine the mechanical zero and a ULN2003 drive with 5–12 VDC power and a maximum output of 500 mA [14] Piece number 6 refers to the engine block. 28BYJ-48 motor with 5 VDC power supply with 4 phases and a 1/64 gearbox [15].

7 User Interface

Figure 7 shows the user interface of the system. The left section visualizes the environment monitored by the radar where the current status of the APs with their respective coverage areas and Wi-Fi operation channel is observed. Each color is associated with a specific channel of the bandwidth. The right section shows the changes of the environment when executing the algorithms of dynamic allocation of channels. In this case, the new channel distribution of the APs is observed when executing the preferential radar algorithm.

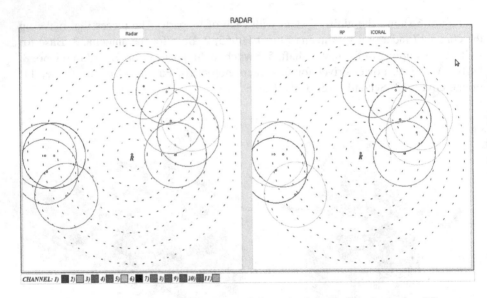

Fig. 7. Scenario of an environment and dynamic channel allocation using RP.

8 Experimental Results

The experimental tests of the radar prototype were carried out at the Pontificia Universidad Javeriana Cali, in the buildings of Lago and Palmas where many APs belonging to different Wi-Fi networks interact. The methodology for conducting the tests consisted of making 10 sweeps in a fixed position of the building and calculating the measure of the accumulated interference. Each method of dynamic allocation was tested under the same conditions. The objective was to characterize them and evaluate their performance based on the same cumulative interference metric with the environment. Next, the results are displayed. Lago building tests:

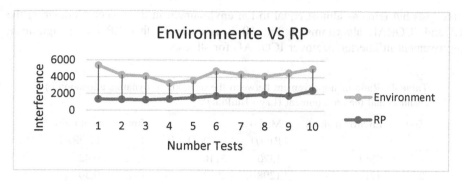

Fig. 8. Comparison of the environment with dynamic assignment of RP channels.

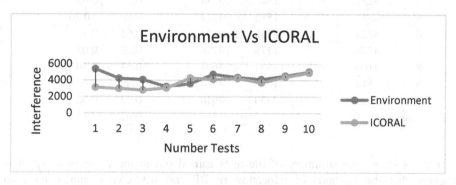

Fig. 9. Comparison of the environment with the dynamic channel assignment ICORAL.

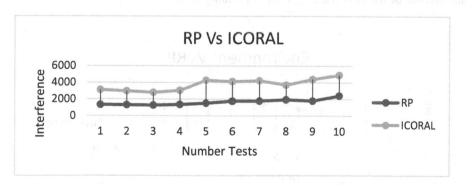

Fig. 10. Comparison of the two methods of dynamic allocation of RP Vs channels ICORAL.

Figures 8 and 9 show the behavior of the cumulative interference with the environment for both methods in the Lago building. The RP algorithm in all tests significantly improves the interference. While the ICORAL algorithm improves in the first three tests but remains almost equal to the environment in the others. Comparing the RP and ICORAL algorithms in Fig. 10 it is observed that RP has a significant improvement in interference over ICORAL for all tests.

Table 4. Ratio of improvement between the methods of dynamic allocation of channels and the environment (Lago Building).

Test	Environment (I)	Methods		Improvement ratio	
		RP (I)	ICORAL (I)	RP	ICORAL
1	5364	1320	3116	0,75	0,42
2	4216	1298	2972	0,69	0,30
3	4064	1258	2792	0,69	0,31
4	3222	1346	3048	0,58	0,05
5	3572	1512	4274	0,58	−0,20
6	4728	1756	4124	0,63	0,13
7	4348	1778	4226	0,59	0,03
8	4104	1934	3698	0,53	0,10
9	4514	1788	4384	0,60	0,03
10	5026	2430	4910	0,52	0,02
Average				0,62	0,12

Table 4 shows the summary of the tests carried out in the Lago building. It is observed that the methods of allocation of RP and ICORAL channels have an improvement of the average environment of 62% and 12% respectively regarding the interference of the environment. Palmas building tests:

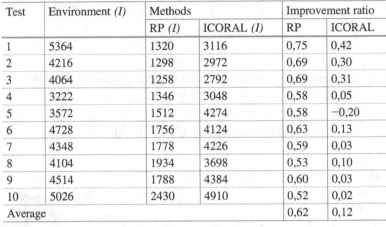

Fig. 11. Comparison of the environment with dynamic allocation of RP channels.

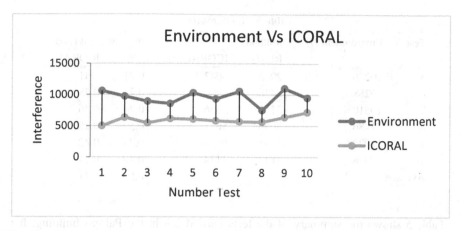

Fig. 12. Comparison of the environment with the dynamic allocation of channels ICORAL.

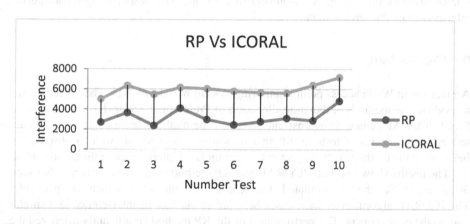

Fig. 13. Comparison of the two methods of dynamic allocation of RP Vs channels ICORAL.

Figures 11 and 12 show the behavior of the accumulated interference with the environment for both methods in the Palmas building. Both RP and ICORAL algorithms greatly improve the interference in all tests. Comparing the RP and ICORAL algorithms in Fig. 13 it is observed that RP has a better behavior in interference with respect to the ICORAL for all the tests.

Table 5. Ratio of improvement between the methods of dynamic allocation of channels and the environment (Palmas Building).

Test	Environment (I)	Methods		Improvement ratio	
		RP (I)	ICORAL (I)	RP	ICORAL
1	10602	2676	4992	0,75	0,53
2	9730	3626	6344	0,63	0,35
3	8918	2344	5460	0,74	0,39
4	8562	4086	6142	0,52	0,28

(continued)

Table 5. (*continued*)

Test	Environment (I)	Methods		Improvement ratio	
		RP (I)	ICORAL (I)	RP	ICORAL
5	10276	2966	6042	0,71	0,41
6	9288	2436	5786	0,74	0,38
7	10510	2760	5656	0,74	0,46
8	7486	3096	5618	0,59	0,25
9	11022	2878	6398	0,74	0,42
10	9498	4814	7174	0,49	0,24
Average				0,66	0,37

Table 5 shows the summary of the tests carried out in the Palmas building. It is observed that the methods of allocation of RP and ICORAL channels have an improvement of the average environment of 66% and 37% respectively regarding the interference of the environment.

9 Conclusions

A radar-based Wi-Fi access point monitoring system was developed. The objective was to evaluate the methods of dynamic allocation of Preferential Radar (RP) and Improved Coral (ICORAL) channels against interference. The methods were completely characterized with a series of tests in different buildings. The tests allowed to observe the behavior of the methods with respect to the accumulated interference of the environment.

The results show that both ICORAL and RP methods improve on average between 30% and 63% the accumulated interference of the environment respectively. The ICORAL algorithm in some cases behaved in the face of interference in a similar way to the environment. The performance of the RP method significantly improved the interference of the environment in all the tests carried out. Comparing the methods, it is observed that the RP method has 33% better performance against interference than the ICORAL method.

References

1. Zhao, Y., Wu, Y., Feng, Y., Zheng, Y., Fang, X.: Dynamic channel selections and performance analysis for high-speed train WiFi network, pp. 1–5. IEEE, Xi'an (2015)
2. CISCO DCA. https://www.cisco.com/c/en/us/td/docs/wireless/controller/technotes/8-3/b_RRM_White_Paper/b_RRM_White_Paper_chapter_0100.pdf. Accessed 12 Jan 2019
3. Wang, B., Wu, W., Liu, Y.: Dynamic channel assignment in wireless LANs, pp. 12–17. IEEE, China (2008)
4. Chang, C.-Y., Chang, C.-T., Huang, P.C.: Dynamic channel assignment and reassignment for exploiting channel reuse opportunities in ad hoc wireless networks, vol. 2, pp. 1053–1057. IEEE, Singapore (2002)

5. Mack, J., Gazor, S., Ghasemi, A., Sydor, J.: Dynamic channel selection in cognitive radio WiFi networks: an experimental evaluation, pp. 261–267. IEEE, Sydney (2019)
6. Mathur, K., Jena, D., Agrawal, S., Baburaj, S., Kondabathini, S., Tyagi, V.: Throughput improvement by using dynamic channel selection in 2.4 GHz band of IEEE 802.11 WLAN, pp. 805–810. IEEE, Prague (2018)
7. Kobayashi, H., Kameda, E., Shinomiya, N.: A matching-based strategy for AP selection in sustainable heterogeneous wireless networks. In: ISCC 2016, pp. 103–107. IEEE (2016)
8. Ribeiro, L., Souto, E., Becker, L.B.: Multi-factor dynamic channel assignment approach for Wi-Fi networks. In: IEEE Symposium on Computers and Communications (ISCC), pp. 1–7 (2018)
9. Ni, C., Gan, C., Ma, X., Wu, C.: Dynamic bandwidth allocation with optimized excess bandwidth distribution and wavelength assignment in multi-wavelength access network. In: International Conference of IEEE Region 10, TENCON 2013, pp. 1–4. IEEE, China (2013)
10. Recommendation ITU-R P.1238-9 (06/2017). https://www.itu.int/dms_pubrec/itu-r/rec/p/R-REC-P.1238-9-201706-I!!PDF-E.pdf
11. Leslye, C., Andrea, J., Marco, M.: Métodos de diseño y cobertura para redes wifi indoor y outdoor, caso UTPL, pp. 1–7. Academia, Ecudor (2013)
12. Pozar, D.: Microwave Engineering, 3rd edn. Wiley, New York (2005)
13. KB ETTUS. https://kb.ettus.com/images/0/06/ettus_research_lp0965_datasheet.pdf
14. Datasheet ULN2003. https://www.electronicoscaldas.com/datasheet/ULN2003A-PCB.pdf
15. Datasheet Stepper Motor. http://robocraft.ru/files/datasheet/28BYJ-48.pdf

Tactical Radio Operator's Combat Readiness as Context Information for Dynamic Spectrum Management Within Military Mobile Ad Hoc Network

Joanna Głowacka$^{(\boxtimes)}$ (iD), Piotr Gajewski(iD), Wojciech Bednarczyk(iD),
Michał Ciołek(iD), Krzysztof Parobczak, and Jarosław Wojtuń(iD)

Military University of Technology, 2, Gen. Sylwestra Kaliskiego Street,
00-908 Warsaw, Poland
joanna.glowacka@wat.edu.pl

Abstract. The Authors present a new approach to a method of dynamic spectrum management within military mobile ad hoc network. They propose that data concerning tactical radio operator's combat readiness be used as context information for spectrum management. The readiness is determined based on monitored vital signs of and interaction with the tactical radio operator. If the operator's inability to operate and protect the tactical radio is identified, actions are taken with the aim to release spectrum resources used by the tactical radio or in order to switch the tactical radio to modes used by search and rescue team.

Keywords: Dynamic spectrum management · DSM · Mobile ad hoc network · MANET · Combat readiness · Context information

1 Introduction

Due to its self-organizing ability and no need of fixed infrastructure, mobile ad hoc networks (MANETs) are of vital importance in military uses. They feature decentralized architecture, meaning that each node can serve as both a terminal and an agent for data transmission. Dynamic conditions of combat operations require modern communication networks to be highly mobile and able to collaborate with other units even if such collaboration has not been expected. This requires quick responses to changes within the network topology for the purpose of maintaining communication among the nodes. Unfortunately, routing updates among the nodes, especially in case of proactive protocols which are recommended for networks operating in destructive environments, e.g. in military environment [1], mean extra burden concerning control traffic.

A dynamic development of radio communication systems which has been observed in recent years calls for introducing more and more effective methods of radio resources control, depending on changing operational conditions. This facilitates optimal use of available frequencies and makes the communication service available to the greatest possible number of users.

© ICST Institute for Computer Sciences, Social Informatics and Telecommunications Engineering 2019
Published by Springer Nature Switzerland AG 2019. All Rights Reserved
A. Kliks et al. (Eds.): CrownCom 2019, LNICST 291, pp. 100–111, 2019.
https://doi.org/10.1007/978-3-030-25748-4_8

During a military action it may often happen that the spectrum assigned at the planning stage of an operation might not be used effectively due to a tactical radio operator's loss of combat readiness. Operator's combat readiness means the ability to carry out certain assignments, such as i.e. tactical radio operation and protection. Assessment of tactical radio operator's combat readiness may be used as context information [2] in MANET spectrum management mechanisms.

The structure of this paper is as follows: first, the authors present current state of affairs concerning dynamic access to spectrum within MANET, which is followed by a description of algorithms used for assessing tactical radio operator's combat readiness; further, the authors discuss the outcomes of applying the proposed solution, which confirm potential capability of its being used for spectrum management in networks featuring opportunistic access to the spectrum.

2 Current Methods of Spectrum Management

More efficient use of the available frequency band is possible thanks to applying a new philosophy of dynamic spectrum management (DSM). Dynamic spectrum management in MANETs may be effected through [3]:

- centralized management which uses a frequency broker featuring implemented spectrum tracking procedures, channel pre-definition and cognitive use of the channels by those network nodes which do not have any features of a cognitive radio,
- opportunistic management, dispersed management within MANET based on cognitive nodes (MANET-CR).

In case of centralized management, tactical radio operation is based on DSM application of coordinated methods of spectrum management in which it is equipped. Coordinated DSM is distributed within essential frequency bands dedicated to DSM, known as dynamic coordinated spectrum access (CSA). Coordinated DSM model uses devices supporting spectrum coordination within a given geographical region in order for them to decide about spectrum access within CSA. Spectrum coordinator collects information acquired from sensors directly from a tactical radio with DSM or from other nodes which sense the environment. The data is processed for the purpose of characterizing radio environment. Thanks to this, coordinators are able to assign free spectrum resources in response to access requests received. This ensures the system operability without causing harmful interference within and outside the network. Requests for assigning transmission resources are sent by tactical radios with DSM to spectrum coordinators by means of a dedicated channel. Spectrum coordinator assigns particular network users with a certain time limit, which is then used for the period of a communication session. Following the session, the channels are released and may be assigned to another system. CSA supports heterogeneous users whose requirements as to a band and operational parameters may cause mutual interference in the remaining users. Coordinated approach is more efficient in spectrum management than any hitherto used method, because thanks to CSA, a licence for a band is granted to each user, as opposed to assigning large portions of static frequency spectrum for use by particular services on a vast geographic area. Moreover, CSAs are assigned to a spectrum coordinator by means of automated processes rather than manual assignment of frequencies as it is in case of classical spectrum management. Such access management results in more flexible session-upon-session operation when the resources are needed.

Opportunistic management involves adaptation of a frequency distribution model, in which sensing of propagation environment is conducted autonomously by a number of tactical radios with DSM, and access to the spectrum is effected based on predefined spectrum management policies (SMPs), for own needs. Tactical radios with DSM, which make use of opportunistic access, identify unused portions of frequency bands in which they can operate without interference with primary users communication or without violating SMP. Tactical radios with DSM which operate within the area can mutually exchange information about the environment and coordinate mutual transmissions without spectrum coordinator support. Autonomous spectrum access (ASA) makes use of a set of frequency ranges which may include a combination of licenced frequency bands, CSA and non-licenced frequency ranges. Thanks to SMPs implemented in ASA, tactical radios with DSM opportunistic model are provided access to the available spectrum when other users are idle. ASA overlaps portions of frequency ranges which are defined by access or band sharing policies. Tactical radios which make use of opportunistic approach to DSM may operate within ASA as long as they observe the band sharing criteria. DSM presents a whole range of technical problems with implementing the method. The first one is a broad range of the spectrum, which may require sensing and describing its parameters. This would require tactical radios to be adjusted to detect broad frequency ranges and to be capable of transmitting and receiving throughout the entire bands. Current broadband antennas and radio technologies are still unfit for implementing that type of DSM in small radios, which could operate within broad frequency ranges. Secondly, DSM-type radios must be able to precisely detect the presence of other users within the band. If a certain band is deemed accessible by a radio, the radio must be able to collaborate with other DSM-type radios within the area in order to fulfil the requirement of not jamming one another. This calls for a development of a set of policies to set out requirements for cooperation of multiple radios within a given area and band.

There are many procedures for DSM use. The most important are the following procedures: "Command and Control" (C&C), "Exclusive Use" (EU), and "Common Spectrum Sharing" (CSS).

In case of "Command and Control" procedure a regulator grants a long-term licence for spectrum use. Such an approach is inflexible and results in unsatisfactory use of the spectrum resources. A slightly more flexible spectrum management method is based on seasonal right to an exclusive use of a frequency band, e.g. one which is not used within a given area. However, in case of long-term exclusivity one cannot talk about dynamic access to the spectrum, thus, neither of the above-mentioned methods belongs to DSA paradigms. Dynamic exclusive use is a method involving the assignment of short-term rights to access a certain frequency band by one user or a cognitive radio network. Another group of spectrum management methods is spectrum sharing with primary and secondary users. This method is based on a detection of a possibility to use the spectrum by cognitive radio devices and on protection of primary users' transmissions against interference generated by secondary users. Spectrum sharing may be effected based on the detection of white spaces in primary user's spectrum and use thereof by the secondary user.

In classical solutions, the above-mentioned spectrum management methods do not take into account a situation in which an operator has lost his/her combat readiness (due to a shot, loss of consciousness or death). In such case we deal with a situation in which a frequency band has been assigned to a tactical radio which is not using it. This may result in failure to assign spectrum resources to new tactical radios within the network. Hence the concept of spectrum management based on information about the life functions of a tactical radio operator.

3   Assessment of Tactical Radio Operator's Combat Readiness

3.1   Data Acquisition

A loss of combat readiness which poses direct threat to a soldier's life or health is most frequently caused by a shot, hyperthermia, hypothermia, a shot accompanied by cardiac arrest and loss of consciousness. Those conditions may cause acute respiratory distress, acute disturbance of consciousness, circulatory disturbance and thermoregulatory disorders. Based on an analysis of chief life- or health threatening conditions of soldiers in combat, the following vital signs have been chosen, which enable assessing health and general condition of a tactical radio operator:

- respiratory rate,
- oxygen saturation (SpO2),
- body temperature,
- blood pressure,
- heart rate.

Assessment of tactical radio operator's vital signs is extremely difficult due to a character of his/her actions which feature a great deal of mobility and a high level of stress. Both these factors significantly impact the values of vital signs monitored. This is why, when assessing vital signs, it is necessary to take account of data concerning current mobility of a person monitored (data acquired from a body position sensor and from a GPS receiver).

Information originating from I/O device which enables interaction with the operator may be an additional source of information about the operator's combat readiness. This device enables the operator to communicate a threat and to verify the operator's identity.

Possible data sources for assessing radio operator's combat readiness have been presented in Fig. 1.

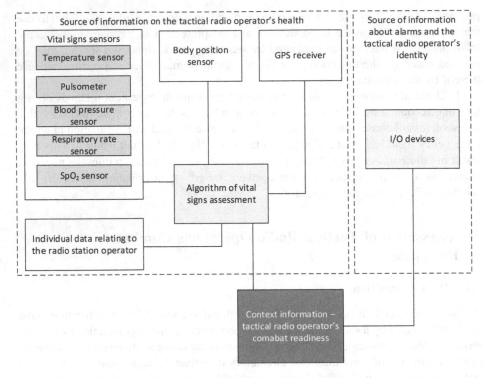

Fig. 1. Data sources for assessing tactical radio operator's combat readiness

3.2   Data Processing

As it has already been mentioned above, harvesting information about vital signs of a soldier in combat is difficult due to continuous mobility of the soldier. In addition, the signal received may be interfered with, which may cause inaccurate reading of the parameters. For vital signs to be assessed correctly, one has to apply an appropriate inference method, which will enable uncertainty modelling.

Inference Method

A method which makes it possible to effectively know true parameters from false ones and which is applied in case of insufficient information, is the Dempster-Shafer theory (DST) [4]. It is also referred to as the evidence theory or the theory of belief functions.

The DST is based on a set of all the elementary hypotheses, referred to as a frame of discernment Θ. In the DST model the frame is composed exclusively of non-overlapping elements, and its sub-sets are assigned a belief mass called a basic belief assignment (BBA), which is often marked as $m()$. BBA must have two properties:

$$m(\varnothing) = 0, \tag{1}$$

$$\sum_{A \subseteq 2^{\Theta}} m(A) = 1, \tag{2}$$

where $m(A)$ – value of basic belief assignment for hypothesis A.

According to the DST, hypotheses are assigned two functions: $Bel(A)$ called belief function and $Pl(A)$ called plausibility function. The belief function enables assessing reliability of clues for A, and the plausibility function assesses reliability of clues against A.

This theory makes uses of a relevant mathematical apparatus in a subjective assessment environment. It makes use of a level of belief and plausibility for the purposes of modelling the fuzzy assessments, and the assignment of belief levels to events and event groups is related with defining a distribution of beliefs.

The Use of the DST

The use of the DST makes it possible to define hypotheses thanks to which a soldier's condition can be assessed. Thanks to developing secondary hypotheses by means of a sum operator, it is possible to formulate imprecise and uncertain hypotheses.

Due to various types of vital signs subject to monitoring, initial assessment should be made separately for each of them. Records of data concerning various parameters may feature frequency of the data occurrence. Thanks to a separate assessment of those parameters there is a possibility to limit the number of secondary hypotheses without the need to use a hybrid model.

When assessing each of the parameters it is possible to formulate primary ("normal", "life-threatening", "serious") and secondary hypotheses ("uncertain serious", "uncertain life-threatening", "uncertain"). A hypothesis is assigned to particular measurements based on an assessment of the range of the value measured, individual features for each operator (age, sex, standard values of particular vital signs) and information on his/her activity. Thanks to a possibility to determine many hypotheses, this algorithm makes it possible, apart from defining the value ranges of a serious condition, to assume predefined critical parameter value ranges as a life-threatening condition. Table 1 presents extreme and alarming value ranges of particular vital signs, which are taken account of when assigning hypotheses to particular measurements.

Basic belief assignment value $m()$ as determined in the DST is set for each hypothesis. It depends on a number of measurements to which a given hypothesis is assigned as well as on the time of observing them. Information concerning the measurement is accepted together with weights which depend on the time of observing a given event, thanks to which a change in a soldier's health condition can be detected more quickly.

$$m(x_1) = \frac{\sum_k w_k n_{1k}}{\sum_k w_k \sum_i^{|2^\Theta|} n_{ik}} \tag{3}$$

where:

$m(x_1)$ – basic belief assignment value based on measurements for hypothesis x_1;

2^Θ = $\{x_1, x_2, \dots , x_N\}$ – a set of all the hypotheses, where $|2^\Theta| = N$;

n_{ik} – number of measurements assessed as x_i in k-th time range of observation;

w_k – weight of measurement for k-th time range of observation

Table 1. Alarming and critical value ranges of vital signs assessment.

		Individual person information – standard values of particular vital signs	Individual person information – age, sex	Lack of individual person information
Pulse	Life-threatening condition	Dependent on individual values	For men: 40÷50/min 120÷(205.8 – (0.685*age))/min For women: 40÷55/min 120÷(205.8 – (0.685*age))/min	40÷50/min 120÷140/min
	Serious condition	<40/min >(205.8 – (0.685*age))/min		<40/min >140/min
Body temperature	Life-threatening condition	Dependent on individual values	35÷36 °C 38÷39.1 °C	
	Serious condition	<35.0 °C >39.1 °C		
SpO$_2$	Life-threatening condition	91÷95% or > 99%		
	Serious condition	<91%		
Arterial pressure	Life-threatening condition	MAP < 75 mmHg SP dependent on individual values	MAP: 60÷75 mmHg SP: 60÷91 mmHg or 219÷249 mmHg	
	Serious condition	MAP < 60 mmHg SP < 75 mmHg or SP > 249 mmHg		
Respiratory rate	Life-threatening condition	Dependent on individual values	9÷12/min 20÷25/min	
	Serious condition	<9/min >25/min		

MAP – mean arterial pressure, the approximate value of which is determined according to [5] as: $MAP = DP + \frac{1}{3}(SP - DP)$, where: DP – diastolic pressure, SP – systolic pressure.

A final assessment of the soldier's condition requires taking account of correlated information on the vital signs monitored. Assessments relating to different vital signs are combined by means of a combination rule. Dempster's rule is a commonly used combination rule. However, it has got multiple shortcomings if the data is conflicted and values of some hypotheses near 0. Those shortcomings have been described in [6] and [7]. The literature abounds in interesting rules of assessment combination, which do not have limitations resulting from Dempster's rule, such as e.g.: disjunctive rule of combination [8, 9], Murphy's rule [10], Smets's rule [11], Yager's rule [12–14], Dubois and Prade's rule [15], Ali, Dutta & Boruah's rule [16].

Based on the analysis of the above-mentioned methods, due to the greatest concentration of basic belief assignment values when combining assessments for basic hypotheses, the Ali, Dutta & Boruah's rule was selected, which is defined by the following correlation:

$$m'(A) = m_1(A) + m_2(A) - m_1(A)m_2(A) \qquad (4)$$

$$m(A) = \frac{m'(A)}{\sum_n m'(A)} \qquad (5)$$

Based on the obtained values of basic belief assignment for all possible hypotheses, the condition of radio operator is classified as:

- serious,
- life-threatening,
- normal.

Node classification follows depending on verified hypothesis. Hypotheses are verified by means of a belief function defined as follows:

$$Bel(X) = \sum_{Y \subseteq X} m(Y), \text{for each } X \subseteq Y \qquad (6)$$

where:

$X \subseteq \Theta$;

Bel: $2^\Theta \to [0, 1]$;

Θ – a set of all basic hypotheses;

2^Θ – a set of all proposals which were created from elements Θ by means of operator $\cup$.

Belief function's value is determined for basic hypotheses. Based on the determination of belief function values a decision is made to accept a given hypothesis.

Algorithm Verification

The accuracy of the operator's condition assessment according to the above-mentioned algorithm was verified for data generated on Hal S3201 adult patient simulator by Gaumard in relation to predicted changes of vital signs parameters under five scenarios:

- a shot,
- hyperthermia,
- a shot accompanied by cardiac arrest,
- hypothermia,
- a shot accompanied by hypothermia,

and for data originating from tests conducted by Centrum Ratownictwa Sp. z. o.o. on 16 healthy volunteers at rest and during activity. The data include values of previously determined vital signs (pulse, body temperature, respiratory rate, saturation, arterial pressure) with one-minute frequency.

The tests confirmed that the combat readiness loss under the five scenarios was detected correctly. In addition, the detection of the condition concerned under each scenario was done before the time the soldier should lose consciousness according to the doctors. In case of testing data originating from healthy volunteers at rest and during activity, in no case has the condition of combat readiness loss been misdefined.

4   Use of Combat Readiness Assessment as Context Information

There are many definitions of context information. A number of them refer to particular cases of use, such as e.g. definitions presented in [17–19]. A more general definition was presented in [20] by Dey and Abowd: *"Context is any information that can be used to characterize the situation of entities (i.e. whether a person, place or object) that are considered relevant to the interaction between a user and an application, including the user and the application themselves. Context is typically the location, identity and state of people, groups and computational and physical objects."* In order to define context, one has to collect adequate amount of information and to analyse it appropriately.

In our case, context information will include data necessary to assess combat readiness of a tactical radio operator. As it has been mentioned above, the information concerns vital signs and interaction with the operator. An appropriate analysis of the data will enable correct assessment of the operator's combat readiness and the detection of a condition which makes it impossible for the operator to handle and protect the ratio. This context may be used in a mechanism of dynamic spectrum management within MANET. If, during a military operation, loss of combat readiness is detected in a tactical radio operator, i.e. if it is detected that he/she has lost control of the tactical radio, it is possible to disconnect the tactical radio from the network, or to switch it to another mode of operation while releasing spectrum resources used by it. This is possible by remote control of the tactical radio by a commander who makes such a decision based on the data received concerning the operator's combat readiness.

5   Research

The goal of research was to verify a possibility of spectrum management within MANET based on information on tactical radio operator's combat readiness.

5.1   Research Environment

The research was conducted with the use of Harris's AN/PRC-117G tactical radios. They are multiband combat-net radios used currently in many regions worldwide, which enable operation within the frequency range from 30 MHz to 2000 MHz. The

AN/PRC-117G tactical radios are capable of operation in narrowband modes – 12,5 kHz or 25 kHz – and wideband modes – up to 5 MHz. Wideband mode is used by adaptive networking wideband (ANW2C) waveform. MANET is made up of tactical radios operating in ANW2C mode. This mode enables simultaneous transmission of data and voice within a radio channel. Both data and voice transmissions are protected by encryption (Type 1, NSA-certified). Data transmission within ANW2C network is based on IPv4 communication protocol. ANW2C mode uses time division multiple access (TDMA) as a medium access method. Maximum network size is limited to 30 nodes. ANW2C network may operate in two modes, namely with a fixed capacity allocation among all the users or in dynamic capacity allocation (DCA) mode.

5.2 Results

The presented method of assessing tactical radio operator's combat readiness may be used to control the work of radio network nodes. In case of AN/PRC-117G tactical radios one can use a set of commands recorded in ASCII format for this purpose. Radio operation may be controlled locally or remotely.

If a tactical radio operator loses combat readiness and radio control as a result of a military action, such a tactical radio may be excluded from the network or switched to another mode of operation, thereby releasing spectrum resources it has hitherto used.

Table 2 presents the results of research which illustrate an average data throughput for ANW2C network depending on a number of active network nodes. The research involved building an ANW2C network composed of 5 AN/PRC-117G tactical radios. One tactical radio was connected to a hypertext transfer protocol (HTTP) server, and the remaining four tactical radios operated as HTTP clients. The research was conducted account taken of two scenarios, i.e. with DCA mode enabled and disabled.

Table 2. 5 MHz ANW2C End-to-end user data throughput [kbps].

Number of active nodes	DCA disabled	DCA enabled
2	1120	2136
3	568	840
4	320	528
5	248	304

As we can see, in both cases an increase in the number of network nodes results in a division of available spectrum among a greater number of users, which translates into a decrease in average data throughput.

6 Summary

The article presents the use of information about tactical radio operator's combat readiness as context information which can be used in dynamic spectrum management mechanisms within MANET. The results obtained confirm a potential capability of the proposed solution to be used for spectrum management within networks which operate based on opportunistic spectrum access.

The solution presented may be used additionally to increase the security of military communications. Monitoring vital signs of a radio operator enables detecting combat readiness loss which results in inability to protect communications. This is particularly important when using Type 1 radios which enable transmitting NATO TOP SECRET information. If such a radio is no longer protected, it must be excluded from the network, including deletion of all the radio settings (encryption keys, mission plans). Moreover, constant monitoring of a radio operator's vital signs enables detecting a case of unauthorised takeover and adversary usage of previously authorised means of communication.

Acknowledgment. This article has been prepared under a project funded by the National Centre for Research and Development under a scientific research programme for state defence and security "Future Technologies for Defence – a Contest for Young Scientists".

References

1. Plesse, T., Lecomte, J., Adjih, C., Badel, M.: OLSR performance measurement in a military mobile ad-hoc network. Ad-Hoc Netw. J. **3**(5), 575–588 (2004). Special Issue on Data Communication and Topology Control in Ad-Hoc Networks
2. Debes, M., Lewandowska, A., Seitz, J.: Definition and implementation of context information. In: Kyamakya, K., Jobmann, K., Kuchenbecker, H.-P. (eds.) Joint Second Workshop on Positioning, Navigation and Communication, WPNC 2005 & first Ultra-Wideband Expert Talk, UET 2005 (2005)
3. Yadav, P., Chatterjee, S.: A survey on dynamic spectrum access techniques in cognitive radio. Int. J. Next-Gener. Netw. **4**, 27 (2012)
4. Shafer, G.: A Mathematical Theory of Evidence. Princeton University Press, Princeton (1976)
5. Klabunde, R.E.: Cardiovascular Physiology Concepts, 2nd edn, p. 97. Lippincott Williams & Wilkins, Philadelphia (2012)
6. Pearl, J.: Reasoning with belief functions: an analysis of compatibility. Int. J. Approx. Reason. **4**, 363–389 (1990)
7. Voorbraak F.: On the justification of Dempster's rule of combination. Artif. Intell. **48**, 171–197 (1991). http://turing.wins.uva.nl/ ∼fransv/#pub
8. Dubois, D., Prade, H.: A set-theoretic view of belief functions. Int. J. Gen. Syst. **12**, 193–226 (1986)
9. Dubois, D., Prade, H.: Representation and combination of uncertainty with belief functions and possibility measures. Comput. Intell. **4**, 244–264 (1988)
10. Murphy, C.K.: Combining belief functions when evidence conflicts. Decis. Supp. Syst. **29**, 1–9 (2000)
11. Smets, P., Kennes, R.: The transferable belief model. Artif. Intell. **66**(2), 191–234 (1994)
12. Smets, P.: Data fusion in the transferable belief model. In: Proceedings of the 3rd International Conference on Information Fusion, 10–13 July 2000, Paris, pp. PS21–PS33 (2000)
13. Yager, R.R.: Hedging in the combination of evidence. J. Inf. Optim. Sci. **4**(1), 73–81 (1983)

14. Yager, R.R.: On the relationships of methods of aggregation of evidence in expert systems. Cybern. Syst. **16**, 1–21 (1985)
15. Yager, R.R.: On the Dempster-Shafer framework and new combination rules. Inf. Sci. **41**, 93–138 (1987)
16. Ali, T., Dutta, P., Boruah, H.: A new combination rule for conflict problem of Dempster-Shafer evidence theory. Int. J. Energy Inf. Commun. **3**(1), 35–40 (2012)
17. Brown, P., Bovey, J., Chen, X.: Context-aware applications: from the laboratory to the marketplace. IEEE Pers. Commun. **4**(5), 58–64 (1997)
18. Ryan, N.S., Pascoe, J., Morse, D.R.: Enhanced reality fieldwork: the context-aware archaeological assistant. In: Gaffney, V., et al. (eds.) Computer Applications in Archaeology. Tempus Reparatum, Oxford (1997)
19. Schmidt, A., Aidoo, K.A., Takaluoma, A., Tuomela, U., Van Laerhoven, K., Van de Velde, W.: Advanced interaction in context. In: Gellersen, H.W. (ed.) HUC 1999. LNCS, vol. 1707, pp. 89–101. Springer, Heidelberg (1999). https://doi.org/10.1007/3-540-48157-5_10
20. Dey, A., Abowd, G.: Towards a better understanding of context and context-awareness. In: Proceedings of the Workshop on the What, Who, Where, When and How of Context-Awareness, Affiliated with the CHI 2000 Conference on Human Factors in Computer Systems, New York (2000)

Dynamic Placement Algorithm for Multiple Classes of Mobile Base Stations in Public Safety Networks

Chen Shen[1] , Mira Yun[2] , Amrinder Arora[1(✉)], and Hyeong-Ah Choi[1]

[1] The George Washington University, Washington DC 20052, USA
{shenchen,amrinder,hchoi}@gwu.edu
[2] Wentworth Institute of Technology, Boston, MA 02115, USA
yunm@wit.edu

Abstract. As new mobile base stations (mBSs) have been constantly developed with various capacities, mobile coverage, and mobility models, the level of heterogeneity in public safety networks (PSNs) has been increasing. Since disasters and emergencies require the ad hoc PSN deployments, dynamic mBS placement and movement algorithm is one of the most important decisions to provide the critical communication channels for first responders (FRs). In this paper, we propose a heterogeneous mBS placement algorithm in an ad hoc public safety network. We define different classes of mobile base stations that have varying performance characteristics and consider three different FRs mobility models. Our proposed algorithm applies the modern clustering technique to deal with the characteristics of different kinds of mBSs.

Keywords: Mobile base station placement ·
Adhoc public safety networks · 5G · LTE

1 Introduction

As Public Safety Networks (PSNs) continue to get more and more coverage, many different classes of mobile base station (mBS) and deployment models are emerging [1–3]. Each of the hardware and deployment models has its own advantages disadvantages. The challenge has quickly shifted from merely meeting the network demand to being able to do that in a cost effective way.

Connectivity and coverage among user equipments (UEs) of some or all first responders (FRs) are the most basic requirements in many PSNs. When the FRs arrive at a disaster site, such as scene of a fire, volcanic eruption, terrorist attack, etc., a PSN must be dynamically deployed to meet the needs of different FRs. Many different deployment mechanisms exist for deploying the base stations. These include, but are not limited to, drones, truck bases, hot air balloons, and being manually established at a location in order to handle the transportation

© ICST Institute for Computer Sciences, Social Informatics and Telecommunications Engineering 2019
Published by Springer Nature Switzerland AG 2019. All Rights Reserved
A. Kliks et al. (Eds.): CrownCom 2019, LNICST 291, pp. 112–125, 2019.
https://doi.org/10.1007/978-3-030-25748-4_9

and installation of the mBSs. It is likely that these mechanisms will continue to evolve over time. For example, in a recent study from AT&T, the concept of the 'Flying Cow' or 'Cell on Wings' used in the extreme hazardous scenario serving the FRs is presented. Therefore, we study the mBSs placement problem from a deployment mechanism independent perspective and generalize these different mechanisms as various classes of transportation models that have their associated movement costs.

The rest of this paper is organized as follows. Section 2 discusses the related work and our contributions. Section 3 outlines the system model and problem statement, e.g. the mBS classes introduction, the UE mobility models, the channel model and the performance metrics. Section 4 describes the proposed method for continuous optimal mobile base station placement solution. Section 5 presents the empirical results regarding the performance of the proposed algorithm, and the comparison to the baseline model. Finally, Sect. 6 summarizes the paper and outlines ideas for future research.

2   Related Work

In this section, we review the recent mBS placement research relevant to cellular networks and at last our contributions are summarized.

In our previous work [4], we proposed algorithms for single class of mBS placement with limited UE mobility models. However, due to different service requirements in PSN scenarios [1] and the advantages of heterogeneous network architecture [5], multiple classes of mBSs have been developed, e.g., the Vehicle Network System (VNS), Cell on Light Truck (CoLT), Cell on Wheels (COW) and System on Wheels (SOW) [6]. The performance of the heterogeneous mBSs of Aerial LTE Base Stations (AeNB) and Portable Land Mobile Unit (PLMU) for PSN communication have been researched in [7]. In this work, our proposed algorithm is designed for dynamic placement of multiple classes of mBSs.

Due to the flexibility and mobility of drones, a growing amount of research has been focused on its applications in cellular network or PSN [8–10]. The channel modeling between drones and UEs in urban setting has been addressed in [11] where a realistic path loss and shadow fading model has been proposed. We apply the same channel model in our work. In [12,13], the energy restricted model for drones has been used. We use the Flying CoW model from AT&T [14] which has unlimited power supply with a thin tether. The dynamic limitation of drones has been considered and analyzed in [15]. But due to the fast development of drones, we apply more flexible restrictions on its dynamic limitation.

UEs in PSN scenarios are often referred as FRs. The studies on mobility models of FRs are very limited. A general study of network performance impact of UE mobility model has been addressed in [16]. In our work, we consider three UE mobility models, including Random Way Point Model (RWP) [15] and two PSN application related models. Generally, the placement of mobile base stations are NP-hard problem [12], where several approaches including greedy, numerical and game theory based are studied in [12,15,17]. We solve this problem with clustering technique.

In this paper, we address the above-mentioned challenges. Our contributions can be summarized as follows:

- We extend the work of mBS deployment with multiple classes of base stations to meet the most critical requirements in PSN scenarios. Different classes of mBSs are associated with corresponding coverage weights, recalculation frequency and relocation threshold.
- Our proposed algorithm applies the modern clustering technique. It solves the base station placement problem from the view of the central control instead of non-cooperative method, which is more suitable in PSN scenarios since most FRs and base stations are deployed with the central control.
- Besides the commonly used RWP model, we also develop two mobility models for PSN scenarios to represent the special characteristics for FRs since they are usually deployed in groups and often have specific working areas.

3   System Model and Problem Statement

In this section, we present the system model and discuss the different mobile base station classes, mobility and channel models and also the quantitative performance metrics that can be used to assess the performance of different solutions.

3.1   Mobile Base Station Classes

Various kinds of mBSs have been developed recently to meet the critical requirement of PSN scenarios. For example, the Emergency Drop Kit from AT&T shows the ability for rapid connectivity during emergencies in rural areas, as well as areas which may be temporarily out of communications [18]. It is designed for a short-term solution until the dedicated deployment arrives. The aerial base stations, especially drone-based ones get a lot of attention and development in recent research. AT&T has proposed their all-weather drone base station (Flying CoW) [14] in 2018 for coverage in extreme conditions. A thin tether is connected between the drone and ground terminal for unlimited power supply and high speed data connection. CoW and SoW have long been used as temporary communication solutions with different capacities due to their flexible mobility and fast deployment. In this paper, we model three kinds of mBSs as in Table 1. Our proposed dynamic placement algorithm can be easily extended for more classes of mBSs.

Table 1. Mobile base station classes

mBS	Drone	CoW	SoW
Capacity	Low	Medium	High
Recalculation frequency	High	Medium	Low
Relocation threshold	Low	Medium	High

3.2   UE Mobility Model

The performance impacts of UE mobility models for ad hoc networks and cellular networks have been analyzed in [16,19]. In this paper, we consider three UE mobility models including the RWP model which has been widely used for general UE mobility modeling. In RWP model, each UE independently selects a random destination inside the Region of Interest (ROI) and moves in a straight trajectory with a constant speed. After reaching the destination, UE pauses for a while before the next move.

In the other two models, more public safety features have been considered. Since most FRs are deployed as a group, in first model, UEs are firstly classified as leaders and non-leaders of the groups. The initial grouping association happens based on UE's role as a leader or following the closest leader. The leaders follow the RWP model and their group members follow the leaders. Figure 1 illustrates this situation. In the second model, a number of Points of Interest(POI) have been initialized in ROI and all the UEs move randomly approaching or around the closest POI depending on the distance between UE and POI, which are shown in Fig. 2. Similar event-driven and role-based (EDRB) mobility models are studied in [20,21]. We acknowledge the more accurate FRs mobility modeling, the better algorithm being able to design and this will be one direction for our future work.

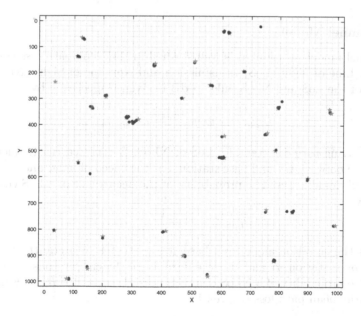

Fig. 1. UE mobility model with leaders in red star (Color figure online)

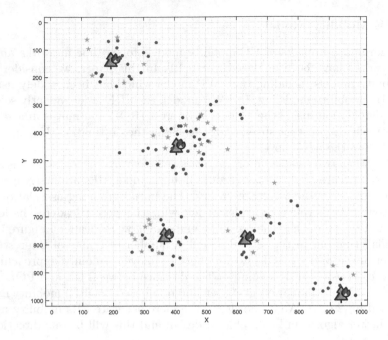

Fig. 2. Screen-shot of UE mobility model with 5 points of interest

3.3   Channel Model

The channel model we adopted is well studied in [11]. The path loss consists of two parts: Line of Sight (LoS) transmission and Non-line of Sight (NLoS). The path loss of the LoS and NLoS links in dB is given respectively by

$$L_{path} = 20log(\frac{4\pi f_c d}{c}) + \eta_{path}$$

where the string *path* can stand for LoS or NLoS, f_c is the carrier frequency, d is the distance between UE and base station, c is the speed of light, and η_{path} is the average additional losses. The probability of the occurrence of a LoS connection is given by:

$$P_{LoS} = \frac{1}{1 + \alpha e^{-\beta(\theta - \alpha)}}$$

where θ is the elevation angle from base station to UE or $\arctan(h/d)$, h is the height of base station, α and β are environment-dependent constants. Consequently, the probability of a NLoS connection is $P_{NLoS} = 1 - P_{LoS}$. Finally, the probabilistic mean path loss is given by

$$L = L_{LoS}P_{LoS} + L_{NLoS}P_{NLoS}$$

To simplify the problem, we assume all the mobile base stations from the same class have the same operational height and transmitting power. The interference from neighboring mBSs under a specific threshold will be neglected.

3.4 Performance Metrics

In this part, we discuss the performance metrics to evaluate our proposed algorithm compared to a baseline static placement method where the same set of mobile base stations are regularly placed in the ROI. The average SINR of all UEs based on the channel and communication model of a specific mobile base stations placement will be the main performance consideration. In order to evaluate the UE at the cell edge, the 5th percentile of SINR will be studied [10]. Since the main objective of base station (BS) placement is to reduce the distance between UE and BS, the UE-to-BS distance will also be addressed. For most drone-based placement problem, the collision avoidance scheme should be investigated, but our proposed clustering algorithm can automatically solve this problem.

4 Heterogeneous Mobile Base Station Placement Algorithm

In this paper, we propose a dynamic placement algorithm for heterogeneous mBSs that employs a variant of K-means++ clustering technique to deal with the characteristics of different kinds of mBSs. The dynamic placement algorithm consists of two parts, static placement of mBS for a specific UE distribution, and the periodical recalculation with mBSs moving threshold. The first part deals with the static situation and the second part makes the process dynamic. In the rest of this section, we describe them separately. To represent the mBS features from Table 1 in the algorithm, for each mBS, we assign three parameters: Capacity Weights C_w, Recalculation Period R_p in seconds and Relocation Threshold R_t in meters.

4.1 Clustering for Static UEs

In our previous work [4], K-means and its variant algorithms have been used for UE clustering by the single class of mBS. The K-means++ algorithm is an improvement with better initialization [22]. We modify the K-means++ with respects of different cluster size to represent the various capacities of mBS classes. The K-means clustering process is a series of UE-mBS association and mBS relocation to the centroid of its associated UEs iteratively. The mBS's capacity C_w will influence the UE association at each iteration by the weighted distance between UE and mBS. Algorithm 1 shows this process.

We show an example of the clustering in Fig. 3. The black dots represent the UEs in the ROI and in this case, total of 8 mBSs are deployed: one SoW (orange truck), two CoW (green car) and five drones (azure drone). The clustering edges are presented by the blue lines. With higher configured capacity weight of SoW, the coverage of the SoW in this example is the biggest. On the other hand, drones and CoW cover with small and medium capacities respectively. The location of mBS is determined by the centroid of its associated UEs.

Algorithm 1. Static UE Clustering with Different Mobile Base Station Capacities

---

Initial mBSs placement;
iter = 1;
while *iter* < *MAX_ITER* **do**
    for $i = 0; i < N_B; i = i + 1$ **do**
        num_UE = 0;
        X = 0;
        Y = 0;
        for $j = 0; j < N_U; j = j + 1$ **do**
            if $U_j \in B_i$ **then**
                $num\_UE = num\_UE + 1$;
                $X = X + U_j.x$;
                $Y = Y + U_j.y$;
            end
        end
        $B_i.x = X \div num\_UE$;
        $B_i.y = Y \div num\_UE$;
    end
    Affiliate UE to mBS in the following loops;
    for $j = 0; j < N_U; j = j + 1$ **do**
        $min_{dist} = Inf$;
        for $i = 0; i < N_B; i = i + 1$ **do**
            $dist = \sqrt{(U_j.x - B_i.x)^2 + (U_j.y - B_i.y)^2}$;
            $dist_w = dist \div B_i.C_w$;
            if $dist_w < min_{dist}$ **then**
                $min_{dist} = dist_w$;
                $U_j.B_{id} = i$;
            end
        end
    end
    iter = iter + 1 ;
end

4.2 Periodic Recalculation for Dynamic UEs

The UE clustering is for the static situation, and we make this re-cluster periodically to adapt to UEs' mobility. The frequency of this recalculation and distance to trigger mBSs' movement depend on the characteristics of mBS classes. For example, the cost of drone's movement should be much less than CoW or SoW, thus the recalculation frequency $(1/R_p)$ of drones should be much higher and the distance threshold (R_t) for movement should be much smaller than CoW and SoW. This process is illustrated in Fig. 4.

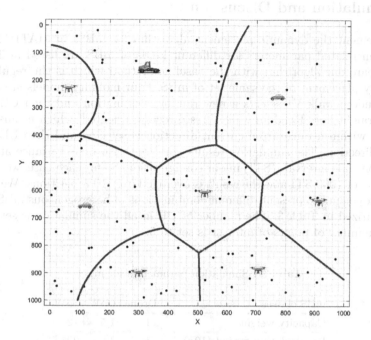

Fig. 3. Static placement of 3 mBS classes: one SoW (truck), two CoW (car) and five drones (Color figure online)

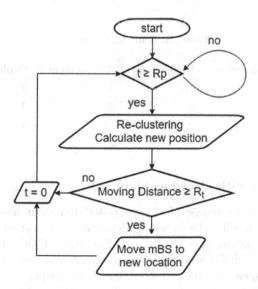

Fig. 4. Flow chart of periodical recalculation of mBS

5   Simulation and Discussion

We implement the dynamic placement algorithm for mBSs in MATLAB. The simulation related parameters for different kinds of mBS are listed in Table 2. We compare our algorithm with the baseline method which is the regular and stationary placement of the same set of mBS. Four mobility models are considered. Random walk V1 selects a new random destination and then UE moves in the straight line. Random walk V2 is direction oriented, which means a new direction within a range from current direction is selected and then UE moves in that direction. The simulation results converge with little variance after 600 simulation time intervals (STI) in our configuration. So for each case, we run the simulation of 1000 STI and the result is averaged over 10 repetitions. We choose four deployments to present different combinations of heterogeneous mBS which is summarized in Table 3. The number of UE in all the simulation is set to 100 and the number of points of interest is set to 5.

Table 2. Simulation parameters for mBS

mBS classes	Drone	CoW	SoW
Capacity weight	1	1.5	2
Recalculation period (10 s)	1	30	90
Relocation threshold (meters)	1	30	50
Height (meters)	30	10	10
Transmit power (watts)	20	30	40

Table 3. Four deployments

	Deployment 1	Deployment 2	Deployment 3	Deployment 4
Drone	5	10	0	0
CoW	2	0	5	0
SoW	1	0	0	3

5.1   Comparison with Baseline Algorithm

We compare the performance of the proposed dynamic heterogeneous mBS placement algorithm with the baseline algorithm in Random Walk V2 model in Sect. 3.2 and deployment 1 in Table 3. The three CDF in Fig. 5 show the SINR, 5th percentile SINR and UE to mBS distance of the two compared algorithms. Generally speaking, the proposed algorithm outperform the baseline one in all the three factors, especially in the 5th percentile SINR.

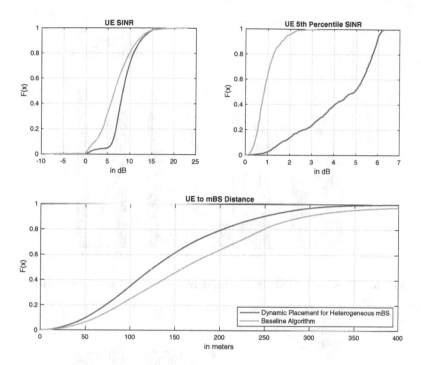

Fig. 5. Performance CDF comparison with baseline algorithm in random walk V2 model with deployment 1

5.2   UE SINR

In Fig. 6, the UE average SINR is compared for different deployments and UE mobility models. In the dimension of UE mobility models, two random walks and the following leader are very similar with slightly better SINR in following leader model. But in the POI model, the first two deployments are much worse than the other two deployments. With the deployments with only CoW or SoW POI provides better performance than the other two models. Because CoW and SoW with much better capacities can be deployed near the POI before UE placements, POI mobility model achieves high performance with Deployment 3 and 4. The second deployment with only drones performs worst in all four mobility models because drones have relatively weak transmission capacity and their moving flexibility is not advantageous when the interesting point location is already defined.

5.3   5th Percentile UE SINR

In order to consider UEs at the cell edge or the worst case in SINR, the 5th percentage SINR of different scenarios is shown in Fig. 7. Similar conclusion can be drawn for the much worse SINR in the first two deployments with drones from the former simulation results. Two observations can be found here: the 5th

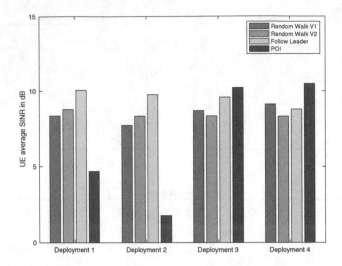

Fig. 6. UE average SINR in various deployments and mobility models

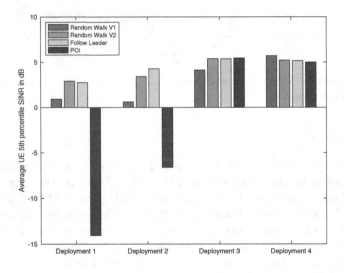

Fig. 7. Average 5th percentile SINR in various deployments and mobility models

percentile SINR in random walk V1 is much lower than in random walk V2, and the first deployment has the lowest value. The first observation is hard to explain since the two random walk models intuitively should perform similarly. But the actual simulation configurations can be the reason. The second observation can draw the conclusion that the only drone deployments can achieve higher SINR for UE at the serving edge due to the high mobility and flexibility of drones.

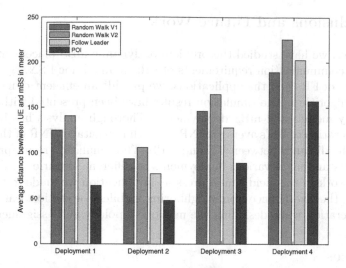

Fig. 8. Average distance between UE and mBS in various deployments and mobility models

5.4 Total mBS-to-UE Distance

Because the normal cell configuration for permanent base station optimization is usually not applicable for mBS, especially the drone-based BSs, the distance between UE and mBS is an important factor. The average UE-to-mBS distance is illustrated in Fig. 8. In the dimension of UE mobility models, the Following Leader and POI models outperform the two random walk models mainly due to the relatively more clustered UE distribution in these models. Otherwise on the dimension of deployments, the one with only drones has the least UE-to-mBS distance and the only CoW and only SoW deployments increase the distance gradually. The UE-to-mBS distance is impacted heavily by the flexibility of mBS.

5.5 Further Discussions

We compare four different deployments in the simulation. In the reality, each kind of mBS should be associated with corresponding cost in operation, which is not considered in our current work. In that case, the optimal deployment should depend on the 'budget' or the mBS' availability in each kind, and the disaster's property.

We use 1000 STI for our simulation, but in reality the disaster scenario and the communication requirements from FRs can vary hugely. For example, in the POI model, the point of interest can move due to the disaster's changing or other factors. But this is out of our consideration in the current work. The simulation result should provide a sight of basic understanding on various deployments and disaster situation.

6   Conclusions and Future Work

In this paper, we have studied the problem of dynamic mBS placement to meet the critical communication requirements of FRs in an ad hoc PSN. By considering the class of FRs and the applications, we provide an efficient dynamic mBS placement algorithm. The simulation results have been presented with different UE mobility models and mBS deployments. Thorough analysis has been done with consideration of UE's average SINR, the 5th percentage SINR of the deployment and the distance between UE and mBS. The simulation result provides in depth understanding of various deployments in different scenarios of FRs.

Future work in this field can address a simplification we made in this work. Specifically, the interference from neighboring mobile base stations can be taken into consideration when designing the network topology and assignment.

References

1. Nationwide Public Safety Broadband Network (NPSBN) QoS Priority and Preemption (QPP) Framework, November 2015. [FirstNet CTO Whitepaper]
2. Rouil, R., Izquierdo, A., Souryal, M., Gentile, C., Griffith, D., Golmie, N.: Nationwide safety: nationwide modeling for broadband network services. IEEE Veh. Technol. Mag. **8**, 83–91 (2013)
3. Li, X., Guo, D., Yin, H., Wei, G.: Drone-assisted public safety wireless broadband network. In: 2015 IEEE Wireless Communications and Networking Conference Workshops (WCNCW), pp. 323–328, March 2015
4. Shen, C., Yun, M., Arora, A., Choi, H.-A.: Efficient mobile base station placement for first responders in public safety networks. In: Arai, K., Bhatia, R. (eds.) FICC 2019. LNNS, vol. 70, pp. 634–644. Springer, Cham (2020). https://doi.org/10.1007/978-3-030-12385-7_46
5. Hu, R.Q., Qian, Y.: Resource Management for Heterogeneous Networks in LTE Systems. Springer, New York (2014). https://doi.org/10.1007/978-1-4939-0372-6
6. (2016). https://firstnet.gov
7. Gomez, K., Hourani, A., Goratti, L., Riggio, R., Kandeepan, S., Bucaille, I.: Capacity evaluation of aerial LTE base-stations for public safety communications. In: 2015 European Conference on Networks and Communications (EuCNC), pp. 133–138, June 2015
8. Li, X.: Deployment of drone base stations for cellular communication without apriori user distribution information. In: 2018 37th Chinese Control Conference (CCC), pp. 7274–7281 (2018)
9. Jiang, F., Swindlehurst, A.L.: Optimization of UAV heading for the ground-to-air uplink. IEEE J. Sel. Areas Commun. **30**, 993–1005 (2012)
10. Merwaday, A., Guvenc, I.: UAV assisted heterogeneous networks for public safety communications. In: 2015 IEEE Wireless Communications and Networking Conference Workshops (WCNCW), pp. 329–334, March 2015
11. Al-Hourani, A., Kandeepan, S., Lardner, S.: Optimal LAP altitude for maximum coverage. IEEE Wirel. Commun. Lett. **3**, 569–572 (2014)
12. Huang, H., Savkin, A.V., Ding, M., Kâafar, M.A.: Optimized deployment of autonomous drones to improve user experience in cellular networks. CoRR, vol. abs/1712.02124 (2017)

13. Zeng, Y., Zhang, R.: Energy-efficient UAV communication with trajectory optimization. IEEE Trans. Wirel. Commun. **16**, 3747–3760 (2017)
14. Pregler, A.: Extreme Connections. AT&T innovation Blog, May 2018. http://about.att.com/innovationblog/extreme_connections
15. Fotouhi, A., Ding, M., Hassan, M.: Flying drone base stations for macro hotspots. IEEE Access **6**, 19530–19539 (2018)
16. Lin, X., Ganti, R.K., Fleming, P.J., Andrews, J.G.: Towards understanding the fundamentals of mobility in cellular networks. IEEE Trans. Wirel. Commun. **12**, 1686–1698 (2013)
17. Ghazzai, H., Yaacoub, E., Alouini, M., Dawy, Z., Abu-Dayya, A.: Optimized LTE cell planning with varying spatial and temporal user densities. IEEE Trans. Veh. Technol. **65**, 1575–1589 (2016)
18. (2018). http://about.att.com/story/firstnet_connected_bubble.html
19. Batabyal, S., Bhaumik, P.: Mobility models, traces and impact of mobility on opportunistic routing algorithms: a survey. IEEE Commun. Surv. Tutor. **17**, 1679–1707 (2015)
20. Nelson, S.C., Harris III., A.F., Kravets, R.: Event-driven, role-based mobility in disaster recovery networks. In: Proceedings of the Second ACM Workshop on Challenged Networks, CHANTS 2007, New York, NY, USA, pp. 27–34. ACM (2007)
21. Hong, X., Gerla, M., Pei, G., Chiang, C.-C.: A group mobility model for ad hoc wireless networks. In: Proceedings of the 2nd ACM International Workshop on Modeling, Analysis and Simulation of Wireless and Mobile Systems, MSWiM 1999, New York, NY, USA, pp. 53–60. ACM (1999)
22. Arthur, D., Vassilvitskii, S.: K-means++: the advantages of careful seeding. In: Proceedings of the 18th Annual ACM-SIAM Symposium on Discrete Algorithms (2007)

Spectrum Analysis Using Semantic Models for Context

Vaishali Nagpure$^{(\boxtimes)}$ (ID), Stephanie Vaccaro, and Cynthia Hood (ID)

Illinois Institute of Technology, Chicago, IL 60616, USA
{vnagpure, svaccaro}@hawk.iit.edu, hood@iit.edu

Abstract. With the ever-increasing demand for spectrum to support wireless innovation, it is critical to understand the fine-grained characteristics of spectrum use in frequency, space and time to facilitate greater spectrum sharing. Contextual information is needed to analyze how the spectrum is being utilized and understand the drivers for spectrum use dynamics. Since human activity often drives spectrum use, understanding this activity can provide significant insight. Analysis of wideband spectrum is extremely time consuming as each band has unique characteristics, domain knowledge and usage drivers. Toward automated analysis, this paper proposes an approach to incorporate contextual information into the analysis utilizing semantic models to capture domain and human activity knowledge. This approach is illustrated through analysis of spectrum measurements of four frequencies licensed to the Chicago White Sox.

Keywords: Spectrum occupancy · Spectrum measurements ·
Semantic modeling · Land Mobile Radio

1 Introduction

Increased access to the radio frequency (RF) spectrum is critical to continued wireless innovation. Spectrum is an extremely valuable natural resource in high demand, yet there is limited understanding about how this resource is being used. Spectrum allocations and licenses provide information about who has the right to utilize the spectrum and how they may use it, but do not provide insight into actual usage. Spectrum measurements play a key role in understanding spectrum use providing information about spectrum utilization in space, frequency and time. In particular, spectrum occupancy measurement studies involve determining the percentage of time that a given frequency band or channel is utilized over a period of time in a given location. These studies can potentially inform spectrum policy and management decisions, and facilitate more efficient spectrum usage through sharing. It is clear that measurement efforts are critical, yet most spectrum measurement datasets appear to be used only by the group that collected them. This lack of sharing along with the expense and effort to collect your own measurements is an impediment to spectrum management research. Although there have been many different types of spectrum measurement campaigns and studies, the technical and regulatory communities struggle to extract the information they need from the existing studies and datasets.

© ICST Institute for Computer Sciences, Social Informatics and Telecommunications Engineering 2019
Published by Springer Nature Switzerland AG 2019. All Rights Reserved
A. Kliks et al. (Eds.): CrownCom 2019, LNICST 291, pp. 126–139, 2019.
https://doi.org/10.1007/978-3-030-25748-4_10

Given our long-term experience collecting spectrum measurements through the IIT Spectrum Observatory [1], we have witnessed first-hand many of the challenges associated with collecting, managing and sharing spectrum measurements. Spectrum measurements are highly complex spatiotemporal data sets that require very specialized domain knowledge to collect, analyze and interpret [2]. Wideband data (as collected at the IIT Spectrum Observatory) spans a large number of frequency bands and each frequency band has it's own unique domain knowledge including physical character-istics, potential transmitters, etc.

Analysis of wideband data begins with parsing the data into frequency bands based on how the spectrum has been allocated. The goal of this type of analysis is typically to determine who is using the spectrum (i.e. emitters) and how they are using it. Although there are bands that are well documented in terms of spectrum allocation (e.g. public safety), it is often challenging to go beyond high-level descriptions such as those found in [3] to determine how a particular band is being utilized in a given geographic area. It can be particularly difficult to get information on government transmitters and since large portions of the spectrum are allocated for either exclusive or shared government use, this is a significant void. In the United States, a subset of potential emitters can be found in the Federal Communication Commission (FCC) Licensee database [4], but the existence of a license in a particular area does not guarantee an active emitter.

One of the challenges of analyzing the data collected in these monitoring efforts is putting the measurements into proper context. Since analysis typically involves iden-tifying both usual and unusual spectrum usage, this generally involves both under-standing how the spectrum is used in the location being measured and identifying potential triggers for changes in usage. Frequently usage can be traced directly or indirectly to human activity. An understanding of the human activity that drives spectrum dynamics can facilitate deeper insights into spectrum usage.

To address these issues, our long-term research goal is to automate analysis of spectrum measurements. This paper describes an approach that utilizes semantic web-based models and tools for ingesting and utilizing domain and human activity infor-mation in machine-readable format. This information can provide the contextual information needed to analyze and understand how the spectrum is being utilized. This paper focuses on the analysis of frequency bands licensed to the Chicago White Sox, a professional baseball team that plays their home games at Guaranteed Rate Field. The main contribution of this paper is to lay the groundwork for automating spectrum analysis. This includes detailing the type of contextual information that is needed to analyze spectrum measurements, identifying machine readable sources for the con-textual information, demonstrating how this information may be collected automati-cally, and using semantic web techniques to model domain knowledge and human activity.

The rest of the paper is organized as follows. Section 2 includes background information about the measurement system along with related work. The contextual information related to spectrum measurement and analysis is described in Sect. 3. Section 4 describes the statistical analysis of the measurements of the Chicago White Sox frequencies studied. The design and implementation of the semantic models is detailed in Sect. 5. The results are highlighted in Sect. 6 and conclusions and future work are discussed in Sect. 7.

2 Background and Related Work

2.1 Measurements

The IIT Spectrum Observatory (IITSO) has been monitoring the 30–6000 MHz radio spectrum of the city of Chicago since mid-2007 from its location on top of the 22 story Tower on IIT's Main campus on the south side of Chicago [1]. The IITSO uses energy detection sensing to capture measurements and has a resolution bandwidth of 3 kHz for all bands. This results in approximately 93 MB of data per day. This data has been used to provide a high-level view of the spectrum occupancy in Chicago [5], but there are limitations to wideband sensing.

In the Land Mobile Radio (LMR) bands where channel bandwidths are narrow (<30 kHz), short transmissions cannot be detected due to the high sweep time. To address this issue, an additional measurement system was deployed. This system measures a subset of the LMR bands utilizing a Tektronix RSA 306 spectrum analyzer. It enables capture of higher resolution data. The band plan and resolution bandwidth were configured as shown in Table 1. This results in approximately 16 GB of data per day.

Table 1. Band plan (Tektronix)

Band plan (MHz)	Resolution bandwidth
100–200	3 kHz
400–460	3 kHz
460–463	1 kHz
463–516	3 kHz
800–900	3 kHz

This paper focuses on the analysis of four frequency bands shown in Table 2. These frequencies have been licensed to the Chicago White Sox and the transmitter address specified in the FCC license database [4] corresponds with Guaranteed Rate Field. Guaranteed Rate Field is the home ballpark for the White Sox and is located approximately .4 miles away from the IIT Tower where the measurement systems are installed. The license database does not provide a description of usage, but this is available from RadioReference.com [6], a crowd sourced radio communications data provider. Radio Reference is utilized and maintained by amateur radio enthusiasts and provides a complete frequency database, trunked radio system information, along with FCC license data.

Table 2. Chicago White Sox frequencies studied

Frequency (MHz)	License	Description
461.2	WPLI617	Parking
462.05	WPXR683	Guest relations operations
464.675	WQDD864	Sportservice concessions
464.95	WPLL482	Medical, guest relations, premium seating

2.2 Related Work

A good survey of spectrum occupancy measurements is given in [7]. Spectrum studies have mainly taken place on the order of days [8, 9], weeks [10–12] months [13] and longer [5]. Globally, there have been a number of spectrum occupancy studies and spectrum surveys conducted over the last decade. Most of these studies are characterized by the fact that they typically collecting wideband data beginning at a few Hz and going up to a 3–5 GHz. These studies focused on spectrum occupancy and as a result have identified substantial opportunities for frequency reuse in many bands, particularly those currently assigned to legacy systems. Analyses have also shown occupancy trends over both short periods of time and seasonal variations occurring weekly and yearly [5].

Occupancy models are broadly classified as time-dimension models, frequency-dimension models, location-dimension models [14–16]. Given knowledge of a particular channel's behavior as described by a set of statistical models, prediction of channel occupancy is possible. These models are limited in their ability to characterize significant dynamics in how spectrum is used.

SpecInsight [12] is an intelligent wideband spectrum sensing and analysis system that learns the characteristics of the signals in each frequency band and adjusts the sensing parameters to maximize detection. SpecInsight classifies usage patterns based on frequency and time attributes and maintains statistics on the timing of pattern occurrences. A detailed overview for the conventional narrowband and wideband spectrum sensing approaches are given in [17] which tackles the trade-off between system performance and practical system implementations very well. In [18], semantic modeling is used to capture configuration, domain knowledge and other potentially relevant information in a way that it can be fused with measurements for analysis and in particular can provide labels for the spectrum data.

3 Contextual Information

The Cambridge English dictionary [19] defines context as "the situation within which something exists or happens, and that can help explain it." Context is necessary to understand how to explain the results of quantitative analysis of spectrum measurements. Spectrum use is a function of frequency, space and time. For a particular location, given how the spectrum is licensed and allocated and how it has been measured, are the results consistent with what is expected? For many bands including LMR, spectrum use is driven by human activity so it is possible to gain additional insight through this lens. Going beyond statistical characterizations to understand how the spectrum is used and when it might or might not be available for sharing is critical for spectrum sharing and dynamic spectrum access.

In this paper, we analyze spectrum measurements of four different frequencies collected at the IIT Tower over a one year period, January 2018–December 2018. We focus on the domain context which includes how the spectrum has been licensed and allocated for use, and the human activity context which includes the drivers of spectrum use. We know that organizational conventions and protocols typically govern how human activity drives spectrum use in the LMR bands. Context is usually included in the narrative part of the analysis. This research proposes to model the contextual information to shift it from being part of the narrative to correlating context with quantitative results with a long-term goal of doing this automatically. As a substantive step in this direction, the machine consumable sources for context information are identified and modeled.

3.1 Domain Context

The domain is considered to be any relevant information about the specific frequency band under study. This includes who the frequency has been licensed to as well as how the licensee allocates the frequency for use in the licensee organization. It also includes characterization and location of the transmitters as specified in the license. In the United States, the FCC licenses spectrum for commercial use. Figure 1 shows part of the FCC database entry for the 461.2 MHz spectrum that has been licensed to the Chicago White Sox. It should be noted that the 461.2 MHz frequency along with other 3 frequencies studied in this paper have been licensed to several entities in the Chicago area, so geographic sharing is already being done. Given the proximity of the IIT Tower to the White Sox ballpark, the spectrum measurements sensed at the IIT Tower are attributed to the White Sox use.

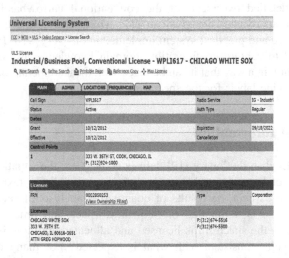

Fig. 1 FCC database entry for White Sox license for 461.2 MHz

Figure 2 shows another part of the database entry with the location of the transmitters. There is a fixed antenna that is used for mobile device communication. The antenna details provided in the license include the emission designator, a code that indicates the frequency bandwidth, modulation type and nature and information type. The emission designator for this antenna is 20K0F3E. Figure 3 shows the decoding of 20K0F3E to indicate a 20 kHz FM signal for analog voice.

Locations Summary

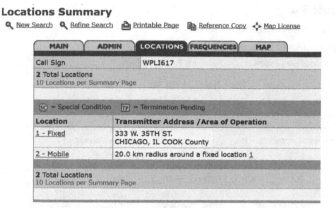

Fig. 2 White Sox 461.2 MHz Transmitter details

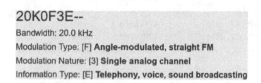

Fig. 3 Emission designator details for 20K0F3E [20]

The FCC database can be used to determine what frequencies the White Sox have licensed, but it doesn't provide information on how these frequencies are being used. Additional information on usage can be obtained from RadioReference.com. RadioReference.com is a crowdsourced wiki that is maintained by radio enthusiasts, many of whom have experience in the field and bring that knowledge into the wiki. RadioReference provides valuable information on how frequencies are being used and the information is logically grouped by the function, organization or venue. For example, RadioReference provides information about how the various sports teams utilized the spectrum they've licensed. Figure 4 shows the RadioReference information about the Chicago White Sox frequencies [21]. There are 16 total frequencies included along with other information including a general description of the functions the frequencies are used for. Fifteen of the frequencies are active and one is deprecated. The report notes that the frequencies were confirmed to be active for the 2016 season and we cross-verified the information with the FCC database to ensure that the frequencies were indeed still active. There is always a concern about the reliability of the sources and correctness of the information gathered. This is an issue for further study, but not one that is unique to spectrum analysis.

Chicago White Sox

■ CARMA Chicago 2010 White Sox frequency report

Conventional frequencies for Guaranteed Rate Field:
Note: these frequencies were confirmed still active during the 2016 season.

Frequency	Input ▫	License	Type	Tone	Alpha Tag	Description	Mode	Tag
463.72500	468.72500	WPLL482	RM		CWS Security	Security & Safety		Security
461.20000	466.20000	WPLI617	RM		CWS Parking	Parking		Business
464.67500	469.67500	WQDD864	RM	223 DPL	CWS Concessn	Sportservice - Concessions		Business
462.05000	467.05000	WPXR683	RM		CWS GstRelations	Guest Relations		Business
464.28750			M	67.0 PL	CWS Ticketng	Ticketing		Business
464.81250			M	466 DPL	CWS Food	Food		Business
464.51250			M	226 DPL	CWS Food	Food		Business
464.75000	469.75000		RM	67.0 PL	CWS Sec old	Security - Main (old)		Deprecated
461.45000	466.45000	WQAU450	RM	67.0 PL	CWS Security	Former Security (expired 2014)		Security
464.65000			M	047 DPL	CWS Ops46455	Operations		Business
456.56250			M	051 DPL	CWS 456.5625	Food-Beverage service		Business
464.83750			M	051 DPL	CWS Janitor	Janitorial		Business
468.66250		WQDD864	M	223 DPL	CWS Concessn	Sportservice - Concessions		Business
463.72500	468.72500	WPLL482	RM		CWS Trades	Trades		Business
464.95000	469.95000	WPLL482	RM		CWS Medical	Medical		Business
464.95000	469.95000	WPLL482	RM		CWS PremSeating	Guest Relations / Premium Seating		Business

Fig. 4   Chicago White Sox frequency report from RadioReference.com

3.2   Human Activity Context

People drive the spectrum use in the LMR bands, including those frequencies licensed to the White Sox. More specifically, the White Sox spectral activity is centered around Guaranteed Rate Field. This is where the baseball games take place, but it also includes offices for day-to-day operations and space for special events. Special events may be public non-baseball events [22] or private events such as weddings [23]. The spectrum is used to support the functions necessary to support the people that come to Guaranteed Rate Field. The specific functions that correspond to the frequencies that we are studying include parking, concessions, guest relations and medical.

Although the White Sox organization operates year round. Baseball is seasonal, with the major league baseball season running from April to October. This means that we would expect much more human activity during the season, especially during game days. This activity at the ballpark includes seasonal workers along with fans who come to watch the games. During the 2018 season, the average attendance at White Sox home games was 20,110 [24].

There is a wealth of information available about each baseball game online. The schedules for each season are published well ahead of the season start. After each game, detailed information about the game including start time, end time, attendance, scoring and any delays is available at Baseball-Reference.com [24]. This study utilized [24] along with a ticket platform, SeatGeek [25]. SeatGeek has an Application Programming Interface (API) so it was easily queryable and reliably provided the planned game days and times. This information had to be verified with the game reports at Baseball-Reference.com to identify games that were rained out or delayed.

4 Statistical Analysis of Spectrum Data

One year of data was analyzed. The spectral occupancy was estimated based on the threshold method. Occupancy thresholds were determined by estimating the noise floor at each frequency. A frequency was considered to be occupied if the noise floor threshold was exceeded. The occupancy was first estimated on an hourly basis. The hourly occupancies were aggregated to provide daily occupancy estimates and then these were further aggregated to provide monthly and yearly estimates. The average occupancy of the four frequencies under study for all of 2018 is shown in Fig. 5.

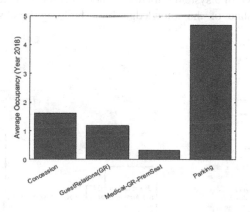

Fig. 5 Average occupancy for the year 2018

Figure 5 shows that there is activity on these frequencies, but they are not highly occupied. Thus there is potential for further sharing beyond the geographic sharing we mentioned earlier. The question is how and when the frequencies are being used.

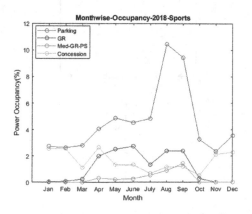

Fig. 6 Month wise occupancy

To get insight into this, we look at the monthly occupancy of each frequency. Figure 6 illustrates the seasonal trends that correspond to baseball season. Three of the four frequencies show increased occupancy during the baseball season. Clearly, further exploration is warranted. To do this, it is necessary to know when events (e.g. games) are taking place at the ballpark.

5 Semantic Models

Semantics can be used to model information in a machine-consumable way that enables reasoning. A key piece of a system utilizing semantics is the knowledge base. Figure 7 shows the current state of the knowledge base building blocks in our prototype application.

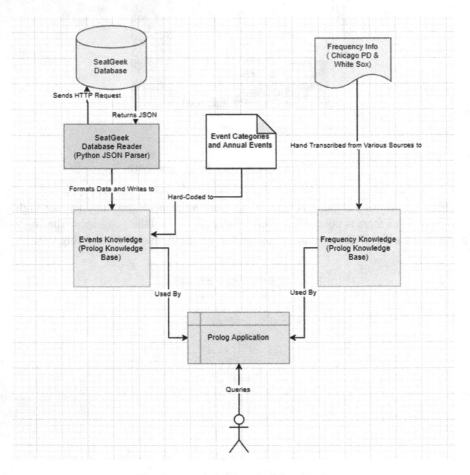

Fig. 7 Knowledge base building blocks

A prototype application was developed to match the frequency occupancy data gathered from our spectrum observatory with frequency specific knowledge and events going on in and around Chicago including White Sox games. In order to do this, Python's http database request abilities and Prolog's knowledge representation and reasoning capabilities were utilized.

Python is used to connect to and query SeatGeek's [25] event database via http request. In response to the query, a JSON file is returned which is then parsed and formatted into prolog readable text. Most of the application including the knowledge base is developed in Prolog. The Prolog knowledge base is split into two files, one for spectrum domain knowledge and one for event knowledge. The spectrum domain knowledge includes models and data for Chicago public safety frequencies, both zone and citywide, as well as White Sox specific frequencies utilized during White Sox games. The human activity domain knowledge captured in this prototype include the models and data for annual holidays, and events pulled from the SeatGeek database. These two knowledge bases are then queried using the main Prolog application.

The knowledge bases can be queried for two types of information: event by date information and frequency information as shown in Fig. 8. Using check Events (Month, Day, Year), a user can give a date in the form of month, day, and year and the knowledge base returns all events occurring on that specified date. Frequencies that might be associated with the event are also returned. In the example query shown in Fig. 8, the Chicago White Sox home game against the Detroit Tigers is identified and the associated White Sox frequencies are identified. Another event scheduled for the same day, Chicago Gourmet is also identified. The frequencies identified for this event include the Chicago Police Department channel for the Zone where this event is located (Zone 4) and the citywide channels used for events.

```
?- checkEvents('September',29,2019).
Got Events: [Detroit Tigers at Chicago White Sox,Chicago Gourmet - Chicago]

***********************************
Event Found: Detroit Tigers at Chicago White Sox
Event Type: outdoor
Time: 2:10pm

Check Channels:
Zones: zone_13
Citywide: citywide_5 | citywide_3 | citywide_1
-----------------------------------
This is a White Sox Game.
The following frequencies are also in use:
-----------------------------------
Frequency: 461.2
Channel Name: parking
Uses: [parking]
Frequency: 464.675
Channel Name: sportservice_concessions
Uses: [concessions]
Frequency: 462.05
Channel Name: guest_relations
Uses: [guest_relations]
Frequency: 464.95
Channel Name: medical
Uses: [medical]
Frequency: 464.95
Channel Name: guest_relations_premium_seating
Uses: [guest_relations_premium_seating]
***********************************
Event Found: Chicago Gourmet - Chicago
Event Type: outdoor
Time: 12:00pm

Check Channels:
Zones: zone_4
Citywide: citywide_5 | citywide_3 | citywide_1
***********************************
true.
```

Fig. 8 Example check events query to knowledge base

Using the command find Freq (INFO), the user can query for frequency information based on a given frequency, frequency name, or frequency usage. Example queries are shown in Fig. 9. The prototype returns all frequency information for a given match in the knowledge base. Using this prototype, we are able to associate human activity with the frequencies that might be impacted, thereby providing context to facilitate analysis of the actual measurements. This can be done with either a date or a frequency as a starting point.

```
?- findFreq(460.125).
Frequency: 460.125
Channel Name: citywide_1
Uses: [major accident,traffic,gangs,public housing,cta]
true.

?- findFreq('wanted flashes').
Frequency: 460.275
Channel Name: citywide_3
Use: [wanted flashes,maintenance,films,admin]
true.

?- findFreq(zone_5).
Frequency: 460.5
Channel Name: zone_5
Uses: [2,21]
true.

?- findFreq(parking).
Frequency: 461.2
Channel Name: parking
Uses: [parking]
true.
```

Fig. 9 Example find Freq queries to knowledge base

6   Results

The White Sox home games were extracted from the knowledge base describe above so the game days and non-game days could be compared. Figure 10 shows an example of gameday occupancy throughout the day and Fig. 10 shows an example of non-game day occupancy.

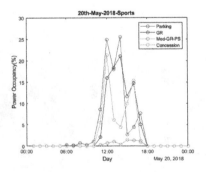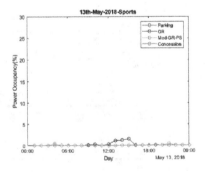

Fig. 10 Occupancy for home game day (left) and non-game day (right)

Figure 10 shows the occupancy of the four frequencies under study on a home gameday. The game details are as follows:

- Date: Saturday, May 20, 2018
- Start Time: 1:10 p.m. CST
- Attendance: 16,829
- Game Duration: 2:28 [24].

Comparison of Fig. 10 illustrates the occupancy increase during game time which is expected. Also, as expected the occupancy increases on the parking channel before the game starts and it is followed by increases on the other channels. This is an example of how human protocols drives spectrum occupancy. It is quite obvious that spectral activity around the parking function will start early as compared to the frequencies used for other functions which become more active once the fans enter the park. Figure 10 shows the non-game day occupancy charts for Sunday, 13th May 2018. There is little spectral activity on a non-gameday, weekend day. The maximum occupancy over the entire day is less than 2% for all four frequencies.

This study has shown that a significant portion of the occupancy on the four channels studied can be attributed to White Sox home game usage. Given the details of the game, the usage is reasonably predictable. We have also discovered that the usage cannot be completely attributed to White Sox games or even other public events. Further research is necessary to determine the patterns of both game-day and non-game day usage. It appears that there is significant opportunity to share this spectrum, particularly on non-game days. Out of four channels, we can see that Medical-Premium Seating-Guest Relations channel is the least utilized, having the highest availability for sharing.

7 Conclusion and Future Work

This paper describes the challenges of spectrum analysis and motivates the need for more in-depth analysis and fine-grained modeling for the purpose of spectrum sharing. Analysis requires a substantial amount of contextual information that includes domain knowledge along with information about human activities. Contextual information can be derived from a variety of different information sources and used to build semantic models that can be used to implement a knowledge base. The goal is to have machine-consumable information that can be used to build and populate semantic models that can be reasoned over.

This paper focuses on the White Sox channels used in Chicago. Semantic models were developed and coded in SWI-Prolog and Python to get detailed insights about events at Guaranteed Rate Field, especially White Sox home gamedays. Some data was pulled automatically through the SeatGeek API, whereas other data was manually entered from online sources. The goal is to automatically pull other relevant information and methods to do this are being investigated. These event (gameday) models allowed us to classify the data into gamedays and non-gamedays for analysis. This is an important step toward automating the analysis.

Ongoing work involves use of machine learning techniques for classifying the data to develop models and also to detect anomalies. As shown, even relatively straight-forward cases like the ones we've explored in this paper are not simple and require significant information to get an accurate understanding of spectrum. Spectrum behavior is challenging to interpret and prediction of usage is driven by many factors such as planned and unplanned events, weather and human protocols. Further development of the semantic models is needed to capture the many different information sources as well as the correlations across time and frequency bands.

This research begins to lay the foundation for intelligent spectrum measurement and monitoring systems that reason over information from a variety of sources, analyze data and situations, and make decisions. The automatic generation of domain, activity and analysis-based metadata opens new avenues for more comprehensive and timely analysis of spectrum measurement data.

Acknowledgements. The authors would like to acknowledge support from the National Science Foundation through NSF 1526638.

References

1. Bacchus, R.B., Fertner, A.J., Hood, C.S., Roberson, D.A.: Long-term, wide-band spectral monitoring in support of dynamic spectrum access networks at the IITtspectrum observatory. In: 3rd IEEE Symposium on New Frontiers in Dynamic Spectrum Access Networks, DySPAN, pp. 1–10. IEEE (2008)
2. Ding, G., Wu, Q., Wang, J., Yao, Y.-D.: Big Spectrum Data: The New Resource for Cognitive Wireless Networking, April 2014. http://arxiv.org/pdf/1404.6508.pdf
3. Kobb, B.Z.: Wireless Spectrum Finder. McGraw-Hill TELECOM, New York (2001)
4. https://wireless2.fcc.gov/UlsApp/UlsSearch/searchLicense.jsp
5. Taher, T.M., Bacchus, R.B., Zdunek, K.J., Roberson, D.A.: Long-term spectral occupancy findings in chicago. In: IEEE Symposium on New Frontiers in Dynamic Spectrum Access Networks (DySPAN), pp. 100– 107. IEEE (2011)
6. https://www.radioreference.com/apps/about/
7. Hoyhtya, M., et al.: Spectrum occupancy measurements: a survey and use of interference maps. IEEE Commun. Surv. Tutor. **18**, 2386–2414 (2016)
8. McHenry, M.A., Steadman, K.: Spectrum occupancy measurements, location 2 of 6: Tyson's square center, vienna, virginia, April 9, 2004. Shared Spectrum Company Report (2005)
9. Wang, Z., Salous, S.: Spectrum occupancy statistics and time series models for cognitive radio. J. Sig. Process. Syst. **62**(2), 145–155 (2011)
10. Sanders, F.H., Lawrence, V.S.: Broadband spectrum survey at Denver, Colorado. US Department of Commerce, National Telecommunications and Information Administration (1995)
11. Islam, M.H.: Spectrum survey in singapore: occupancy measurements and analyses. In: 3rd International Conference on Cognitive Radio Oriented Wireless Networks and Communications, 2008, CrownCom 2008, pp. 1–7. IEEE (2008)
12. Shi, L., Bahl, P., Katabi, D.: Beyond sensing: Multi-GHz realtime spectrum analytics. In: Proceedings of the 12th USENIX Symposium on Networked Systems Design and Implementation, NSDI 2015, pp. 159–172 (2015)

13. Petrin, A., Steffes, P.G.: Analysis and comparison of spectrum measurements performed in urban and rural areas to determine the total amount of spectrum usage. In: International Symposium on Advanced Radio Technologies, pp. 9–12 (2005)
14. Chen, Y., Oh, H.S.: A survey of measurement-based spectrum occupancy modeling for cognitive radios. IEEE Commun. Surv. Tutor. **18**(1), 848–859 (2016)
15. Łopatka, J., Malon, K., Kryk, M.: Hybrid model of radio channels occupancy prediction for dynamic spectrum access. In: IEEE-2018 Baltic URSI Symposium (URSI), 09 July 2018
16. López-Benítez, M., Casadevall, F.: An overview of spectrum occupancy models for cognitive radio networks. In: Casares-Giner, V., Manzoni, P., Pont, A. (eds.) NETWORK-ING 2011. LNCS, vol. 6827, pp. 32–41. Springer, Heidelberg (2011). https://doi.org/10.1007/978-3-642-23041-7_4
17. Ali, A., Hamouda, W.: Advances on spectrum sensing for cognitive radio networks: theory and applications. IEEE Commun. Surv. Tutor. **19**(2), 1277–1304 (2017)
18. Nagpure, V., Hood, C., Vaccaro, S.: Semantic Models for Labeling Spectrum Data. In: IFIP International Conference on Artificial Intelligence Applications and Innovations, pp 3–12 (2018)
19. https://dictionary.cambridge.org/us/dictionary/english/context
20. https://fccid.io/Emissions-Designator/20K0F3E
21. https://wiki.radioreference.com/index.php/Illinois_Sports#Chicago_White_Sox
22. https://www.facebook.com/events/summers-end-chicago-food-truck-fest-at-labagh-woods/285694112161786/
23. https://www.mlb.com/whitesox/ballpark/meeting-and-event-spaces
24. https://www.baseball-reference.com/
25. https://www.seatgeek.com/build

Margin-Based Active Online Learning Techniques for Cooperative Spectrum Sharing in CR Networks

K. Praveen Kumar[✉], Eva Lagunas, Shree Krishna Sharma,
Satyanarayana Vuppala, Symeon Chatzinotas, and Björn Ottersten

Interdisciplinary Centre for Security, Reliability and Trust University of Luxembourg,
Luxembourg City, Luxembourg
praveen.korrai@uni.lu

Abstract. In this paper, we consider a problem of acquiring accurate spectrum availability information in the Cooperative Spectrum Sensing (CSS) based Cognitive Radio Networks (CRNs), where a fusion center collects the sensing information from all the sensing nodes within the network, analyzes the information and determines the spectrum availability. Although Machine Learning (ML) techniques have been recently applied to enhance the cooperative sensing performance in CRNs, they are mostly supervised learning based techniques and need a significant amount of labeled data, which is difficult to acquire in practice. Towards relaxing this requirement of large labeled data of supervised learning, we focus on Active Learning (AL), where the fusion center can query the label of the most uncertain cooperative sensing measurements. This is particularly relevant in CRN environments where primary user behavior changes in a quick manner. In this regard, we briefly review the existing AL techniques and adapt them to the considered CSS based CRNs. More importantly, we propose a novel margin based active on-line learning algorithm that selects the instance to be queried and updates the classifier by using the Stochastic Gradient Descent (SGD) technique. In this approach, whenever an unlabeled instance is presented, the proposed AL algorithm compares the margin of instance with a threshold to decide whether it should query a label or not. Supporting results based on numerical simulations show that the proposed method has significant advantages on classification and detection performances, and time-complexity as compared to state-of-the-art techniques.

Keywords: Active learning · Cooperative spectrum sensing ·
Cognitive radio network

1 Introduction

Due to the extensive proliferation of wireless applications and services, the wireless data traffic increases at an alarming rate and it is estimated to increase 10

© ICST Institute for Computer Sciences, Social Informatics and Telecommunications Engineering 2019
Published by Springer Nature Switzerland AG 2019. All Rights Reserved
A. Kliks et al. (Eds.): CrownCom 2019, LNICST 291, pp. 140–153, 2019.
https://doi.org/10.1007/978-3-030-25748-4_11

folds in the next few years as compared to the current data traffic [1,2]. Though this increasing demand for wireless services has made the usable radio frequency spectrum a scarce and expensive resource, field trails for channel measurements have illustrated that most of the time the radio spectrum is under-utilized by the licensed users, also known as primary users (PUs) [3]. In this context, Opportunistic Spectrum Access (OSA) is envisaged as a candidate technology to mitigate the spectrum scarcity by improving the utilization of spectrum [4]. In interweave CR networks (CRNs), the secondary users (SUs) can access the spectrum of the PUs opportunistically when the PU transmission is detected to be idle. Therefore, an efficient spectrum sensing mechanism is essential for the realization of efficient OSA.

In the above context, various spectrum sensing methods have been extensively discussed in the literature [5,6]. In this paper, we concentrate on energy detection (ED), which is one of the most widely accepted method as it is by far the cheapest and simplest option. However, due to the noise effect, the sensing efficiency of the ED method may be significantly degraded in the lower SNR regime [7]. Further, when the SUs are distributed in distinct locations, due to shadowing and severe multi-path fading, the hidden PU problem may occur. In this respect, the cooperative spectrum sensing (CSS) mechanism has been proposed to improve the sensing reliability by jointly processing the sensing information in a fusion center [8].

Machine learning techniques applied to CSS for CRNs have received significant research attention recently, e.g. [9–11]. In the context of interweave CR, the spectrum sensing problem reduces to a binary classification on the channel availability. Two major categories of learning methods to train the classifier are supervised and unsupervised learning methods. In supervised learning, first, the feature vectors (i.e., data) should be labeled prior to training. Next, the feature vectors and its corresponding labels are fed into the classifier for training. Support vector machine (SVM) is one of the most popular supervised learning methods. SVM was exploited for spectrum sensing in [9,10], showing its superior performance as compared with the conventional ED based spectrum sensing. In contrast to the supervised learning algorithms, unsupervised learning algorithms do not require training (i.e., no labeled data). Examples of unsupervised learning applied to CSS can be found in [10,11], where [11] has adopted a linear fusion rule for CSS and utilized the linear discriminant analysis to obtain linear coefficient weights, and [10] explored techniques such as Gaussian Mixture Models (GMM) and K-means clustering.

In the CSS problem, labeling data means knowing whether the PU is active or not at a particular time instant. In such a case, unlabeled data is abundant (i.e., spectrum sensing measurements) but labeling is challenging, as it requires feedback from the PUs to the SUs. To address this problem, Active Learning (AL) have been recently proposed, allowing the learning algorithms to dynamically query instances for labeling. AL has been applied in [12,13] to learn the interference channel between the PU and SU in an underlay CRN. In addition, the PUs behavior is not static, i.e., it may change from active to inactive at any time and vice-versa. In order to react faster to PU activity changes, in this paper, we investigate and analyze the applicability of online AL in ED-based CSS.

First, we consider the perceptron based online AL technique [14] (also known as CBGZ algorithm), and we apply it to the proposed CSS problem. In this approach, when the classifier receives the energy measurements from the distributed SUs, it first predicts the availability of the channel from the energy instance, and then utilizes a random sampling approach to decide whether the label for such instance should be queried or not (i.e., if PU feedback is needed or not). If yes, the classifier acquires the true label from the system and follows the standard perceptron approach to update the classifier. Subsequently, the online passive aggressive AL (PAAL) algorithm [15] is considered for the proposed CSS problem. This algorithm follows the similar idea of CBGZ algorithm for querying, but utilizes an efficient PA learning strategy to update the classifier. However, the aforementioned AL algorithms depend on the uncertainty sampling strategy for querying, which is not accurate. To address this, we propose a margin-based online AL algorithm with reduced threshold and Stochastic Gradient Descent (SGD) update. In this algorithm, margin-based condition is utilized, which compares the margin of incoming instance with a threshold to decide whether it should query the class label or not. Finally, we compare and illustrate the performance of all the aforementioned algorithms through numerical simulations, and also we compare them with the conventional supervised and unsupervised counterparts.

The rest of the paper is structured as follows. Section 2 presents the system model. We discuss the ML and AL based frameworks for the CSS, and propose a margin-based AL algorithm for CSS in Sect. 3. The numerical evaluations of proposed algorithms are discussed in Sect. 4. Finally, conclusions are drawn in Sect. 5.

2   System Model

We consider an interweave CRN, consisting of a PU, and N number of SUs where each SU is indexed by $m = 1, 2, 3, ..., N$. The SUs collect energy measurements which are sent to a fusion center which jointly analyzes them and determines the channel availability. The PU is placed at the coordinate d^{PU} and the mth SU is placed at the coordinate d_m^{SU} in the two-dimensional (2D) space. Further, all channels are considered to experience path loss, and independent and identically distributed (i.i.d.) Nakagami-m fading. Therefore, the channel gain h_m between the PU and the mth SU is expressed as

$$h_m = PL(|d^{PU} - d_m^{SU}|)F_m \qquad (1)$$

where $PL(r) = r^{-\beta}$ represents the path loss respective to distance between the PU and SU with the path loss exponent β, and F_m is the Nakagami-m fading component.

We assume the PU activity to vary over time in a random fashion, with 0.5 of probability of being in active state. Let S denotes the state of PU. If the PU is in active state, then $S = 1$, and otherwise $S = 0$. Thus, the availability of channel at the tth instant is written as

$$Y_t = \begin{cases} +1, & \text{if } S = 1 \\ -1, & \text{if } S = 0 \end{cases} \tag{2}$$

where $Y_t = +1$ represents the unavailability of channel and $Y_t = -1$ represents the availability of channel.

We assume the cooperative sensing approach as shown in Fig. 1, in which each SU first determines the energy level of the signal transmitted by PU and reports it to a fusion center or classifier. Further, based on the sensed energy levels reported by all SUs, the fusion center or classifier estimates the availability of spectrum.

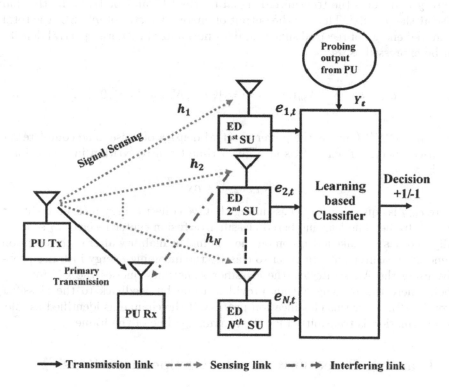

➡ **Transmission link**   ╌╌➤ **Sensing link**   ╌·➤ **Interfering link**

Fig. 1. Learning based cooperative spectrum sensing model

In order to perform the energy detection, SUs require to compute the energy level of primary transmitted signal for a certain period of time. The frame structure of the SU is depicted in Fig. 2. Each frame consists the deciding time τ (i.e., a combination of sensing time (τ_s), classification time (τ_c), and reporting time (τ_r)) and a transmission duration ($T - \tau$). Let W be the assigned bandwidth for the signal transmission and τ_s be the spectrum sensing duration; then, the energy detector senses $W\tau_s$ complex samples during the sensing time. Therefore, the ith signal sample sensed by the mth SU can be expressed as

$$X_m(i) = Sh_m y(i) + Z_m(i) \tag{3}$$

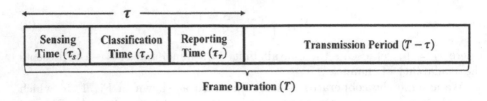

Fig. 2. Frame structure of SU transmission

where $y(i)$ denotes the transmitted signal by the PU, and $Z_m(i)$ is the thermal noise at the mth SU. Thus, at the output of energy detector of mth SU, the total estimated energy of received signal samples normalized by noise spectral density can be expressed as [8]

$$e_{m,l} = \frac{2}{\eta} \sum_{i=1}^{W\tau_s} |X_m(i)|^2, \quad m = 1, 2, .., N, \text{and } l = 1, 2, ..., L \tag{4}$$

where $\eta = \mathbb{E}\{|Z_m(i)|^2\}$ is the power spectral density of noise. The complete set of sensed energies from N SUs at a given time t can be written in vector form as follows,

$$\mathcal{E}_t = \begin{bmatrix} e_{1,t}\, e_{2,t} \cdots e_{N,t} \end{bmatrix}^T \tag{5}$$

where $e_{m,t}$ is equivalent to (4) assuming that we collect energy measurements at each tth frame. The learning based classifier or fusion center receives $\mathcal{E}_t$ sequentially and has to make a decision on the channel availability after each reception of energy measurements. In order to classify the incoming energy instance properly, using the AL strategies, the classifier sometimes queries the PU for true label, where the probing output provides a one bit feedback to the classifier given by (2). After this classification process, if the channel is identified as idle, the information is transmitted for the remaining time of the frame.

3   Learning Based Frameworks for Spectrum Sensing

ML based frameworks have already been adapted for CSS applications in the literature [8] and it has been shown that supervised learning algorithms are efficient for these applications. However, in supervised learning tasks, a huge data set with labels is required to train the ML model. The gathering of data and labeling for it is very time-consuming and costly in these specific applications. To address this problem, in this section, we develop online AL algorithms based CSS schemes to predict the channel availability. First, we exploit the margin based online AL algorithms such as CBGZ [14] and PAAL [15] and we adapt them to fit in the proposed CSS problem. Next, we propose a reduced threshold and SGD update based online AL algorithm for CSS.

3.1 The CBGZ Algorithm for CSS

In an attempt to minimize the number of requested labeled queries and so the feedback from the PU, the CBGZ algorithm makes use of an improved Perceptron-type of algorithm, which considers a margin-based filtering rule. The application of CBGZ in the problem at hand is as follows. After receiving the energy measurement vector $\mathcal{E}_t$, the algorithm queries the PU for the label with a probability of $d/(d+|\hat{b}|)$ [14], where d is a constant and $\hat{b}$ represents the uncertainty of the received instance, and it is computed as a margin of the received measurement instance with respect to the present hypothesis. Next, the algorithm's prediction is compared with the feedback received by the PU and if does not match, the standard Perceptron update process is invoked. The summary of the CBGZ algorithm for CSS is presented in Algorithm 1.

Algorithm 1. The CBGZ Algorithm for CSS

   Input: d (Smoothing parameter)
   Initialization: $w_t = 0$;
   for $t = 1, 2, 3, ..., M$ **do**
   Receive $\mathcal{E}_t$
   Set $f_t = w^T \mathcal{E}_t$
   Predict $\hat{y} = sign(f_t)$
   Flip a coin with $P(Heads) = \frac{d}{d+|\hat{b}|}$
      if Heads **then**
         Receive Y_t
         if $Y_t \neq \hat{y}$ **then**
            $w_{t+1} = w_t + \eta Y_t \mathcal{E}_t^T$
            $w_{t+1} = w_t$
         end if
      end if
   end for

3.2 Passive Aggressive AL for CSS

In this section, we consider the PAAL algorithm for the proposed CSS problem. As in CBGZ, this algorithm consists of two stages: (i) querying process and (ii) classifier updating process. In the first stage, the algorithm follows the same approach of margin-based AL.

First, the algorithm estimates the uncertainty or margin of the incoming energy instance using the current classifier i.e., $p_t = w_t.\mathcal{E}_t$. Next, it draws a Bernoulli random variable $Z_t \in \{0, 1\}$ with probability $C/(C + |p_t|)$ [15] to decide whether the label should be requested or not. If $Z_t = 0$, the class label is not queried and the classifier is not updated. Otherwise, the class label of the incoming energy instance is queried at the PU, and unlike the previous margin-based AL algorithm that updates the classifier only whenever a misclassification occurs, the PAAL algorithm updates the classifier whenever the value of loss function is not zero. Then, the classifier updates as $w_{t+1} = w_t + \xi_t Y_t \mathcal{E}_t$, where ξ_t is the step-size estimated as [15]

$$\xi_t = \begin{cases} l_t(w_t)/(\|\mathcal{E}_t\|^2 + 1/(2A)) \ ; \text{PAA-3} \\ \min(C, l_t(w_t)/(\|\mathcal{E}_t\|^2 \quad ; \text{PAA-2} \\ l_t(w_t)/(\|\mathcal{E}_t\|^2) \qquad ; \text{PAA-1} \end{cases} \tag{6}$$

The summary of the PAAL algorithm for CSS is provided in Algorithm 2.

Algorithm 2. PAA algorithm for CSS

Input: $A > 0$ (Penalty Parameter); $C > 1$ (Smoothing Parameter)
Initialization: $w_t = 0$
for $t = 1, 2, 3, ..., M$ **do**
Receive $\mathcal{E}_t$
Predict $\hat{y}_t = sign(w^T \mathcal{E}_t)$
Draw a Bernoulli random variable $Z_t \in \{0, 1\}$ w.p $\frac{C}{C+|p_t|}$
    if $Z_t = 1$ **then**
        Query the label $Y_t \in \{-1, +1\}$
        compute the loss $l_t(w_t) = max(0, 1 - Y_t w_t^T \mathcal{E}_t)$
        $w_{t+1} = w_t + \xi_t Y_t \mathcal{E}_t$
    else
        $w_{t+1} = w_t$
    end if
end for

3.3 Proposed Margin-Based Method with Reduced Threshold and SGD Update

Both PAAL and CBGZ algorithms utilize a computationally complex uncertain sampling strategy for querying. Hence, in order to avoid the uncertainty sampling step, in this section, we propose a threshold based querying strategy algorithm for CSS. The proposed algorithm comprises of two steps: (i) selection of instance, and (ii) Update process of the classifier. The margin of $\mathcal{E}_t$ is computed as $W_t^T \mathcal{E}_t$, and that is compared with the existing threshold C_t. If the margin is lower than the existing threshold, then the classifier w_t is updated using the SGD step. Finally, we determine the upper bound of estimation error, which is used in the update of threshold C_t. The complete procedure is summarized as Algorithm 3.

Computation of Threshold (C_t): Assume w_* is an optimal classifier with v error rate. Here, our objective is to find the condition that for a given energy instance, the classifier w_t should provide the same prediction as the optimal one w_*. This condition reveals if a certain energy instance is not useful to our classifier, and thus if it can be excluded by the threshold.

For each iteration t, consider ϕ_t is the angle of w_t and w_*. For every energy instance $\mathcal{E}_t$, $w_*(\mathcal{E}_t)$ and $w_t(\mathcal{E}_t)$ predict the same sign if the margin of the instance $|w_t^T \mathcal{E}_t|$ is greater than the angle of classifier ϕ_t. If $|w_t^T \mathcal{E}_t| \leq \phi_t$, for instance $\{\mathcal{E}_t : sign(h_t(\mathcal{E}_t)) \neq sign(h_*(\mathcal{E}_t))\}$. Therefore, we have to make $C_t \geq \phi_t$.

The upper bound for ϕ_t can be expressed as

$$P\{|w_t^T \mathcal{E}_t| \leq \phi_t\} \leq b_a P\{\mathcal{E}_t : sign(h_t(\mathcal{E}_t)) \neq sign(h_*(\mathcal{E}_t))\} \tag{7}$$

where b_a is a constant and P is the instances distribution. We have $P\{E_t : sign(h_t(\mathcal{E}_t)) \neq sign(h_*(\mathcal{E}_t))\} \leq 2v + Er_t$ and $\phi_t \leq b_c P\{|w_t^T \mathcal{E}_t| \leq \phi_t\}$. Finally, by using equations, we get $\phi_t \leq B(2v + Er_t)$. where $B = b_a b_c$. Therefore, we set the threshold to be $C_t = B(2v + Er_t)$.

Classifer Update using SGD: SGD updates the classifier as

$$w_{t+1} = w_t - \epsilon_t \frac{\partial}{\partial w} l(w^T \mathcal{E}_t, Y_t) \tag{8}$$

where $l(w^T \mathcal{E}_t)$ is the hinge loss. After some mathematical manipulations, we obtain the following SGD update policy,

$$w_{t+1} = \begin{cases} w_t + \frac{1}{\sqrt{t}} Y_t \mathcal{E}_t, & \text{if } Y_t w^T \mathcal{E}_t < 1/\sqrt{t}, \\ w_t & \text{otherwise.} \end{cases} \tag{9}$$

Algorithm 3. Reduced Threshold with SGD update for CSS

Input: B; v (Error)
Initialization: $w_t = 0$; $C_t = 1$;
for $t = 1, 2, 3, ..., M$ **do**
Receive $\mathcal{E}_t$
Predict $sign(w^T \mathcal{E}_t)$
Receive Y_t
  if $|w^T \mathcal{E}_t| < C_t$ **then**
    if $Y_t w^T \mathcal{E}_t < \gamma$ **then**
      $w_{t+1} = w_t + \frac{1}{\sqrt{t}} Y_t \mathcal{E}_t$
    end if
  end if
$Er_t = \frac{\log(t)}{\sqrt{t}}$
$C_{t+1} = B(2v + Er_t)$
end for

---

4 Numerical Evaluations

To show the benefits of the proposed scheme, in this section, we evaluate the empirical performance of the considered online AL algorithms for CSS tasks in CRNs and we compare them with state-of-the art ML algorithms such as SVM and K-means in terms of miss-classification rate and time-complexity.

4.1 Simulation Parameters

We consider a wireless network with a PU that is located at coordinates (500 m, 0) and nine SUs involving in CSS that are located in a 3-by-3 grid topology in the area of 2000 m × 2000 m as depicted in Fig. 3. Further, we assign 5 MHz bandwidth W for each transmission, each frame duration is 100 ms, sensing duration τ_s is assumed as 10 ms, the PL exponent is 3, and the additive white Gaussian noise (AWGN) is considered.

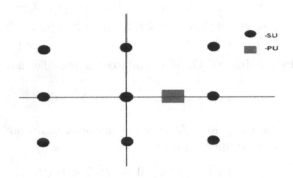

Fig. 3. Example of network model with $N = 9$ SUs

In the online AL algorithms, there are some other parameters such as optimal error v and constant B in Algorithm 3, smoothing parameter C and penalty parameter A in Algorithm 2, learning parameter η and constant d to compute the probability of sampling in Algorithm 1, which are all estimated through the cross validation procedure. We ran the experiment over a total of 1024 frames. All the experiments were performed and averaged over 5 independent Monte-Carlo runs, each with a random PUs behavior resulting in a different permutation of the energy dataset within the 1024 instances.

For the experiments related to the passive learning SVM algorithm, we used 1024 energy instances for the training and 1024 energy instances for the testing. For the experiments related to the passive learning K-means algorithm, we computed the classifier assuming 1024 non-labeled energy instances and next we proceeded to the testing phase with 1024 different energy instances.

In the testing of online AL algorithms, all energy instances come one by one in a sequential manner. If the incoming energy instance is identified as a most informative, then the querying process is executed to obtain the true class label. The latter is assumed to be provided by the PUs in the form of 1-bit feedback. After receiving the true channel availability, that energy instance is added to training set. The same process is continued till to the end of the experiment.

4.2 Results and Discussions

Figure 4 illustrates the performance of SVM, K-means and online AL based CSS schemes in terms of mis-classification rate. From the results, it is observed that the CSS schemes based online AL algorithms outperform the conventional SVM and K-means based CSS schemes. This is due to fact that whenever the misclassification occurs, the considered online AL algorithms update the classifier according to the margin of the incoming energy instance. This classifier updating process helps to reduce the further classification errors. Also, we notice from the results in Fig. 4 that the CSS scheme based on the proposed online AL algorithm with reduced threshold and SGD update outperforms the online PAAL algorithms and the state-of-the art CBGZ AL algorithm based CSS schemes. The proposed algorithm constantly shows almost 10 % of lower misclassification rate compared to PAAL algorithms and more than 15 % of lower misclassification rate compared to CBGZ algorithm.

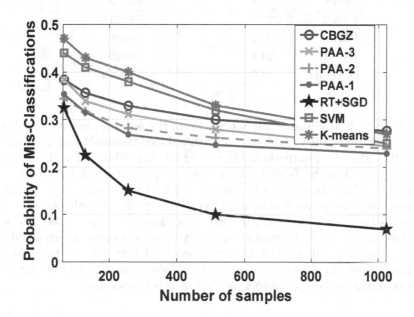

Fig. 4. Evaluation of CSS schemes based on on-line AL methods and conventional ML methods.

We also provide the detailed comparison of the supervised, unsupervised and online AL algorithms in Tables 1 and 2. As can be seen form Table 1, both the supervised and unsupervised algorithms SVM and K-means are interestingly illustrating the almost same performance using the size of 1024 data samples for training and for the classifier computation, respectively. However, the time-complexity of K-means algorithm is very high as compared to the SVM algorithm. Also, from Table 2, it is observed that the proposed RT+SGD AL algorithm shows better performance compared to other AL algorithms in terms of both misclassification and run time.

In Fig. 5, we show the performance of the proposed online AL based CSS schemes in terms of probability of detection when 2×2 SUs (i.e., 4 SUs) and

Table 1. Percentage of misclassifications and time complexity

AL algorithms	Percentage of misclassifications	Run time (s)
SVM	25.9	0.0531
K-means	27.1	0.0953

Table 2. Percentage of misclassifications and time complexity of online AL algorithms

AL algorithms	Percentage of misclassifications	Run time (s)
CBGZ	27.4	0.0423
PAA-3	25.7	0.0311
PAA-2	25.2	0.0301
PAA-1	24.1	0.0295
SGD+RT	7.5	0.0046

3×3 SUs (i.e., 9 SUs) cooperate in CSS. From the obtained results, we observe that the proposed online AL algorithm with reduced threshold and SGD update (SGD+RT) algorithm outperforms the other AL algorithms with almost 3 % of detection rate. Also, in Table 3, we provide the percentage of misclassification and the time duration to identify the busy or idle PU channel state using the AL classifiers with 4 and 9 SUs in cooperation for CSS. We also observe from the results that with 9 cooperating SUs, the detection rate of the AL based algorithms improves by increasing the number of instances, as expected. For example, after 200 energy instances the proposed CSS scheme based on the SGD+RT algorithm achieves almost +7 % of detection rate with 9 number of SUs cooperation as compared to 4 number of SUs cooperation.

Table 3. Percentage of misclassifications and time complexity of 2×2 and 3×3 SUs cooperation

AL algorithms	Percentage of misclassifications		Run time (s)	
	2×2	3×3	2×2	3×3
CBGZ	32.7	27.4	0.0411	0.0423
PAA-1	26.4	24.1	0.0290	0.0295
SGD+RT	13.3	7.5	0.0038	0.0046

We show the performance comparison of AL algorithms based CSS schemes in terms of probability of false alarm (P_{fa}) in Fig. 6. From the results, it is observed that the proposed RT+SGD algorithm shows the less false alarm rate as compared to the other two algorithms.

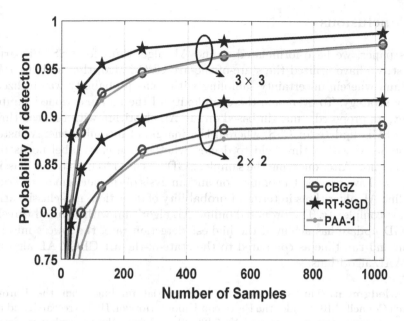

Fig. 5. Probability of detection of CSS schemes based on online AL methods

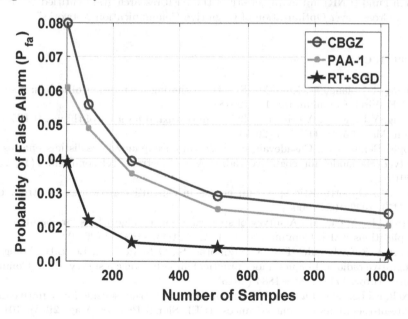

Fig. 6. Probability of false alarm of CSS schemes based on online AL methods

5   Conclusions

In this paper, we have formulated online AL algorithms for CSS. In particular, first, we have utilized the state-of-the-art CBGZ algorithm and the PAAL algorithm, wherein uncertainty sampling with some probability was utilized as a query strategy. To overcome the complexity of the aforementioned sampling strategy, we proposed a margin-based online AL algorithm with reduced threshold and SGD update for CSS, wherein the margin of incoming energy instance was compared with a threshold to decide whether the class label needs to be queried or not. After querying the sample, SGD was used to update the classifier. Finally, we have provided a comparison and analysis of the performance of different online AL algorithms in terms of probability of detection, mis-classifications, and time complexity. The proposed online AL algorithm with reduced threshold and SGD update has achieved the highest detection rate, the lowest misclassification and run time as compared to the state-of-the art CBGZ AL algorithm and PAAL algorithms.

Acknowledgment. This work has received partial funding from the European Research Council (ERC) under the European Union's Horizon H2020 research and innovation programme (grant agreement No 742648), and from the Luxembourg National Research Fund (FNR) in the framework of the AFR research grant entitled "Learning-Assisted Cross-Layer Optimization of Cognitive Communication Networks".

References

1. Zhang, L., Liang, Y., Xiao, M.: Spectrum sharing for internet of things: a survey. IEEE Wirel. Commun. **99**, 1–8 (2018)
2. Wang, Y.E., et al.: A primer on 3GPP narrowband internet of things. IEEE Commun. Mag. **55**(3), 117–123 (2017)
3. Lopez-Benitez, M., Casadevall, F.: Spectrum occupancy in realistic scenarios and duty cycle model for cognitive radio. Adv. Electron. Telecommun. **1**(1), 26–34 (2010)
4. Lee, Y.: Opportunistic spectrum access in congnitive networks. Electron. Lett. **44**(17), 1022–1024 (2008)
5. Yucek, T., Arslan, H.: A survey of spectrum sensing algorithms for cognitive radio applications. IEEE Commun. Surv. Tutor. **11**(1), 116–130 (2009)
6. Sharma, S.K., Bogale, T.K., Chatzinotas, S., Ottersten, B., Le, L.B., Wang, X.: Cognitive radio techniques under practical imperfections: a survey. IEEE Commun. Surv. Tutor. **17**(4), 1858–1884 (2015)
7. Axell, E., Leus, G., Larsson, E.G., Poor, H.V.: Spectrum sensing for cognitive radio: state-of-the-art and recent advances. IEEE Signal Process. Mag. **29**(3), 101–116 (2012)
8. Saifan, R., Jafar, I., Al-Sukkar, G.: Optimized cooperative spectrum sensing algorithms in cognitive radio networks. Comput. J. **60**(6), 835–849 (2017)
9. Zhang, D., Zhai, X.: SVM-based spectrum sensing in cognitive radio. In: International Conference on Wireless Communications, Networking and Mobile Computing (WICOM), Wuhan, China (2011)

10. Thilina, K.M., Choi, K.W., Saquib, N., Hossain, E.: Machine learning techniques for cooperative spectrum sensing in cognitive radio networks. IEEE J. Sel. Areas Commun. **31**(11), 2209–2221 (2013)

11. Choi, K.W., Hossain, E., Kim, D.I.: Cooperative spectrum sensing under a random geometric primary user network model. IEEE Trans. Wirel. Commun. **10**(6), 1932–1944 (2011)

12. Tsakmalis, A., Chatzinotas, S., Ottersten, B.: Interference constraint active learning with uncertain feedback for cognitive radio networks. IEEE Trans. Wirel. Commun. **16**(7), 4654–4668 (2017)

13. Tsakmalis, A., Chatzinotas, S., Ottersten, B.: Constrained Bayesian active learning of interference channels in cognitive radio networks. IEEE J. Sel. Topics Signal Process. **12**(1), 6–19 (2018)

14. Cesa-Bianchi, N., Gentile, C., Zaniboni, L.: Worst-case analysis of selective sampling for linear-threshold algorithms. In: Advances in Neural Information Processing Systems Conference (2004)

15. Lu, J., Zhao, P., Hoi, S.C.H.: Online passive-aggressive active learning. Mach. Learn. **103**(2), 141–183 (2016)

On the Feasibility of a Secondary Service Transmission over an Existent Satellite Infrastructure

Luciano Barros Cardoso da Silva[1]([⊠]), Tarik Benaddi[2], and Laurent Franck[3]

[1] IMT Atlantique, LabSTICC, Toulouse, France
luciano.barroscardosodasilva@imt-atlantique.fr
[2] IMT Atlantique, LabSTICC, DEOS, Toulouse, France
tarik.benaddi@imt-atlantique.fr
[3] Airbus Defence and Space, Toulouse, France
laurent.franck@airbus.com

Abstract. In this paper, we present a realistic use case in order to investigate the feasibility of a secondary service transmission over an existent satellite infrastructure. By introducing the overlay cognitive radio paradigm towards satellite communications, we compute a theoretical achievable data rate greater than 16 kbps for the secondary service, which is suitable for most M2M applications. Using simulation results, we show that this can be achieved while preserving the primary service performance. In addition, a system design framework is discussed in order to dimension such systems.

Keywords: Satellite communications · Cognitive radio ·
Overlay paradigm · Dirty paper coding ·
Machine-to-Machine application

1 Introduction

Machine-to-Machine (M2M) communications are one of the central use cases in the upcoming new fifth generation (5G) mobile network [1] as they play a major role in the internet of things (IoT). [2] predicts the deployment of around 1 million devices per km^2 like sensors/actuators, vehicles, factory machines, beacons *etc.* In this new machine-type communication environment and since available radio spectrum is today a scarce resource (cf [3] for example), one of the main faced challenges concerns the design of efficient spectrum utilization and a better coordination between legacy and future services.

Within this context, despite of the continuous technological developments in terrestrial networks, the satellite communications systems are still relevant in the modern telecommunications world. This affirmation can be sustained especially today, since the demand for the rising new services has experienced a significant growth, supported by the unique characteristics such as multicast

© ICST Institute for Computer Sciences, Social Informatics and Telecommunications Engineering 2019
Published by Springer Nature Switzerland AG 2019. All Rights Reserved
A. Kliks et al. (Eds.): CrownCom 2019, LNICST 291, pp. 154–167, 2019.
https://doi.org/10.1007/978-3-030-25748-4_12

and broadcasting capabilities, mobility aspects, global reach, besides the ability to cover and connect green space and hostile environments [4]. In this sense, the use of satellite for M2M applications provides to the end-users connectivity anytime, anywhere, for any media and device.

As a counterpoint, to meet these increasingly challenging requirements while keeping the competitiveness facing the terrestrial technologies, the satellite segment needs to push the boundaries in the direction to more and more efficient technical solutions. In this sense, the search for power and bandwidth efficiency as well as the actual trend to low complexity systems are of the upmost importance. It is within this framework that the terrestrial cognitive radio techniques have also attracted the attention for satellite applications. Supported by the recent developments in the space qualified Software Defined Radios (SDR) [5], and also by the maturity of concepts such as flexible [6] and hosted payloads, these techniques have become feasible whereby some relevant research have been developed, resulting in a more smart spectrum management.

In a nutshell, the cognitive user (CU), unlicensed or less prioritized to operate in a specific spectrum band, senses the environment around it and adapts its transmission as a function of the interference, by adjusting the frequencies, waveforms and protocols in order to access the licensed primary user (PU) spectrum efficiently. Without going into further details, three paradigms classifies the CU operation [7]:

- *interweave*, where the CU transmits opportunistically into the spaces not currently used by the PU;
- *underlay*, where the CU adjusts its parameters according to the PU signal characteristics in order to transmits simultaneously while respecting an interference power threshold;
- *overlay*, where the CU has the noncausal knowledge about the PU signal and message and, by using judiciously chosen coding and signal processing techniques, is able to use simultaneously the PU channel (same frequency, time and polarization), without deteriorating this latter.

The first two schemes were well studied in [8,9]. In this paper, we investigate the third scheme.

The main reason to propose the overlay paradigm for satellite communications lies in the feasibility of transmitting both unlicensed and licensed services simultaneously towards its respective terminals. We emphasizes that, due the priority among users, the superposition coding strategies is required [10], unlike the technical solutions adopted for the broadcast channel. In practice, this method enables the addition of a secondary service above legacy infrastructure of the primary service, instead of using a dedicated satellite or constellation.

This paper presents a practical scenario for the techniques previously exposed in the recent publications [11,12], which concern the design of the overlay paradigm transmission towards satellite communication systems. By using the concepts and the framework well characterized by these references, this work extends the previous analysis focusing on the feasibility of a low data rate

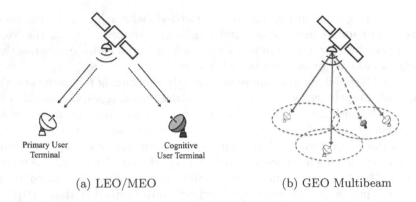

(a) LEO/MEO   (b) GEO Multibeam

Fig. 1. Satellite scenarios (Color figure online)

transmission. In this sense, a practical use case is investigated, which considers commercial off-the-shelf (COTS) parts [13] and assumes realistic link budget parameters in its evaluation. The discussions and results contained herein could be seen as part of a "preliminary phase" of an engineering process plan [14].

2   Overlay Model Description

The following scenarios are provided as examples where the overlay CR techniques might be applied to satellite communications. In the first case, presented in the Fig. 1a, an ordinary LEO/MEO satellite provides two different services towards different terminals. In this context, a single licensed user PU takes priority over the added unlicensed CU. The interference presented at both terminals should be mitigated by properly designed CU encoder, without any changes in the PU transmission chain.

In the same way, the GEO multibeam satellite is illustrated in the Fig. 1b. In this case, considering the frequency reuse, the CU is able to transmits by using, for instance, the determined blue frequency (or polarization) into the red spot footprint, as far as the interference among adjacent beams is resolved. It is worth noting also that all possible different PU transmissions, represented by several blue spots, should be taken into account in the interference mitigation design. By this way, the total satellite capacity could be increased as well as the spectrum resources better managed.

Equally suitable for both scenarios, the interference model with side information, adapted from [10], is presented in Fig. 2. Assuming that the signals are onboard the satellite, the cognitive encoder has full and noncausal knowledge about each PU i-th signal and message, which addresses the main overlay paradigm requirement. In this sense, the encoded cognitive signal X_c^n is function of both primary and cognitive messages $m_{p,i}$ and m_c.

Without loss of generality, considering the i-th PU and the added CU, the channel gains $|h_{yx,i}|$ (from the transmitter x to the receiver y) are defined by

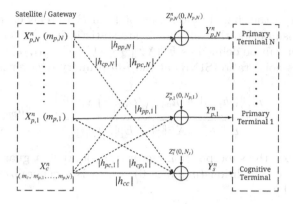

Fig. 2. Overlay model.

the direct paths ($|h_{cc}|$ and $|h_{pp,i}|$), and the interfering paths ($|h_{pc,i}|$ and $|h_{cp,i}|$) losses (the Fig. 2 summarizes these notation). In our context, these gains are computed as function of the each transmission link budget.

The following equations describes the output of the channel, where n refers to the n-*th* symbol:

$$Y_{p,i}^n = |h_{pp,i}|X_{p,i}^n + |h_{pc,i}|X_c^n + Z_{p,i}^n \tag{1}$$

$$Y_s^n = |h_{cc}|X_c^n + \sum_{i=1}^N |h_{cp,i}|X_{p,i}^n + Z_s^n, \tag{2}$$

Based on the fact that the terminals may be located in different geographical sites, the Gaussian noise component $Z_{p,i}^n$ (resp. Z_s^n) is assumed to follow the normal law $\mathcal{N}(0, N_{p,i})$ (resp. $\mathcal{N}(0, N_s)$). Also, the power constraints to be satisfied are $E[\|X_{p,i}^n\|^2] = P_{p,i}$ and $E[\|X_c^n\|^2] \leq P_c$, respectively.

Finally, since each PU has the same transmission priority, we highlight that the interference among them could be solved by precoding techniques as proposed, for instance, in DVB-S2X standard [15]. Under this assumption, this work provides a design method to permit a secondary service transmission without affecting the PU transmission performance.

3 Enabling Techniques

3.1 Superposition Strategy

The purpose of the superposition technique is to ensure that the signal-to-noise ratio (SNR) at each PU receiver is not decreased in the presence of interference. To accomplish this goal, the CU shares part of its power to relay each PU. Based on that operation, the CU transmitted signal is given by:

$$X_c^n = \hat{X}_c^n + \sum_{i=1}^N \sqrt{\alpha_i \frac{P_c}{P_{p,i}}} X_{p,i}^n, \tag{3}$$

where $\alpha_i \in [0, 1]$ is the shared power fraction from P_c to relay each PU message.

Under the assumption that all signals are statistically independent, the new power constraint can be defined as $E[\|\hat{X}_c^n\|^2] \leq (1 - \sum_{i=1}^{N} \alpha_i)P_c$. The signal-to-interference-plus-noise ratio (SINR) at the i-th primary receiver is given by:

$$SINR_{P,i} = \frac{E\left[\|(|h_{pp,i}| + |h_{pc,i}|\sqrt{\alpha_i \frac{P_c}{P_{p,i}}})X_{p,i}^n\|^2\right]}{E[\||h_{pc,i}|\hat{X}_c^n\|^2] + E[\|Z_{p,i}^n\|^2]} = \frac{\|h_{pp,i}\|^2 P_{p,i}}{N_{p,i}} \tag{4}$$

In this context, the superposition factor $\alpha_i \in [0, 1]$ that guarantees Eq. (4), for the interference condition ($|h_{pc,i}| > 0$), which is a generalized form of [10, Eq. 14], is given by:

$$\alpha_i = \left(\frac{|h_{pp,i}|\sqrt{P_{p,i}}\left(\sqrt{N_{p,i}^2 + \|h_{pc,i}\|^2 P_c(N_{p,i} + \|h_{pp,i}\|^2 P_{p,i})} - N_{p,i}\right)}{|h_{pc,i}|\sqrt{P_c}(N_{p,i} + \|h_{pp,i}\|^2 P_{p,i})}\right)^2 \tag{5}$$

By inspection of Eq. (3), we emphasize that the CU transmission is feasible only if the condition $\sum_{i=1}^{N} \alpha_i < 1$ is satisfied. By this assumption, note that the CU data rate should be decreased when aggressive frequency reuse scenarios are considered.

3.2 Dirty Paper Coding

Once the superposition factors are computed and the CU partially shares its power to relay each PU signal, the next step is to design $\hat{X}_c^n$ efficiently, in order to minimize the PU interference. The optimal strategy uses the theoretical results presented by Costa [16]. On the assumption that the interference is non-causally known at transmitter, a transmitter-based interference presubtraction can be implemented, without any power increase, reaching the AWGN capacity. In this sense, according to the main concept of this technique, the CU adapts its waveform, instead of cancel, as a function of the channel interference.

By rearranging the Eq. (2) and considering the superposition, we have:

$$Y_s^n = |h_{cc}|\hat{X}_c^n + \sum_{i=1}^{N} \left(|h_{cp,i}| + |h_{cc}|\sqrt{\alpha_i \frac{P_c}{P_{p,i}}}\right)X_{p,i}^n + Z_s^n. \tag{6}$$

Without loss of generality, given that the signals in Eq. (6) are statistically independents each other, the implemented model considers a single Gaussian distributed PU constellation, in respect to the total interfering power received at CU terminal. In addition, in order to simplify the notation through this paper, the Eq. (6) is normalized by the direct path attenuation factor $|h_{cc}|$. Thus, the signal at CU receiver is given by:

$$Y_s^n = \hat{X}_c^n + \underbrace{\left(b + \sqrt{\alpha \frac{P_c}{P_p}}\right)X_p^n}_{S^n} + Z_s^n, \tag{7}$$

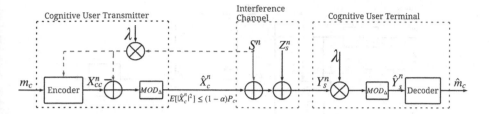

Fig. 3. Proposed DPC encoder.

where the factor b represents the normalized interfering path and S^n represents the total channel interference.

The Fig. 3 presents the basic diagram of the DPC encoder. In this configuration, assuming low and intermediate SNR regime, the partial interference presubtraction (PIP) is implemented [16]. In this way, the signal $\hat{X}_c^n$ is designed as:

$$\hat{X}_c^n = \left[X_{cc}^n - \lambda S^n\right] \text{MOD}_\Delta, \tag{8}$$

where X_{cc}^n is the coded signal and the factor λ, to be properly chosen, controls the fractioned interference to be presubtracted. Also, MOD_Δ is the complex-valued modulo operation. The modulo amplitude is defined by $\Delta = \sqrt{M}d_{min}$, where M is the number of points of the square QAM constellation and d_{min} the minimum intersymbol distance.

Concerning the historical perspective for practical DPC implementations, the use of Tomlinson-Harashima precoding (THP) for intersymbol interference (ISI) cancelling was firstly introduced by Erez et al. in [17]. Subsequently, Eyuboglu and Forney [18] developed the trellis precoding technique (TP), also recovering partially the so-called shaping loss by the trellis shaping (TS) technique [19]. Finally, the TP technique for multiuser interference, presented in [20] further developed in [21], formed the basis of our implemented DPC encoder.

The Fig. 4 details the practical DPC encoder implemented in this work. Three separated gains can be reached by this system:

- the coding gain, which is realized by the coset select code C_c, specified G_c;
- the shaping gain, generated by the shaping code C_s, according to the TS technique [19] for multiuser interference;
- the precoding gain, which mitigates the interference by the presubctraction combined with the shaping metric and modulo operation.

This work implements the following branch metric Eq. (9), where the precoder selects the proper region sequence with minimum average energy to steer the scaled interference sequence λS, taking also account the modulo operation.

$$\left\| \left[X_{cc}^n - \lambda S^n\right] \text{MOD}_\Delta \right\|^2. \tag{9}$$

The Fig. 5 presents the scatter plot of the signals corresponding to the encoder processing. The 16-QAM and further 256-QAM expanded constellations signal

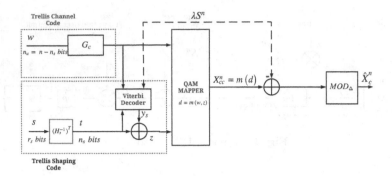

Fig. 4. Proposed DPC encoder [12]

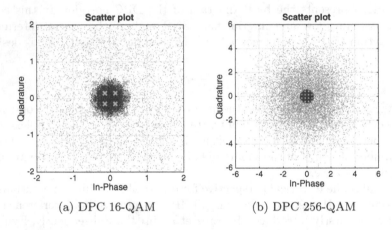

(a) DPC 16-QAM   (b) DPC 256-QAM

Fig. 5. Scatter Plot of Signal Constellation at CU Transmitter, considering $10.log(Pp/Pc) = 9.5$ dB and $\alpha = 0.85$, according with the use case of this paper (X_{cc}^n in green "x", λS^n in red dots and $\hat{X}_c^n$ in blue dots). (Color figure online)

X_{cc}^n is shown superposed by the gaussian distributed scaled interference λS^n, considering the study case presented in the next section. The transmitted signal $\hat{X}_c^n$, after presubtraction and modulo operation, is also illustrated by the blue dots. We highlight that, despite of the large amount of interference, the transmitted signal is confined into the expanded constellation, respecting the power constraint of the DPC theory. In addition, we can note that further expansion can be done to confine the total interference [12]. However, the complexity of the system will be increased.

At the receiver, the reverse operations are done. In this sense, at the decoder input, the signal is given by:

$$\hat{Y}_s^n = \left[(\hat{X}_c^n + S^n + Z^n)\lambda \right] \mathrm{MOD}_\Delta. \tag{10}$$

Finally, the choice of the λ is done with the goal to maximize the equivalent CU signal to noise ratio (SNR) in Eq. (10). The λ factor can be optimized and is given by [16]:

$$\lambda = \frac{(1-\alpha)P_c}{(1-\alpha)P_c + E[|Z_s^n|^2]}. \tag{11}$$

3.3   CU Transmitted Power

By reason of the TP operation, a power reduction of $\hat{X}_c^n$ is obtained, which impacts directly both links performance as follows:

- For the PU, by inspection of Eq. (4), it is noticed that this power reduction will result in an increase of the SINR at the PU receiver (in comparison to the case with no interference). This occurs because the superposition factor α in Eq. (5) does not take into account the shaping gain induced by TP. Consequently, the PU link would present a better BER performance than that required;
- Concerning the CU, it is also noted that under these conditions, we obtain $E[|X_c^n|^2] < P_c$. This means that the CU is not operating in its total power capacity. Consequently, this will lead to an under utilization of the available resources on the satellite.

The proposed method for controlling the CU output power, presented in [12], allows to reach the power equality $E[|X_c^n|^2] = P_c$ (or equivalently, $E[|\hat{X}_c^n|^2] = (1-\alpha)P_c$) as a function of the shaping gain. This is obtained by an adequate scaling of the minimum distance d_{min} of the cognitive transmitted constellation. In short, according to this method, we define the power of the baseline, without considering the shaping operation (i.e. uniformly distributed constellation points), as:

$$P_\oplus = \frac{2^R}{6}d_{min}'^2, \tag{12}$$

where R is the data rate in bits per two dimensions. In addition, the shaping gain is given as follows:

$$\gamma_s = \frac{P_\oplus}{(1-\alpha)P_c}. \tag{13}$$

As a consequence, we define the scaled minimum distance d_{min}', such that the available power after the shaping operation is equals to $(1-\alpha)P_c$, as:

$$d_{min}' = \sqrt{\frac{[(1-\alpha)P_c]6\gamma_s}{2^R}}. \tag{14}$$

Based on this proposed design procedure exposed in this section, the trellis-shaped based DPC encoder using a M-QAM expanded constellation at the transmission rate of $R_{cu} = 2$ bits/symbol was implemented for the proposed use case of the next section. Further analyses and the complete characterization of the performance as well as the distortions evolved can be find in [12]. In the next section, we investigate the feasibility in a realistic application.

3.4 Practical System Analysis

As a matter of system engineering, the design for the CU payload could either be a standalone system (implemented by a dedicated transmission chain and antenna) or a shared transmitter (by using the same transponder and antenna as the PU). In this latter configuration, notice that more caution should be taken into account when the transmission of both signals inputs the same high power amplifier (HPA). In fact, this practice should be avoided since this implementation may induce higher nonlinear distortions, particularly in terms of AM/AM and AM/PM conversions [22].

Moreover, at the receiver side (PU and CU), two design solutions could be adopted: (i) by the deployment of geographically separated receiving sites for each user, in this way reducing the interference due to the attenuation at the interfering paths, or (ii) by using the same receiving station with two dedicated demodulators and decoders. In this last case, the attenuation of the interfering and direct paths are the same (i.e. $|h| = |h_{pp,i}| = |h_{cc}| = |h_{pc,i}| = |h_{cp,i}|$). In fact, it increases the interference of both links and, as a consequence, requires higher value of the superposition factor α (in other words, reducing the secondary service data rate). On the other hand, we reduce the infrastructure costs, since the hardware is further simplified by utilizing the same earth station to receive also the CU signal.

Deepening the vision on the techniques described, we point out that, due to the superposition, the bit rate of the secondary service might be very low with respect to the primary. However, this practice generates two implementation problems: (i) in the DPC presubtraction technique, the same symbol rate for both signals is considered in order to be able to compute the Eq. (8) and (ii) in the superposition technique, the interference generated by the CU signal would appears as spikes in the PU bandwidth, which makes the usual interference model unrealistic in this case. In order to avoid both constraints, we can think of the implementation of the chirp spread spectrum technique [23] at CU transmission. In this sense, the DPC encoder can correctly perform its operation and the CU receiver can demodulated at a more flexible transmitted data rate.

To improve the whole system performance, the channel estimation techniques could be realized at the terminals end through a link feedback, for instance, according to the DVB-S2X standard [15]. By these features, the superposition factor α, which depends directly on the channels conditions, as well as the λ, which depends on SNR, can be periodically updated, changing the achievable secondary service data rate and, as consequence, optimizing CU performance.

4 Realistic Use Case

In order to investigate the system feasibility, we adopted a scenario where a Cubesat at a height of 600 km with same orbital parameters as [24], using COTS parts, transmits both signals (primary and cognitive) from the same satellite antenna towards a single earth station, which is equipped with two dedicated demodulators. In this sense, the channel attenuations are the same and defined as $|h|$. In this study, just the downlink is considered.

The main specification for the PU signal are: output power of $1W$ [13], operating frequency of 2200 MHz (downlink band assigned for Earth Exploration Satellite Service), bit rate of 3.4 Mbps, BER specified to 10^{-5} and coded QPSK modulation with FEC ($R = 1/2$). The following Table 1 presents the link budget of PU without secondary service addition.

Table 1. Primary user link budget (QPSK Coded FEC R = 1/2).

Frequency (MHz)	2200
Throughput rate (Mbps)	3.4
Transmitted power (mW)	1000
Satellite carrier EIRP (dBm)	38.3
Free space loss (dB)	-162.2
Depointing loss (dB)	-10
E. S. Antenna max gain - 5 m, eff 50 % (dBi)	38.2
System noise temperature (K)	130
C/N0 (dB-Hz)	81.8
Eb/N0 (dB)	16.5
Demodulation losses (dB)	-6
Eb/N0 required (dB) – for BER = 1E-5	7
Margin (dB)	3.5

It is worth noting that a conservative margin for demodulation losses of 6 dB is assumed in order to cover the impairments of the communication chain. The overall link margin is about 3.5 dB, as required by the targeted BER.

The principle behind this design strategy was to use part of the power remaining in this margin to transmit the CU signal. Therefore, we defined that 900 mW were allocated for PU transmission (which still maintain the recommended link margin of 3 dB) and 100 mW were used for CU. The next Fig. 6 presents the overlay model considering this use case.

The powers transmitted and received are provided considering the realistic link budget parameters. By computations according the Table 1, the channel attenuations equal $\|h\|^2_{dB} = -125.7$ dB. In this condition, the interference-to-noise ratio (INR) and the link degradation D by interference are given by:

$$INR = \frac{I_{pc}}{N_p} = 4.47; \tag{15}$$

$$D(dB) = 10.log(1 + INR) = 7.38 \ dB. \tag{16}$$

The CR overlay techniques are employed to mitigate both link degradation. Firstly, considering all parameters, the CU performs the superposition strategy

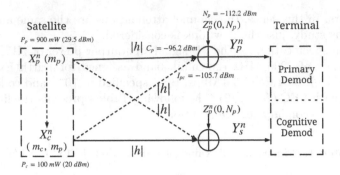

Fig. 6. Overlay model considering realistic satellite link budget.

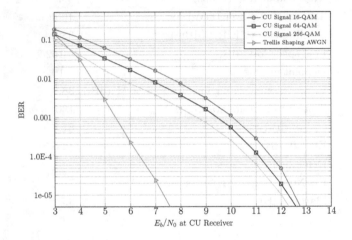

Fig. 7. BER CU, considering $10.log(Pp/Pc) = 9.5$ dB and $\alpha = 0.85$, according with the use case.

and, by the Eq. (5), the factor $\alpha = 0.85$ is evaluated. This value guarantees a $SINR$ of 16 dB at PU. We highlight that, thanks to the superposition strategy and the power controlling design of the DPC encoder (see results in [12]), the PU maintains the same performance as in absence of the CU interference.

From this point, a simulation for CU link is realized considering the whole CU channel interference (see Eq. (7)), which is composed by the PU signal and the CU shared power in the superposition.

By taking the link parameters and the CU BER curve presented in Fig. 7, we can now compute the link budget, presented at Table 2. It is important to note that, by reason of superposition, just 15 % of the power originally allocated for CU is used for its own transmission. However, despite of the low power received, the sensitivity of the receiver is in line with the specifications usually attributed for small satellites links (i.e. sensitivity threshold of -118 dBm [25]).

We emphasize that all conservative margins are still being considered in order to guarantee the performance.

Table 2. Cognitive user link budget.

Transmitted power (mW)	100 * 0.15
Satellite carrier EIRP (dBm)	20.1
Free space loss (dB)	−162.2
Depointing loss (dB)	−10
Ground station antenna max gain - 5 m, eff 50 % (dBi)	38.2
System noise temperature (K)	130
C (dBm)	−113.9
C/N0 (dB-Hz)	63.6
Demodulation losses (dB)	−6
Eb/N0 required (dB) – for BER = 1E-3	
DPC 16-QAM	10
DPC 64-QAM	9.4
DPC 256-QAM	8.5
Eb/N0 required (dB) – for BER = 1E-5	
DPC 16-QAM	12.5
DPC 64-QAM	12.25
DPC 256-QAM	12
Margin (dB)	3
Minimum Bit Rate (kbps) – DPC 16-QAM	
BER = 1E-3	28.8
BER = 1E-5	16.2
PAPR	
DPC 16-QAM	5.0
DPC 64-QAM	4.67
DPC 256-QAM	4.79

It is worth noting that different M-QAM schemes present the peak to average power ratio (PAPR) in the same order of magnitude. Also, the further expansion impacts directly in the BER performance, even when the total interference is not confined inside the expanded constellation (as in this case). As pointed in [11], by practical effects, the SNR here does not consider the defined effective noise in Eq. (11). It could suggest that DPC performs better than AWGN in low SNR, which is not the case. As an important result of this feasibility analysis, we assure that the bit rate of 16 and 28 kbps can be reached for the secondary service, as a function of the BER specified, which depends on the service application.

5   Conclusion

This paper investigated the feasibility of a low data rate secondary service transmission over a primary user infrastructure. A realistic scenario was presented with COTS parts and the different techniques were implemented to resolve the interference of both links. As a result, we obtained the same performance for the PU as in absence of the CU operation (AWGN channel). Concerning the secondary service, we have reached a data rate greater than 16 kbps, which is suitable for most of M2M applications.

As a drawback, we point out an increase of the output power of the satellite due to the intrinsic signal correlation presented at the superposition technique. In the described scenario, the total power transmitted is about 32 dBm instead of 30 dBm specified. In this case, the transmitted antenna should be properly designed in order to consider this output power.

Considering future works, we will seek another proof of concept by means of SDR implementation. In addition, the control of PAPR by shaping operation, typically on satellite communication, will be also investigated.

Acknowledgment. This work was supported by National Council for Scientific and Technological Development (CNPq/Brazil) and by National Institute for Space Research (INPE/Brazil).

References

1. ITU-R: Emerging trends in 5G/IMT2020. Geneva Mission Briefing Series, September 2016
2. Huawei: 5G network architechture - a high-level perspective. White paper, December 2016
3. United States radio spectrum frequency allocations chart. United States Department of Commerce (2016). https://www.ntia.doc.gov/files/ntia/publications/january-2016-spectrum-wall-chart.pdf
4. Minoli, D.: Innovations in Satellite Communications and Satellite Technology: The Industry Implications of DVB-S2X, High Throughput Satellites, Ultra HD, M2M, and IP. Wiley, Hoboken (2015)
5. Maheshwarappa, M.R., Bowyer, M., Bridges, C.P.: Software defined radio (SDR) architecture to support multi-satellite communications. In: IEEE Aerospace Conference, pp. 1–10 (2015)
6. Porecki, N., Thomas, G., Warburton, A., Wheatley, N., Metzger, N.: Flexible payload technologies for optimising Ka-band payloads to meet future business needs. In: Proceedings of the 19th Ka Broadband Communications, Navigation and Earth Observation Conference, pp. 1–7 (2013)
7. Biglieri, E.: An overview of cognitive radio for satellite communications. In: IEEE First AESS European Conference on Satellite Telecommunications (ESTEL), pp. 1–3 (2012)
8. Sharma, S.K., Chatzinotas, S., Ottersten, B.: Cognitive radio techniques for satellite communication systems. In: IEEE 78th Vehicular Technology Conference (VTC Fall), pp. 1–5 (2013)

9. Álvarez-Díaz, M., Neri, M., Mosquera, C., Corazza, G.: Trellis shaping techniques for satellite telecommunication systems. In: IEEE International Workshop on Satellite and Space Communications, pp. 148–152 (2006)
10. Jovicic, A., Viswanath, P.: Cognitive radio: an information-theoretic perspective. IEEE Trans. Inf. Theory **55**(9), 3945–3958 (2009)
11. da Silva, L.B.C., Benaddi, T., Franck, L.: Cognitive radio overlay paradigm towards satellite communications. In: 2018 IEEE International Black Sea Conference on Communications and Networking (BlackSeaCom), pp. 1–5 (2018)
12. da Silva, L.B.C., Benaddi, T., Franck, L.: A design method of cognitive overlay links for satellite communications. In: IEEE 9th Advanced Satellite Multimedia Systems Conference and the 15th Signal Processing for Space Communications Workshop (ASMS/SPSC), pp. 1–6 (2018)
13. ISIS high data rate S-band transmitter mission specifications (online). https://www.isispace.nl/wp-content/uploads/2016/02/isis-communication-systems-brochure-v2-compressed.pdf
14. DVB: DVB-doc A-172 white paper on the use of DVB-S2X for DTH applications DSNG and professional services broadband interactive services and VL-SNR applications (2015)
15. E.-M. ECSS: ST-10C: Space project management-project planning and implementation (2009)
16. Costa, M.: Writing on dirty paper (corresp.). IEEE Trans. Inf. Theory **29**(3), 439–441 (1983)
17. Erez, U., Shamai, S., Zamir, R.: Capacity and lattice strategies for canceling known interference. IEEE Trans. Inf. Theory **51**(11), 3820–3833 (2005)
18. Eyuboglu, M.V., Forney, G.D.: Trellis precoding: combined coding, precoding and shaping for intersymbol interference channels. IEEE Trans. Inf. Theory **38**(2), 301–314 (1992)
19. Forney, G.: Trellis shaping. IEEE Trans. Inf. Theory **38**(2), 281–300 (1992)
20. Yu, W., Varodayan, D.P., Cioffi, J.M.: Trellis and convolutional precoding for transmitter-based interference presubtraction. IEEE Trans. Commun. **53**(7), 1220–1230 (2005)
21. Sun, Y., Xu, W., Lin, J.: Trellis shaping based dirty paper coding scheme for the overlay cognitive radio channel. In: IEEE 25th Annual International Symposium on Personal, Indoor, and Mobile Radio Communication (PIMRC), pp. 1773–1777 (2014)
22. Maral, G., Bousquet, M.: Satellite Communications Systems: Systems, Techniques and Technology. Wiley, Hoboken (2011)
23. Proakis, J.G., Salehi, M.: Digital Communications, vol. 4. McGraw-Hill, New York (2001)
24. Barbarić, D., Vuković, J., Babic, D.: Link budget analysis for a proposed cubesat earth observation mission. In: 2018 41st International Convention on Information and Communication Technology, Electronics and Microelectronics (MIPRO), pp. 0133–0138. IEEE (2018)
25. Arias, M., Aguado, F.: Small satellite link budget calculation. In: ITU Symposium and Workshop on Small Satellite Regulation and Communication Systems (2016)

Interpolation-Based Interference Rejection Combining for Black-Space Cognitive Radio in Time-Varying Channels

Sudharsan Srinivasan$^{(\boxtimes)}$, Sener Dikmese, and Markku Renfors

Electrical Engineering, Faculty of Information and Communication Sciences,
Tampere University, Tampere, Finland
{sudharsan.srinivasan, sener.dikmese,
markku.renfors}@tuni.fi

Abstract. In this paper, we investigate multi-antenna interference rejection combing (IRC) based black-space cognitive radio (BS-CR) operation in time-varying channels. The idea of BS-CR is to transmit secondary user (SU) signal in the same frequency band with the primary user (PU) such that SU's power spectral density is clearly below that of the PU, and no significant interference is inflicted on the PU receivers. We explore the effects of interpolation and mobility on the novel blind IRC technique which allows such operation mode for effective reuse of the PU spectrum for relatively short-distance CR communication. We assume that both the PU system and the BS-CR use orthogonal frequency division multiplexing (OFDM) waveforms with common numerology. In this case the PU interference on the BS-CR signal is strictly flat-fading at subcarrier level. Sample covariance matrix-based IRC adaptation is applied during silent gaps in CR operation. We propose an interpolation-based scheme for tracking the spatial covariance in time-varying channels, demonstrating significantly improved robustness compared to the earlier scheme. The performance of the proposed IRC scheme is tested considering terrestrial digital TV broadcasting (DVB-T) as the primary service. The resulting interference suppression capability is evaluated with different PU interference power levels, silent gap durations, data block lengths, and CR device mobilities.

Keywords: Black-space cognitive radio · Underlay CR · IRC ·
Interference rejection combining · Multi-antenna system · Receiver diversity ·
Mobility · OFDM · DVB-T

1 Introduction

Cognitive radios (CRs) are intended to operate in radio environments with a high level of interference and, simultaneously, produce negligible interference to the primary users (PUs) [1–3]. CR studies in the past have been focusing on opportunistic whitespace scenarios where the unused spectrum is dynamically identified and used. Also underlay CR operation has received some attention. Here the idea is to transmit in wide frequency band with low power-spectral density, typically using spread-spectrum techniques [4]. Black-space CR (BS-CR), where a CR deliberately transmits

© ICST Institute for Computer Sciences, Social Informatics and Telecommunications Engineering 2019
Published by Springer Nature Switzerland AG 2019. All Rights Reserved

A. Kliks et al. (Eds.): CrownCom 2019, LNICST 291, pp. 168–179, 2019.
https://doi.org/10.1007/978-3-030-25748-4_13

simultaneously along the primary signal in the same time-frequency resources without causing objectionable interference has received limited attention [5–8]. BS-CR systems effectively reuse the spectrum over short distances. It can operate with limited spectrum resources and can be used without any additional spectrum sensing.

As discussed in our previous paper [9], one of the major requirements for CR operation is to minimize the interference to the primary transmission system. In BS-CR this is reached by setting the CR transmission power at a small-enough level. The most important factor that enables such a radio system is that stronger interference is easier to deal with as compared to weaker interference [10], if proper interference cancellation techniques are utilized. Previous studies from information theory provide theoretically achievable bounds for such cognitive radios [11].

Multi-antenna systems allow for spatio-temporal signal processing, which do not only improve the detection capability of the receiver but also improve performance in fading multipath channels with interference. Various methods of interference cancellation can be found in [12–17] and the references therein. All other detection algorithms except the multi-user detector perform sub-optimally [12].

The interference rejection combining (IRC) receivers have the significant advantage in comparison to the other receivers in multi-user scenarios that they do not need detailed information about the interfering signals, such as modulation order and radio channel propagation characteristics. For CR scenarios, IRC receivers in general are simple and desirable compared to optimum detectors. IRC techniques are widely applied for mitigating co-channel interference, e.g., in cellular mobile radio systems like LTE-A [18]. The use of multiple antennas in CRs has been studied earlier, e.g., in [17]. Our initial study on this topic in highly simplified scenario with suboptimal algorithms was in [19], but to the best of our knowledge, IRC has not been applied to BS-CR (or underlay CR) elsewhere. In this current study we develop the ideas that we presented in [9] under more practical situations and study the performance of our algorithms in more details. Notably, the scheme studied in [9] was found to be very sensitive to mobility, because its performance is critically affected by errors in spatial covariance estimation. Here in this work we extent our previous studies on the effects of mobility and propose a scheme to improve the quality of the covariance estimation with time-varying channels using interpolation between sample covariance-based estimates.

In this paper we consider BS-CR operation in the terrestrial TV frequency band, utilizing a channel with an on-going relatively strong TV transmission. The PU is assumed to be active continuously. If the TV channel becomes inactive, this can be easily detected by each of the CR stations in the reception mode. Then the CR system may, for example, continue operation as a spectrum sensing based CR system. In our case study, we focus on the basic scenario of IRC based multi-antenna CR receiver with co-channel interference generated by a single PU transmitter. The performance of such a system under different interference levels, mobilities, frame structures, and modulation orders is studied, and the improved robustness obtained through covariance interpolation is highlighted.

The rest of the paper is organized as follows: In Sect. 2, the BS-SC scenario and proposed IRC scheme are explained. The system model and IRC solution are formulated in Sect. 3. Section 4 presents the simulation setup and performance evaluation results. Finally, concluding remarks are presented Sect. 5.

2 IRC-Based Black-Space Cognitive Radio Scenario

In our basic scenario, illustrated in Fig. 1, we consider a CR receiver using multiple antennas to receive data from a single-antenna cognitive transmitter. The CR operates within the frequency band of the PU, and the PU power spectral density (PSD) is very high in comparison to that of the CR. The primary transmission is assumed to be always present when the CR system is operating. The primary transmitter generates a lot of interference to the CR transmission, which operates closer to the noise floor of the primary receiver, and due to this, the primary communication link is protected. We consider frequency reuse over relatively small distances, such as an indoor CR system. The multi-antenna configuration studied here is that of single-input multiple output (SIMO). Other configurations, involving also transmit diversity in the CR link are also possible, but they are left as a topic for future studies.

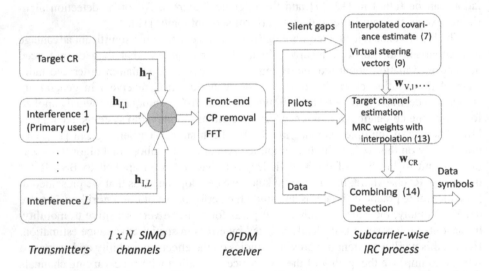

Fig. 1. Modified black-space CR system mode with silent gap interpolation. The related equation numbers are indicated.

Here the PU is a cyclic prefix orthogonal frequency division multiplexing (CP-OFDM) based DVB-T system [20]. The CR system is also an OFDM based multi-carrier system using the same subcarrier spacing and CP length as the primary system. Thus, it has the same overall symbol duration. The CR system is assumed to be synchronized to the primary system in frequency and in quasi-synchronous manner also in time. The CP length is assumed to be sufficient to absorb the channel delay spread together with the residual offsets between the two systems observed at the CR receiver. Consequently, the subcarrier-level flat-fading circular convolution model for spatio-temporal channel effects applies to the target CR signal and to the PU interference signal as well. Then the IRC process can be applied individually for each subcarrier. Since the CR receiver observes the PU signal at very high SINR level, synchronization task is not particularly difficult and low-complexity algorithms can be utilized. Considering short-range CR scenarios, the delay spread of the CR channel has a minor effect on the overall channel delay spread to be handled in the time alignment of the two systems. Basically, if all CR stations are synchronized to the PU, they are also synchronized with each other. In addition to these, we assume that the secondary user is mobile within a given range and the effects of mobility on the system performance are studied.

Both the primary and the CR systems use QAM subcarrier modulation, but usually with different modulation orders. The received CR signal consists of contributions from both the desired CR communication signal and the primary transmission signal, the latter one constituting a strong interference. Our proposed scheme includes two phases in the CR system operation as described in our previous work [9]. The spatial characteristics of the PU interference are modeled using multiantenna sample covariance matrix, which is estimated during silent gaps in the CR transmission, independently for each active subcarrier. No explicit channel estimation of the PU channel is required. The CR channel is estimated from the partial IRC signals, from which the PU interference has been effectively suppressed.

In this work we study how channel fading with mobility affects the performance. We consider adaptation of the IRC process using interpolation of the sample covariance matrices between the silent gaps. This is expected to improve the performance, allowing to increase the data block length between silent gaps, thus reducing the related overhead in throughput.

3   IRC for Black-Space Cognitive Radio

Based on the OFDM model mentioned above, subcarrier-wise detection is considered with flat-fading process to get rid of the challenge of frequency selectivity in the IRC process.

In the SIMO configuration, the CR is assumed to have N receiver antennas and L different interference sources are assumed. Based on this model, the signal received by the CR can be formulated for each active subcarrier as follows:

$$\mathbf{r} = \mathbf{h}_T x_T + \sum_{l=1}^{L} \mathbf{h}_{I,l} x_{I,l} + \boldsymbol{\eta}. \tag{1}$$

Here x_T is a transmitted subcarrier symbol and $\mathbf{h}_T$ is the target channel vector with N receiver antennas in the CR, $x_{I,l}$ is the lth interfering signal, and $\mathbf{h}_{I,l}$ is the channel vector for the lth interferer. Finally, $\boldsymbol{\eta}$ is the additive white Gaussian noise (AWGN) vector. In this generic system model, it is assumed that the PU is the dominant interferer, and the other interference sources are, e.g., other CR systems introducing co-channel interference at relatively low power level.

3.1   Stage 1: Covariance Matrix Estimation and IRC Process

As it is illustrated in Fig. 1, the interference minimizing IRC weights are obtained during the silent period. Due to that, Eq. (1) can be modified during silent gaps of CR operation as

$$\mathbf{r} = \sum_{l=1}^{L} \mathbf{h}_{\mathrm{I},l} x_{\mathrm{I},l} + \boldsymbol{\eta}. \tag{2}$$

Here it is assumed that only interference and noise are present during the silent period in the signal observed by the CR. Linear combiner is used for the signals from different antennas with a weight process in detection as follows:

$$y = \mathbf{w}^{H} \mathbf{r}, \tag{3}$$

where y is the detected signal, $\mathbf{w}$ is the weight vector with N elements, and superscript H denotes the Hermitian (complex-conjugate transpose).

Determining the optimum weight values is an optimization problem which can be solved with the linear minimum mean-squared error (LMMSE) criterion [21] that aims to minimize the mean-squared error with respect to the target signal x_{T},

$$J = E\left[\left| x_{\mathrm{T}} - \mathbf{w}^{H} \mathbf{r} \right|^{2} \right]. \tag{4}$$

When knowledge of the covariance matrix is calculated/known, interference rejection combining (IRC) can be applied. For the covariance matrix, two cases can be considered: (i) perfect channel state information and (ii) sample covariance based approaches.

Perfect Channel Information Case
Assuming that the channel vectors form the interferers are perfectly known, the noise plus interference covariance matrix can be calculated as

$$\Sigma_{\mathrm{NI}} = \sum_{l=1}^{L} P_{l} \mathbf{h}_{\mathrm{I},l} \mathbf{h}_{\mathrm{I},l}^{H} + P_{\mathrm{N}} \mathbf{I}, \tag{5}$$

where P_l is the variance of interferer l, P_{N} is the noise variance that can be obtained from the SNR and $\mathbf{I}$ is the identity matrix of size NxN. Assuming that the channel vector for the target signal is known, the conventional LMMSE solution for the weight vector is

$$\mathbf{w} = \Sigma_{\mathrm{NI}}^{-1} \mathbf{h}_{\mathrm{T}} \left(\mathbf{h}_{\mathrm{T}}^{H} \Sigma_{\mathrm{NI}}^{-1} \mathbf{h}_{\mathrm{T}} + 1/P_{\mathrm{T}} \right)^{-1} \tag{6}$$

where P_{T} is the target CR signal power and unit noise variance is assumed.

In the BS-CR scenario with a single dominant interferer, the estimation of the PU channel is relatively straightforward if the CR knows PUs pilot structure. However, in case of multiple interferers, the channel vectors of all interferers should be estimated, which becomes quite challenging. Furthermore, the target channel cannot be estimated before the interference cancellation. Therefore, the perfect channel information case serves mainly as an ideal reference in performance comparisons.

Sample Covariance Based Case Without PU Channel Information
It is difficult to have the prefect channel state information on the CR receiver side. Alternatively, the joint interference and noise covariance matrix can be estimated by the sample covariance matrix of the received signal in the absence of the target transmission, i.e., during the silent gaps as

$$\bar{\Sigma}_{\text{NI}} = \sum_{m=1}^{M} \mathbf{r}(m)\mathbf{r}(m)^{\text{H}}. \tag{7}$$

Here m is the OFDM symbol index and M is the observation length in subcarrier samples, which is chosen equal to the length of the silent gap.

Linear Interpolation for Interference Covariance Tracking with Mobility
Regarding the mobility aspects, there are significant differences in the effects of PU transmitter mobility, CR transmitter mobility, and CR receiver mobility. If the PU transmitter is stationary and CR receiver is stationary, the mobility of CR transmitter is easier to handle, because the dominating PU interference is stationary, and the variations in the noise and interference covariance matrix are only due to the co-channel CR interferes. However, even in this case, radio environment of the CR receiver may vary due to movement of people or vehicles nearby. Therefore, some tolerance to mobility is required also in such scenarios, at least with pedestrian mobilities. The mobility of PU transmitter or CR receiver make the dominant interference time-varying, and in the BS-CR scenario, the CR link performance is very sensitive to quality of the PU interference covariance matrix estimate. Therefore, it is important to investigate these mobility effects and consider enhanced schemes for tracking the interference covariance with mobility.

While considering the sample covariance-based approach, we assume that the fading effect during each silent gap is small enough to be neglected. Then we apply linear interpolation for the covariance matrix elements when calculating the weight vectors for the data symbols between two consecutive silent gaps.

There are two key parameters in this process, the silent gap length and the data block length between two consecutive gaps. Increasing the gap length improves the performance with low mobility but degrades the performance with higher mobility and increases the overhead in throughput. Increasing the data block length increases the throughput but degrades the performance with mobility. These tradeoffs are investigated through simulations in Sect. 4 of this paper.

IRC Process
As indicated in the model shown in Fig. 1, the CR channel cannot be estimated before the step of the interference cancellation. Hence, we apply the IRC process with N orthogonal virtual steering vectors. For this operation, following unit vectors are applied as the virtual steering vectors,

$$\begin{aligned}
\mathbf{h}_{\text{V},1} &= [1, 0, 0, 0, \ldots, 0]^{\text{T}} \\
\mathbf{h}_{\text{V},2} &= [0, 1, 0, 0, \ldots, 0]^{\text{T}} \\
&\cdots \\
\mathbf{h}_{\text{V},N} &= [0, 0, 0, 0, \ldots, 1]^{\text{T}}
\end{aligned} \tag{8}$$

The obtained weight vectors are normalized as follows:

$$\mathbf{w}_{V,n} = \bar{\Sigma}_{NI}^{-1}\mathbf{h}_{V,n} \Big/ \left\|\bar{\Sigma}_{NI}^{-1}\mathbf{h}_{V,n}\right\|. \tag{9}$$

In this case the weighted output signals

$$y_n = \mathbf{w}_{V,n}^{H}\,\mathbf{r}, \qquad n = 1, 2, \ldots, N \tag{10}$$

have equal noise variances. This normalization provides optimum performance for the following maximum ratio combining (MRC) stage.

3.2   Stage 2: Target Channel Estimation with Linear Interpolation and MRC Combining

In the second stage, data symbols of the CR link are transmitted together with the training/pilot symbols. The IRC process targets to cancel the interference from all of the weighted output signals, while the MRC stage combines these signals with maximum SNR at the output. The N channel coefficients for the weighted output signals can be estimated using the pilot symbols as follows:

$$\hat{h}_{V,n} = y_n/p = \mathbf{w}_{V,n}^{H} \cdot \mathbf{r}/p, \qquad n = 1, 2, \ldots, N, \tag{11}$$

where p is the transmitted pilot symbol value.

Linear Interpolation for Channel Estimation

In the traditional pilot-based channel estimation process, it is required to use efficient interpolation techniques, such as linear interpolation or quadratic interpolation, based on the channel information at pilot sub-carrier symbols. The performance of linear interpolation technique is better than the piecewise-constant interpolation methods [22, 23]. Due to that, linear interpolation is considered in our study due to its simplicity. Generally, with a 2D pilot structure, the channel estimate for a data symbol is obtained by 2D linear interpolation between the three closet pilot symbols in time and frequency. The MRC weights for a data symbol are then calculated as

$$\mathbf{w}_{MRC} = [\hat{h}_{V,1}\,\hat{h}_{V,2}\ldots\hat{h}_{V,N}]^{T} \Big/ \sqrt{\sum_{k=1}^{N}\left|\hat{h}_{V,k}\right|^{2}}, \tag{12}$$

where $\hat{h}_{V,n}$, $n = 1, \ldots, N$, denote the corresponding interpolated channel estimates.

3.3   Stage 3: Combining for Detection

After the MRC weights are calculated, the effective weight vectors for CR can be obtained as,

$$\mathbf{w}_{CR} = \begin{bmatrix} \mathbf{w}_{V,1} & \mathbf{w}_{V,2} & \cdots & \mathbf{w}_{V,N} \end{bmatrix} \cdot \mathbf{w}_{MRC}$$

$$= \sum_{n=1}^{N} \hat{h}_{V,n} \mathbf{w}_{V,n} \Big/ \sqrt{\sum_{k=1}^{N} \left| \hat{h}_{V,k} \right|^2}. \tag{13}$$

In the final stage, the equalized data symbols are calculated by maximum ratio combining the N samples obtained by applying the virtual steering vectors,

$$\hat{d} = \mathbf{w}_{CR}^{H} \cdot \mathbf{r}. \tag{14}$$

It is enough to calculate and use this weight vector $\mathbf{w}_{CR}$, instead of separately applying the MRC weights on the samples obtained by the weight vectors $\mathbf{w}_{V,n}$.

4 Performance Evaluation

The simulations are carried out for the system setup explained in Sect. 2. The carrier frequencies of CR and PU are the same and it is here set to 700 MHz, which is close to the upper edge of the terrestrial TV frequency band. The modulation order used by CR varies between 4QAM, 16QAM, and 64QAM. The pilot symbols are binary and have the same power level as the data symbols. The primary transmitter signal follows the DVB-T model with 16QAM modulation, 8 MHz bandwidth, and CP length of 1/8 times the useful symbol duration, i.e., 28 μs. The IFFT/FFT length is 2048 for both systems. The DVB-T and CR systems use 1705 and 1200 active subcarriers, respectively. ITU-R Vehicular A channel model (about 2.5 μs delay spread) is used for the CR system and Hilly Terrain channel model (about 18 μs delay spread) for PU transmission. The CR receiver is assumed to have four antennas, and uncorrelated 1×4 SIMO configurations are used for both the primary signal and the CR signal.

The number of spatial channel realizations simulated in these experiments is 300. The ratio of CR and PU signal power levels at the CR receiver (referred to as the signal to interference ratio, SIR) is varied. The lengths of the OFDM symbol frame and silent gap for interference covariance matrix estimation are also varied (expressed in terms of CP-OFDM symbol durations). A very basic training symbol scheme is assumed for the CR: training symbols contain pilots in all active subcarriers and the spacing of training symbols is 8 OFDM symbols. Frame length is selected in such a way that training symbols appear as the first and last symbol of each frame, along with other positions. Channel estimation uses linear interpolation between the training symbols. We have tested the BS-CR link performance with SIR values of $\{-10, -20, -30\}$ dB using silent gap durations of $\{8, 16, 32\}$ OFDM symbols, and data block lengths of $\{17, 25, 33, 41\}$ OFDM symbols.

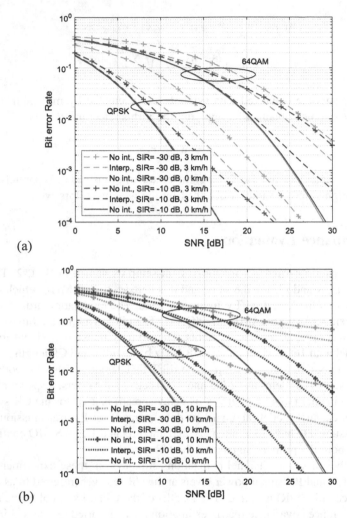

Fig. 2. Performance of QPSK and 64QAM BS-CR systems for SIR $\in \{-10, -30\}$ dB with or without covariance matrix interpolation. (a) 0 and 3 km/h CR receiver mobilities. (b) 0 and 10 km/h CR receiver mobilities. Silent gap length of 16 symbols, and OFDM frame length of 17 symbols.

Figure 2 shows the impact of covariance matrix interpolation on the BS-CR link performance. Here the data block length and gap duration are fixed to 17 and 16 OFDM symbols, respectively. This choice provides performance that is no more than 1 dB from the configuration reaching 1% or 10% BER with lowest SNR, among the tested configurations with even higher overhead. We can see that covariance interpolation provides significant improvement of robustness in time-varying channels. Focusing on the 1–10% BER region, the performance with interpolation at 10 km/h mobility clearly exceed the performance at 3 km/h without interpolation. However, for the 64QM case with −30 dB SIR, this is true only for BER of 10% or higher, due to the high error floor at very low SIR and high mobility. We can also see that with stationary channel, the performance is practically independent of the SIR (comparing with [9] the performance is slightly improved by fine-tuning the used algorithms).

In Fig. 3, the effect of silent gap duration is tested with 16 QAM modulation, SIR of −20 dB, and data block length of 17. The overhead in data rate is about 58% and 44% for gap lengths of 16 and 8, respectively. The shorter gap length results in about 1.5 dB performance loss in the 1–10% range in stationary case and about 1.8–3.5 dB loss with 10 km/h mobility, compared to the gap length of 16. With 20 km/h mobility, the corresponding loss is about 2.2 dB at 10% BER, but longer gap leads to higher error floor, and the performance with shorter gap becomes better for BER below 3%.

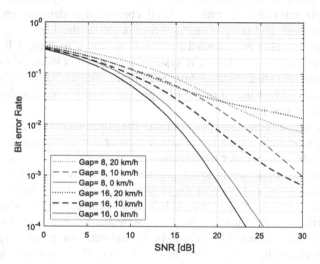

Fig. 3. Performance of QPSK and 16QAM BS-CR systems for SIR = −20 dB with covariance matrix interpolation for CR receiver mobilities of 0, 10 km/h, and 20 km/h. Silent gap length of 8 or 16 symbols and OFDM frame length of 17 symbols.

5   Conclusion

The performance of black-space CR transmission links in the presence of strong interferences and mobility was investigated using spatial covariance interpolation between silent gaps. The interference rejection capability of IRC using multiple receive antennas for various modulation orders under varying mobility and channel setups was studied. It was found that the IRC performs very well in the basic SIMO-type BS-CR scenario when stationary channel model is applicable, e.g., in fixed wireless broadband scenarios. However, the scheme is rather sensitive to the fading of the PU channel, e.g., due to people moving close to the CR receiver. Due to the strong interference level, the interference cancellation process is affected by relatively small errors in the covariance matrix estimate. For covariance estimation, the silent gap length in the order of 16–32 OFDM symbols provides optimum performance with stationary channels, but even with 3 km/h mobility, the performance degrades greatly when considering SIR levels below -10 dB. The data block length should be of the same order or less, which leads to high overhead due to the silent gaps. Covariance interpolation was shown to greatly improve the robustness with time-varying channels, such that good link performance can be obtained with up to 20 km/h mobility at 700 MHz carrier frequency. This indicates that the proposed BS-CR scheme could be feasible at below 6 GHz frequencies with pedestrian mobilities. However, there is a significant tradeoff between link performance and overhead in data rate due to the silent gaps.

In the basic TV black-space scenario, there is only one strong TV signal present in the channel, in agreement with our assumption about the primary interference sources. DVB-T system allows also single-frequency network (SFN) operation and the use of repeaters to improve local coverage. In both cases, the primary transmissions can be seen as a single transmission, with a spatial channel that depends on the specific transmission scenario, and the proposed scheme is still applicable.

The scheme can also be extended to scenarios where multiple CR systems are operating in the same region. If all CR systems are time-synchronized to the PU and they are at a relatively small distance from each other, they are also synchronized with each other, and could be handled by the IRC process as additional interference sources following the model of Eq. (1).

In future work, it is important to optimize the silent gap and data block lengths along with the modulation order to maximize throughput with given PU interference level and mobility. Lower-order modulations are more robust to errors in covariance estimation, allowing significantly lower gap and training overhead than higher order modulation. Complexity reduction of the covariance interpolation and IRC process is also an important topic for further studies. It is also worth to consider adaptation of the IRC process without silent gaps after the first one required for the initial solution. This would help to reduce the related overhead in throughput. One possible approach is to do this in a decision-directed manner: first estimating the covariance matrix in the presence of the target signal and then cancelling its effect based on detected symbols and estimated target channel. In future studies, also the effect of antenna correlation will be taken into consideration.

Acknowledgment. This work was supported in part by the Finnish Cultural Foundation.

References

1. Yucek, T., Arslan, H.: A survey of spectrum sensing algorithms for cognitive radio applications. IEEE Commun. Surv. Tutor. **11**(1), 116–129 (2009)
2. Zeng, Y., Liang, Y.C., Hoang, A.T., et al.: A review on spectrum sensing for cognitiveradio: challenges and solutions. EURASIP J. Adv. Signal Process. **2010**, 381465 (2010). https://doi.org/10.1155/2010/381465
3. Dikmese, S., Srinivasan, S., Shaat, M., et al.: Spectrum sensing andresource allocation for multicarrier cognitive radio systems under interference and powerconstraints. EURASIP J. Adv. Signal Process. **2014**, 68 (2014). https://doi.org/10.1186/1687-6180-2014-68
4. Wyglinski, A.M., et al.: Cognitive Radio Communications and Networks: Principles and Practice. Academic Press, Cambridge (2010)

5. Selén, Y., Baldemair, R., Sachs, J.: A short feasibility study of a cognitive TV black space system. In: Proceedings of the IEEE PIMRC 2011, Toronto, ON, pp. 520–524 (2011)
6. Rico-Alvariño, A., Mosquera, C.: Overlay spectrum reuse in a broadcast network: covering the whole grayscale of spaces. In: Proceedings of the IEEE DySPAN 2012, WA, pp. 479–488 (2012)
7. Wei, Z., Feng, Z., Zhang, Q., Li, W.: Three regions for space-time spectrum sensing and access in cognitive radio networks. IEEE Trans. Veh. Technol. **64**(6), 2448–2462 (2015)
8. Beyene, Y., Ruttik, K., Jantti, R.: Effect of secondary transmission on primary pilot carriers in overlay cognitive radios. In: Proceedings of the CROWNCOM 2013, Washington, DC, pp. 111–116 (2013)
9. Srinivasan, S., Renfors, M.: Interference rejection combining for black-space cognitive radio communications. In: Moerman, I., Marquez-Barja, J., Shahid, A., Liu, W., Giannoulis, S., Jiao, X. (eds.) CROWNCOM 2018. LNICST, vol. 261, pp. 200–210. Springer, Cham (2019). https://doi.org/10.1007/978-3-030-05490-8_19
10. Carleial, A.B.: A case where interference does not reduce capacity. IEEE Trans. Inf. Theory **21**, 569–570 (1975)
11. Devroye, N., Mitran, P., Tarokh, V.: Achievable rates in cognitive radio channels. IEEE Trans. Inf. Theory **52**, 1813–1827 (2006)
12. Verdu, S.: Multiuser Detection. Cambridge University Press, Cambridge (1998)
13. Winters, J.: Optimum combining in digital mobile radio with cochannel interference. IEEE Trans. Veh. Technol. **2**(4), 539–583 (1984)
14. Liaster, J., Reed, J.: Interference rejection in digital wireless communication. IEEE Signal Process. Mag. **14**(3), 37–62 (1997)
15. Klang, G.: On interference rejection in wireless multichannel systems. Ph.D. thesis, KTH, Stockholm, Sweden (2003)
16. Beach, M.A., et al.: Study into the application of interference cancellation techniques. Roke Manor Research Report 72/06/R/036/U, April 2006
17. Bakr, O., Johnson, M., Mudumbai, R., Ramchandran, K.: Multi antenna interference cancellation techniques for cognitive radio applications. In: Proceedings of the IEEE WCNC (2009)
18. Cheng, C.C., Sezginer, S., Sari, H., Su, Y.T.: Linear interference suppression with covariance mismatches in MIMO-OFDM systems. IEEE Trans. Wirel. Commun. **13**, 7086–7097 (2014)
19. Srinivasan, S., Dikmese, S., Menegazzo, D., Renfors, M.: Multi-antenna interference cancellation for black space cognitive radio communications. In: Proceedings of the 2015 IEEE Globecom Workshops, San Diego, CA, pp. 1–6 (2015)
20. Ladebusch, U., Liss, C.A.: Terrestrial DVB (DVB-T): a broadcast technology for stationary portable and mobile use. In: Proceedings of the IEEE, vol. 94, pp. 183–193, January 2006
21. Haykin, S.: Adaptive Filter Theory, 4th edn. Prentice-Hall, Upper Saddle River (2001)
22. He, C., Peng, Z., Zeng, Q., Zeng, Y.: A novel OFDM interpolation algorithm based on comb-type pilot. In: 5th International Conference on Wireless Communications Networking and Mobile Computing, WiCom 2009, pp. 1–4 (2009)
23. Cimini, L.J.: Analysis and simulation of a digital mobile channel using orthogonal frequency division multiplexing. IEEE Trans. Commun. **33**(7), 665–675 (1985)

Cooperative Delay-Constrained Cognitive Radio Networks: Throughput Maximization with Full-Duplex Capability Impact

Ali Gaber[1(✉)], El-Sayed Youssef[1], Mohamed R. M. Rizk[1], Mohamed Salman[2], and Karim G. Seddik[3]

[1] Department of Electrical Engineering, Alexandria University, Alexandria, Egypt
`aligaber@alexu.edu.eg`, `dr.e.a.youssef@gmail.com`, `mrmrizk@ieee.org`
[2] ECEE Department, University of Colorado, Boulder, CO, USA
`mohamed.salman@colorado.edu`
[3] ECNG Department, American University in Cairo, Cairo, Egypt
`kseddik@aucegypt.edu`

Abstract. In this paper, we study the problem of maximizing the secondary user (SU) throughput under a quality of service (QoS) delay requirement of the primary user (PU). In addition, we investigate the impact of having a full-duplex capability at the SU on the network performance, compared to the case of a half-duplex SU. We consider a cooperative cognitive radio (CR) network with multipacket reception (MPR) capabilities at the receiving nodes. In our proposed system, the SU not only exploits the idle time slots (i.e. when PU is not transmitting) but also chooses between cooperating or sharing the channel with the PU probabilistically. We formulate our optimization problem maximizing the SU throughput under a PU delay constraint; we optimize over the SU transmission modes' selection probabilities. The resultant optimization problem turns out to be a non-convex quadratic constrained quadratic programming (QCQP) optimization problem, which is, in general, an NP-hard problem. An efficient approach is devised to solve it and characterize the stability region of the network under a delay constraint on the PU. Numerical results, surprisingly, reveal that the network performance with a full-duplex SU is not always better than that of a half-duplex SU. In fact, we show that a full-duplex capability at the SU can adversely affect the stability performance of the network especially if the channel condition between the SU and the destinations is weaker than that between the PU and the destinations.

1 Introduction

Cooperative communication techniques have gained a lot of interest over the years due to the important key role they play in wireless communications [1]. In cooperative communications, intermediate nodes capture the source transmitted packets and contribute via cooperatively relaying them to the destination. In

© ICST Institute for Computer Sciences, Social Informatics and Telecommunications Engineering 2019
Published by Springer Nature Switzerland AG 2019. All Rights Reserved
A. Kliks et al. (Eds.): CrownCom 2019, LNICST 291, pp. 180–194, 2019.
https://doi.org/10.1007/978-3-030-25748-4_14

[2], the authors showed that a reasonable enhancement in the stable throughput region of the network can be achieved as a result of the existence of a cooperative relay node. On the other hand, cognitive radio (CR) has emerged as a powerful leading technology in alleviating the scarcity of available spectrum resources, which are relatively under-utilized. CR achieves efficient utilization of the spectrum while maintaining some primary users (PUs) QoS [3]. Recently, researchers started to integrate cooperative communications into CR networks, e.g., [4], by allowing the secondary users (SUs) to serve as relays for the PUs. As a result, the SUs use their available resources to transmit their own data as well as PUs' packets. It was shown that cooperation could, in general, enhance both the SU and the PU throughputs.

The authors in [5] and [6] have derived the stability region for a cooperative CR network consisting of one PU, one SU (with two queues, one for its own data packets and the other one to relay the PU packets), and a common destination. The authors in [5] considered a model in which the SU transmits only when the PU is inactive with firm priority in favor of the PU relaying queue. Assigning firm priority to the relaying queue can degrade the SU throughput, especially when the PU average packet delay becomes much smaller than the target delay constraint. To overcome this problem, the authors in [6] presented a model in which the SU can serve either its own data queue or the relaying queue according to some probability assigned to each queue (i.e., the SU randomizes the service between its own data queue and the relaying queue whenever it has access to the channel). In [5] and [6], optimizing the SU throughput and the average packet delay experienced by the PU and SU were studied, respectively, subject to network stability constraints. It was shown that cooperation is beneficial only when the channel between the SU and the common destination is better than that between the PU and the destination (same restriction as in [2]). Recently, the authors in [7] modified the scheme presented in [6] by allowing the SU to have an extra queue (battery queue). The authors then derived the stability region of the network subject to some energy harvesting constraints on the SU.

Another important aspect of our proposed framework is to consider the case when the SU has a full-duplex capability and characterize its effect on the system stability region. Most of the previous works on cooperative CR networks have assumed that the users are all half-duplex; this is because full-duplex communications were considered infeasible in the past due to the effect of self-interference. Recently, the authors in [8] presented feasible approaches that can achieve a drastic self-interference suppression. Using these approaches, the authors in [9] showed that the existence of a full-duplex relay can enhance the users' throughput. In [10], the model presented in [9] was extended assuming a full-duplex SU that has its own data queue in addition to the relaying queue, unlike [9] where the relay node is assumed not to have data of its own; the SU throughput was maximized subject to some stability constraints of the network. Note that the recent works presented in the context of cooperative CR, e.g., [6,7] and [11] consider only a half-duplex SU to simplify the analysis for their models.

Our main contributions in this paper are as follows.

- We propose a delay-aware scheme that enables us to maximize the SU throughput subject to a QoS delay requirement of the PU. Our optimization problem turns out to be a non-convex quadratic constrained quadratic programming (QCQP) optimization problem which is NP-hard in general. An efficient approach is proposed to solve our optimization problem to characterize the stability region. The importance of our proposed framework arises from the emergence of numerous real-time applications that should be supported by cooperative CR networks e.g., video streaming, gaming, and other multimedia applications; these applications demand high throughput with strict delay requirements. This aspect is mostly neglected in all of the above-cited cooperative CR works [5–7,9,10], which mainly focused on enhancing certain performance subject to some network stability and/or energy harvesting constraints with no delay provisioning.
- We study the impact of having the full-duplex capability at the SU on the network performance from a queuing-theoretic perspective. The full-duplex capability enables the SU to decode the PU transmission while simultaneously transmitting over the channel. We, unexpectedly, show that having a full-duplex capability at the SU is not always beneficial. It can adversely affect the stability performance of the network and, in some cases, provide strictly inferior performance compared to the system with a half-duplex SU.

It is worth mentioning that optimizing network performance under PU delay constraints have been considered before in [11] for a different cooperation policy and for a simpler network configuration. More precisely, [11] considered only a half-duplex SU which simplifies the stability and delay analysis for their network. Moreover, [11] considered only the case in which the SU accesses the channel when the PU is sensed to be inactive (similar to [5,6]). This, in effect, reduces the number of optimization parameters, and hence, simplifies the optimization problem.

The rest of the paper is organized as follows. The system model is described in Sect. 2. Section 3 describes our cooperation policy and presents the delay and stability analysis. The problem formulation and the solution approach are presented in Sect. 4. The numerical results are presented in Sect. 5. Finally, conclusions are drawn in Sect. 6.

2   System Model

In this paper, we consider a full-duplex cooperative CR network, shown in Fig. 1, consisting of one PU (p), one SU (s) and two different destinations (d and d'). The SU is a full-duplex node that can simultaneously receive a packet from the PU and transmit a packet to d or d'. We also assume that the primary destination (d) has MPR capability, i.e., it can simultaneously decode multiple packets received from the PU and the SU relaying queue.

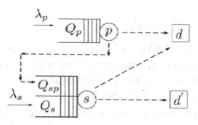

Fig. 1. The system model

Time is divided into slots; each slot has a fixed time duration, and for simplicity of presentation, we assume that every packet transmission takes only one slot. The SU is assumed to have two infinite queues Q_s and Q_{sp}. The queue Q_s is used to store the SU arriving packets, whereas the queue Q_{sp} is used to store the relayed packets received from the PU. The PU is assumed to have only one infinite queue Q_p used to store its arriving packets. The arrivals at the PU and the SU are considered to be Bernoulli processes with rates λ_p and λ_s, respectively. The arrival processes at both users are assumed to be independent.

The wireless channel between any two nodes (m, n), where $m \in \{p, s\}$ is the transmitting node and $n \in \{s, d, d'\}$ is the receiving node, where $m \neq n$, is modeled as a stationary Rayleigh flat-fading channel. The channel gain is denoted by h_{mn}, where $\mathbb{E}(|h_{mn}|^2) = \rho_{mn}^2$. The channel gains, h_{mn}'s, are assumed to be constant within any given time slot but vary independently from one time slot to another. All users (nodes) are exposed to independent complex additive white Gaussian noise with unit variance and zero mean. The transmitted power from the PU or the SU is fixed and is denoted by P.

In our proposed scheme, we will have four different cases (modes of operation) for packet transmissions. The first case is when either the PU or the SU transmits alone, i.e, the non-transmitting node remains idle. In this case, the probability of successful decoding is given by:

$$f_{mn} = \mathbb{P}\{R < \tfrac{1}{2}\log_2(1 + P|h_{mn}|^2)\} = \exp\left(-\frac{2^{2R} - 1}{P\rho_{mn}^2}\right), \tag{1}$$

where $mn \in \{pd, ps, sd', sd\}$ and R is the transmission rate.

The second case is when the SU transmits a packet from Q_s simultaneously with the PU, i.e., each user is causing interference on the other user transmission. In this case, we let v_{mn}^I denote the probability that node n successfully decodes the packet transmitted from node m by considering the interference caused by node I transmission as noise, where $(mn, I) \in \{(pd, s), (sd', p)\}$.

The third case is when the SU transmits a relayed packet from Q_{sp} simultaneously with the PU, i.e., both nodes transmit primary packets to node d. In this case, the primary destination attempts decoding both transmissions by using its MPR capability (using successive interference cancellation). The probability of successful transmission, in this case, is given by

$$g_{mn}^I = u_{mn}^I + (1 - u_{mn}^I)v_{mn}^I, \tag{2}$$

where $(mn, I) \in \{(pd, s), (sd, p)\}$ and u_{mn}^I is the probability that node n successfully decodes both packets transmitted from nodes m and I. The derivations of u_{mn}^I and v_{mn}^I are omitted due to space limitations.

The fourth case occurs when the SU exploits its full-duplex capability, i.e., it attempts decoding a primary packet while transmitting either a secondary or primary packet. Let f_{sd}^{dup} denote the probability that s decodes a primary packet from p in the full-duplex mode, and is given by

$$f_{ps}^{dup} = \mathbb{P}\left\{ R < \frac{1}{2}\log_2\left(1 + \frac{P|h_{ps}|^2}{1 + Pg}\right) \right\} = \exp\left(-\frac{(2^{2R} - 1)(1 + Pg)}{P\rho_{ps}^2}\right), \tag{3}$$

where the scalar gain $g \in [0, 1]$ represents the effectiveness of self-interference cancellation [9,12]. For example, if $g = 0$ then self-interference is perfectly suppressed, and if $g = 1$ then no self-interference cancellation is considered. The details of the techniques utilized for self-interference cancellation are beyond the scope of this paper (the interested reader is referred to [8] and references therein). Note that to consider the case where the SU is a half-duplex node, we set $f_{sd}^{dup} = 0$ not $g = 1$. As $g = 1$ corresponds to the case of full-duplex mode with no self-interference cancellation which is different from the half-duplex mode.

3   Cooperation Policy and System Analysis

In this section, our proposed cooperation policy is introduced. Then, we provide our proposed scheme delay and stability analysis. Note that we assume ACK/NACK packet transmission from s, d', and/or d at the end of each time slot. These ACK/NACK packets are assumed to be received error-free at all nodes. For simplicity of presentation, the SU is assumed to be able to perfectly sense the existence or the absence of the PU. As a result of this sensing process, the cooperation policy can be split into two cases as follows:

3.1 Q_p Has Packets (An Active Primary User; A Busy Time Slot)

In this case, the SU could select one of two access decisions as follows.

- The SU decides to access the channel, causing interference to the PU, by sending a packet from Q_s or Q_{sp} with probabilities p_s^{busy} or p_{sp}^{busy}, respectively (unlike [5,6,11], where the SU always refrains from accessing the channel upon detecting a PU transmission). This, in turn, should result in better utilization of the PU channel. In this case, the SU can still decode the PU packet if d fails to decode it using its full-duplex capability. Note that, for a half-duplex SU, the SU will not be able to help the PU in this case.

- The SU decides to refrain from sending any packets and only listen to the PU transmission to help to relay it if the primary destination (d) does not succeed in decoding it. This occurs with probability $1 - p_{sp}^{busy} - p_s^{busy}$.

It should be noted that the PU is the owner of the spectrum, therefore, it does not have to provide any provisions to the SU; the PU just sends the packet on the top of Q_p provided that Q_p is nonempty. As a result, three possible scenarios would emerge:

- If d decodes the received packet successfully, then Q_p drops the packet irrespective of the decision taken by the SU.
- If the PU packet is decoded successfully by the SU, and at the same time d could not decode it, then the packet will be dropped from Q_p and stored at Q_{sp}.
- If both s and d were not able to decode the primary packet, then the PU keeps the packet in Q_p to attempt re-sending it in the next time slot.

3.2 Q_p Is Empty (An Inactive Primary User; An Idle Time Slot)

In this case, the SU could select one of two access decisions as follows.

- The SU transmits one packet from Q_{sp} with probability p_{sp}^{idle}. This packet will be dropped from Q_{sp} if the SU receives an ACK from d.
- The SU transmits one packet from Q_s with probability $p_s^{idle} = 1 - p_{sp}^{idle}$. This packet will be dropped from Q_s if the SU receives an ACK from the secondary destination (d').

Our model involves interaction among different queues. The stability and delay analysis of more than two interacting queues is, in general, a complex problem [13]. As a result, we resort to the dominant system approach to decouple the interaction among the system queues. This approach was used before in different contexts, e.g., [14], to derive sufficient stability conditions of a system of interacting queues. In our dominant system, the SU is assumed to send dummy packets when it chooses to transmit a packet from an empty queue. This has the effect of decoupling the interaction between the queues by decoupling the service rate for any queue from the number of packets in other queues. It is obvious that dominant system stability implies original system stability (as transmitting dummy packets can only degrade the system performance). Also, the average delay experienced by the packets in the dominant system is an upper bound on that of the original system.

Next, we present the details of the stability analysis of our system followed by the delay analysis. Loynes' theorem [15] is applied to examine the stability of each queue in the system. Loynes' theorem states that a queue is stable as long as its average arrival rate is strictly less than its average service rate if both the arrival and the service processes are stationary. It should be noted that the PU packets can be served by two queues, Q_p and Q_{sp}, and this should be taken into consideration while calculating the PU delay.

According to the scenarios illustrated above in Sects. 3.1 and 3.2, the PU packets' departure rate from Q_p can be expressed as [1]

$$\mu_p = p_s^{busy}(v_{pd}^s + (1 - v_{pd}^s)f_{ps}^{dup}) + p_{sp}^{busy}(g_{pd}^s + (1 - g_{pd}^s)f_{ps}^{dup})$$
$$+(1 - p_s^{busy} - p_{sp}^{busy})(f_{pd} + (1 - f_{pd})f_{ps}). \qquad (4)$$

For Q_p to be stable, its arrival rate λ_p should be less than μ_p, i.e.,

$$\lambda_p < p_s^{busy}(v_{pd}^s + (1 - v_{pd}^s)f_{ps}^{dup}) + p_{sp}^{busy}(g_{pd}^s + (1 - g_{pd}^s)f_{ps}^{dup})$$
$$+(1 - p_s^{busy} - p_{sp}^{busy})(f_{pd} + (1 - f_{pd})f_{ps}). \qquad (5)$$

For Q_{sp}, the service rate, μ_{sp}, can be given as

$$\mu_{sp} = \frac{\lambda_p}{\mu_p}p_{sp}^{busy}g_{sd}^p + (1 - \frac{\lambda_p}{\mu_p})p_{sp}^{idle}f_{sd}, \qquad (6)$$

where $\frac{\lambda_p}{\mu_p}$ is the probability that Q_p is nonempty. Moreover, the arrival rate at Q_{sp}, λ_{sp}, is given by

$$\lambda_{sp} = \frac{\lambda_p}{\mu_p}\{p_s^{busy}f_{ps}^{dup}(1 - v_{pd}^s) + p_{sp}^{busy}f_{ps}^{dup}(1 - g_{pd}^s)$$
$$+(1 - p_s^{busy} - p_{sp}^{busy})f_{ps}(1 - f_{pd})\}. \qquad (7)$$

To satisfy the Q_{sp} stability requirement, λ_{sp} should be less than μ_{sp}, i.e.,

$$\lambda_p < \frac{C}{A - B + C}\mu_p, \qquad (8)$$

where A, B, and C are given by

$$A = \{p_s^{busy}f_{ps}^{dup}(1 - v_{pd}^s) + p_{sp}^{busy}f_{ps}^{dup}(1 - g_{pd}^s)$$
$$+ (1 - p_s^{busy} - p_{sp}^{busy})f_{ps}(1 - f_{pd})\},$$
$$B = p_{sp}^{busy}g_{sd}^p, C = p_{sp}^{idle}f_{sd}.$$

From (5) and (8), it can be easily shown that the PU arrival rate should satisfy the following condition for the system to be stable.

$$\lambda_p < \min\{\mu_p, \mu_r\}, \qquad (9)$$

where μ_p is as in (4) and μ_r is given by

$$\mu_r = \frac{C}{A - B + C}\mu_p. \qquad (10)$$

---

[1] Note that throughout the analysis presented in this paper, we consider a dominant system in which the SU transmits dummy packets if it selects to transmit from an empty queue. This has the effect of decoupling the service rates of each queue from the state of other queues.

Finally, for Q_s, the service rate μ_s can be shown to be given by

$$\mu_s = \frac{\lambda_p}{\mu_p} p_s^{busy} v_{sd'}^p + \left(1 - \frac{\lambda_p}{\mu_p}\right)(1 - p_{sp}^{idle}) f_{sd'}. \tag{11}$$

To satisfy the Q_s stability requirement, the arrival rate λ_s should be less than μ_s, i.e.,

$$\lambda_s < \frac{\lambda_p}{\mu_p} p_s^{busy} v_{sd'}^p + \left(1 - \frac{\lambda_p}{\mu_p}\right)(1 - p_{sp}^{idle}) f_{sd'}. \tag{12}$$

This completes our stability region characterization.

If a PU packet is delivered to its destination d via the SU, the packet will experience two queuing delays; the delay in Q_p and the delay in Q_{sp}. As a result, the average delay experienced by the PU packet can be expressed as follows:

$$D_p = \tau T_p + (1 - \tau)(T_p + T_{sp}) = T_p + (1 - \tau)T_{sp}, \tag{13}$$

where the average delays at Q_{sp} and Q_s are represented by T_{sp} and T_s, respectively. And τ is given by

$$\tau = \frac{(1 - p_s^{busy} - p_{sp}^{busy}) f_{pd} + p_s^{busy} v_{pd}^s + p_{sp}^{busy} g_{pd}^s}{\mu_p}, \tag{14}$$

and it represents the possibility that the PU packet is decoded successfully by d conditioned on that it was dropped from Q_p. The queues Q_{sp} and Q_s are discrete time M/M/1 queues with Bernoulli arrival processes and Geometric service rates. Therefore, by applying Pollaczek-Khinchine [16] and Little's law, T_p and T_{sp} can be expressed as

$$T_p = \frac{1 - \lambda_p}{\mu_p - \lambda_p}, \quad T_{sp} = \frac{1 - \lambda_{sp}}{\mu_{sp} - \lambda_{sp}}. \tag{15}$$

4   Problem Formulation and Solution Approach

In this section, we introduce our optimization problem. Our objective is to maximize the SU throughput λ_s subject to a PU QoS delay requirement. As a result, our optimization problem can be written, in an epigraph form [17, Chapter 4], as follows:

$$\begin{aligned}
\max_{p_s^{busy}, p_{sp}^{busy}, p_{sp}^{idle}, \lambda_s} \quad & \lambda_s \\
\text{subject to} \quad & \lambda_s \le \mu_s, \\
& D_p \le \phi, \\
& 0 \le p_s^{busy} + p_{sp}^{busy} \le 1, \\
& 0 \le p_{sp}^{idle} \le 1, \\
& p_i^{busy} \ge 0 \quad i \in \{s, sp\},
\end{aligned} \tag{16}$$

where the stability of Q_s is guaranteed by the first constraint, while the PU QoS delay requirement is guaranteed by the second constraint. Introducing a PU delay constraint is stricter than a PU queue stability constraint and, hence, implies the stability of the primary queue, i.e., Q_p length is guaranteed not to grow to infinity. Consequently, there is no need for having an extra stability constraint of Q_p.

It should be noted that D_p and μ_s, given in (13) and (11), are non-convex functions in the optimization parameters, which renders the overall optimization problems to be non-convex. Next, we go through a number of steps to solve this non-convex optimization problem. Note that, due to space limitations, we just provide a concise description of our proposed approach to solve the above non-convex optimization problem.

First, it should be noted that if we fix μ_p (make it constant) then μ_s from (11) becomes a linear function of the optimization parameters. Hence, the first constraint in the optimization problem in (16) becomes convex. Moreover, if we fix μ_p then the delay constraint, which is the second constraint in our optimization problem, becomes a quadratic function of the optimization parameters in the form of $\boldsymbol{p}^T \boldsymbol{A} \boldsymbol{p} + \boldsymbol{c}^T \boldsymbol{p} + d \leq 0$ where $\boldsymbol{p}$ is a vector that contains all the optimization parameters. Unlike the first constraint, the second one is, unfortunately, non-convex because the matrix $\boldsymbol{A}$ is not a positive semi-definite matrix as will be explained later. Hence, by fixing μ_p, we convert the first constraint to be linear and the second to be non-convex quadratic, while the rest of the constraints are already linear constraints. This form of optimization problems is called non-convex QCQP optimization problems. To solve this problem, we use the feasible point pursuit successive convex approximation (FPP-SCA) algorithm presented in [18], which solves the problem by linearizing the non-convex parts of the delay constraint. We use it, due to its advantages, illustrated in [18], over the other methods that can be used to solve QCQP optimization problems.

To find the optimum μ_p, we iterate over all possible values of μ_p and for each value we solve a non-convex QCQP optimization problem. For each value of μ_p, we solve for the maximum stable λ_s given μ_p. Finally, we find the maximum value of λ_s over all feasible μ_p's to be our solution. The values of feasible μ_p's range from λ_p to λ_p^m, which is the maximum feasible value of λ_p. The minimum value of μ_p is λ_p to guarantee the stability of the PU queue. The value of λ_p^m can be easily calculated by setting $p_{sp}^{idle} = 1$ and $p_s^{busy} = 0$ in (9) and optimizing only over p_{sp}^{busy} (which can be done easily via one-dimensional numerical search).

Based on the above, and for a given μ_p, we can easily see from (11) that the first constraint becomes a linear function in the optimization parameters. Define a 4-D vector $\boldsymbol{p} = [p_s^{busy}, p_{sp}^{busy}, p_{sp}^{idle}, \lambda_s]^T$; the second constraint can now be rewritten as $\boldsymbol{p}^T \boldsymbol{A} \boldsymbol{p} + \boldsymbol{c}^T \boldsymbol{p} + d \leq 0$ where $\boldsymbol{A}$, $\boldsymbol{c}$, and d can be directly obtained from (6), (7), and (13). It can be readily seen that this constraint is a non-convex quadratic constraint since $\boldsymbol{A}$ is an indefinite matrix. As mentioned above, we use FPP-SCA algorithm, presented in [18] which linearizes the non-convex parts of the constraints as shown next.

Using eigenvalue decomposition, we can express the matrix A, which is an indefinite matrix, as $A = A^+ + A^-$, where $A^+ \succeq 0$ and $A^- \preceq 0$. For any $z \in R^{4 \times 1}$, we have

$$(p - z)^T A^- (p - z) \leq 0, \tag{17}$$

$$p^T A^- p \leq 2 z^T A^- p - z^T A^- z. \tag{18}$$

With the aid of the above inequalities, the quadratic non-convex constraint $(p^T A p + c^T p + d \leq 0)$ can be replaced by the following convex constraint:

$$p^T A^+ p + 2 z^T A^- p + c^T p + d \leq z^T A^- z, \tag{19}$$

which relaxed the non-convex part of the constraint to a linear one. Now, our non-convex problem is converted into a convex one. Finally, we rewrite the optimization problem in (16) to that given in Algorithm 1, where $x = [0, 0, 0, -1]^T$, $b = [1, 1, 0, 0]^T$ and v, u, m, n can be obtained from (4).

Algorithm 1 finds the feasible solution that maximizes the throughput of the SU under a QoS delay constraint on the PU. Note that a slack variable (s) is used to ensure the feasibility of the approximated problem, and a penalty (Λ) is used to guarantee that the slack is mildly used.

Algorithm 1

For $\mu_p = \lambda_p : \delta : \lambda_p^m$
Initialization: set $i = 0$ and $z_0 = 0$.
Repeat
1. solve

$$\lambda_s(\mu_p) \quad = \min_p \ x^T p + \Lambda s$$

$$\text{s. t.} \quad v^T \alpha + u = \mu_p,$$

$$m^T p + n \leq 0,$$

$$p^T A^+ p + 2 z_i^T A^- p + c^T p + d \leq z_i^T A^- z_i + s,$$

$$0 \leq b^T p \leq 1,$$

$$0 \preceq p \preceq 1,$$

$$0 \leq s.$$

2. Let p_k^* denote the optimal p obtained at the i-th iteration, and set $z_{i+1} = p_i^*$.
3. Set $i = i + 1$.
until convergence
Return the maximum $\lambda_s(\mu_p)$.

To draw the stability region (λ_p versus λ_s) with the delay constraint on the PU, we vary λ_p from zero to λ_p^m and obtain the maximum stable throughput of the SU for each λ_p using Algorithm 1. Then, we get the convex hull for the obtained values. Note that we already know that the stable throughput point

$(\lambda_p, \lambda_s) = (0, f_{sd'})$ is in the stable region. This point corresponds to the case when $\lambda_p = 0$, and hence, the SU is free to transmit its own packets in all time slots achieving its maximum stable throughput $f_{sd'}$.

5   Numerical Results

In this section, we illustrate the impact of having a full duplex SU on the stability region under a PU delay constraint. We compare the stability region for three different configurations (a) full-duplex SU with perfect self interference cancellation $(g = 0)$, (b) full-duplex SU with no self interference cancellation $(g = 1)$, and (c) half-duplex SU such that $f_{ps}^{dup} = 0$. This comparison is shown for four different channel scenarios, i.e., different sets of channel gains variances ρ_{pd}^2 and ρ_{sd}^2, while fixing the rest of the simulation parameters. The fixed parameters are: $P = 10$, $R = 1$, $\rho_{p,s}^2 = 0.6$, $\rho_{s,d'}^2 = 0.8$, and $\rho_{p,d'}^2 = 0.3$.

For the first channel scenario, shown in Fig. 2, we choose high channel gains between $p - d$ and $s - d$ such that $\rho_{pd}^2 = 0.8$ and $\rho_{sd}^2 = 0.9$. Since there is already a good channel between the PU and its destination, most of the packets are successfully transmitted from the PU to d, and hence, the three configurations achieve almost the same maximum stable λ_p (relaying will not be that beneficial in this case). For small λ_p, the SU does not play an important role in delivering the PU packets, and hence, all the three configurations have a close performance. While, for high λ_p, the SU full-duplex capability helps to have faster delivery of the PU packets, allowing the SU to exploit more resources to send its own packets. Thus, as shown in Fig. 2, the stability region achieved by the full-duplex configuration is larger than that of the half-duplex.

In the second scenario, we choose low channel gain between $p - d$ $(\rho_{pd}^2 = 0.04)$ and high channel gain between $s - d$ $(\rho_{sd}^2 = 0.9)$. In this scenario, the SU plays an important role acting as a relay to deliver the PU packets because of the bad $p - d$ channel. It is clear from Fig. 3 that the full-duplex capability significantly improves the stability region compared to that of the half-duplex. The rationale behind this observation is that the full-duplex capability allows the SU to capture more PU packets in the relaying queue to account for the bad $p - d$ channel. In fact, the SU relaying rate dominates the service rate of the PU. On the other hand, the half-duplex capability reduces λ_{sp} as shown in (7) because the SU has to listen more to the PU transmissions, as a result of the bad $p - d$ channel, which reduces the opportunities available for the SU to send its own packets.

In the previous scenarios, we considered a good channel between s and d, which in turn allows the SU to deliver the relayed packets easily to d. Hence, the full-duplex capability was always advantageous and it enlarged the stability region. In the following two scenarios, we consider two other possibilities in which the SU suffers from bad channel to d.

In the third scenario, we choose high channel gain between $p - d$ $(\rho_{pd}^2 = 0.8)$ and low channel gain between $s - d$ $(\rho_{sd}^2 = 0.04)$. Figure 4 shows that the three configurations achieve almost the same maximum stable λ_p because there is already a good channel between the PU and its destination and most of the

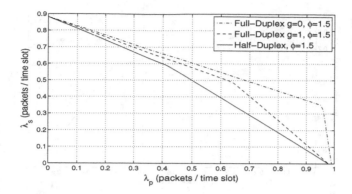

Fig. 2. High channel gain p-d and high channel gain s-d.

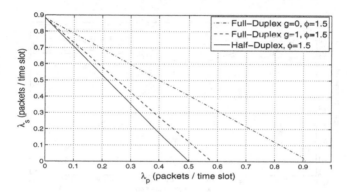

Fig. 3. Low channel gain p-d and high channel gain s-d.

packets are successfully transmitted from the PU to the destination on the direct channel. However, none of the three configurations is strictly better over the whole feasible range of λ_p. For small λ_p, the half-duplex configuration provides the best performance, while for large λ_p the full-duplex configuration is the best. The rationale behind this is that for small PU arrival rate λ_p, Q_{sp} is dominating the stability conditions; refer to (9). Hence, having only half-duplex capability reduces the arrival rate of Q_{sp} (7) which, in turn, preserves the stability of Q_{sp} for a wider range of λ_p than that for the full-duplex case. As we increase λ_p, it is clear that full-duplex is becoming the best configuration. This happens because, for large λ_p, Q_p is dominating the stability conditions; refer to (9). Consequently, having the full-duplex capability at the SU increases the service rate of Q_p (4) which, in turn, preserves the stability of Q_p for larger values of λ_p.

In the fourth scenario, we choose low channel gain between $p-d$ ($\rho_{pd}^2 = 0.04$) and low channel gain between $s-d$ ($\rho_{sd}^2 = 0.04$). Figure 5 shows that the half-duplex configuration is strictly better than the full-duplex configuration. For this scenario, when the SU has a full-duplex capability, most of the PU packets have to be delivered to the destination through the relying queue Q_{sp} due to the good

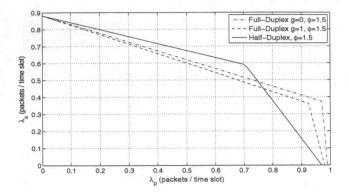

Fig. 4. High channel gain p-d and low channel gain s-d.

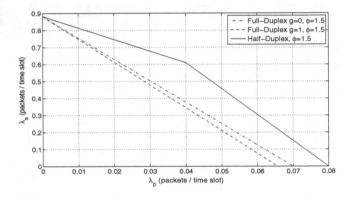

Fig. 5. Low channel gain p-d and low channel gain s-d.

channel condition between p and s and bad channel condition between p and d. However, the SU cannot relay these packets due to the bad channel condition $s - d$ causing congestion of the packets at the relaying queue Q_{sp}. On the other hand, having only half-duplex SU reduces the accumulation of packets at Q_{sp} by reducing the arrival rate of Q_{sp} which, in turn, enlarges the stability region. This clearly shows that having a full-duplex capability at the SU is not always beneficial.

Finally, Fig. 6 compares the performance of our adopted FPP-SCA algorithm, presented in Algorithm 1, to the exact results obtained from the exhaustive grid-search based solution. This is performed, without taking the convex hull, for the case where the channel $p - d$ is low and that between $s - d$ is high. Figure 6 Demonstrates that an almost identical performance is achieved by both for the entire range of λ_p. Hence, this shows the efficacy of the adopted FPP-SCA algorithm.

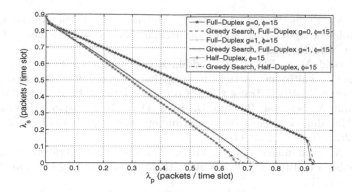

Fig. 6. Comparison between our adopted FPP-SCA algorithm and the greedy search algorithm.

6   Conclusion and Future Work

We have studied cooperative CR networks with the target of maximizing the SU throughput subject to a PU QoS delay constraint. We have also studied the impact of having a full-duplex SU on the network performance compared to having a half-duplex SU. We formulated an optimization problem to maximize the SU throughput subject to a PU delay constraint, which was shown to be non-convex. We proposed to solve the problem by iterating over a set of non-convex QCQP optimization problems and using the FPP-SCA algorithm to solve each iteration. Unexpectedly, our numerical results have revealed that having a full-duplex capability at the SU is not always beneficial and it can adversely affect the stability region of the network in some scenarios. In our future work, we will consider comparing the performance of our proposed cooperation policy to that of the other cooperation and no-cooperation policies presented in the same context. In addition, we will investigate the impact of the SU sensing errors on our system performance. We will also consider finding the optimal trade-off between the PU delay and the SU throughput.

References

1. Tao, X., Xu, X., Cui, Q.: An overview of cooperative communications. IEEE Commun. Mag. **50**(6), 65–71 (2012)
2. Sadek, A.K., Liu, K.R., Ephremides, A.: Cognitive multiple access via cooperation: protocol design and performance analysis. IEEE Trans. Inf. Theory **53**(10), 3677–3696 (2007)
3. Wang, B., Liu, K.R.: Advances in cognitive radio networks: a survey. IEEE J. Sel. Topics Sig. Process. **5**(1), 5–23 (2011)
4. Letaief, K.B., Zhang, W.: Cooperative communications for cognitive radio networks. Proc. IEEE **97**(5), 878–893 (2009)
5. Rong, B., Ephremides, A.: Cooperative access in wireless networks: stable throughput and delay. IEEE Trans. Inf. Theory **58**(9), 5890–5907 (2012)

6. Ashour, M., El-Sherif, A.A., ElBatt, T., Mohamed, A.: Cognitive radio networks with probabilistic relaying: stable throughput and delay tradeoffs. IEEE Trans. Commun. **63**(11), 4002–4014 (2015)
7. Abd-Elmagid, M.A., ElBatt, T., Seddik, K.G., Ercetin, O.: Stable throughput of cooperative cognitive networks with energy harvesting: finite relay buffer and finite battery capacity. IEEE Trans. Cogn. Commun. Netw. **4**(4), 704–718 (2018)
8. Sabharwal, A., Schniter, P., Guo, D., Bliss, D.W., Rangarajan, S., Wichman, R.: In-band full-duplex wireless: challenges and opportunities. IEEE J. Sel. Areas Commun. **32**(9), 1637–1652 (2014)
9. Pappas, N., Ephremides, A., Traganitis, A.: Stability and performance issues of a relay assisted multiple access scheme with MPR capabilities. Comput. Commun. **42**, 70–76 (2014)
10. ElAzzouni, S., Ercetin, O., El-Keyi, A., ElBatt, T., Nafie, M.: Full-duplex cooperative cognitive radio networks. In: 2015 13th International Symposium on Modeling and Optimization in Mobile, Ad Hoc, and Wireless Networks (WiOpt), pp. 475–482. IEEE (2015)
11. Elmahdy, A.M., El-Keyi, A., ElBatt, T.A., Seddik, K.G.: Optimizing cooperative cognitive radio networks performance with primary QoS provisioning. IEEE Trans. Commun. **65**(4), 1451–1463 (2017)
12. Ramirez, D., Aazhang, B.: Optimal routing and power allocation for wireless networks with imperfect full-duplex nodes. IEEE Trans. Wirel. Commun. **12**(9), 4692–4704 (2013)
13. Tsybakov, B.S., Mikhailov, V.A.: Ergodicity of a slotted aloha system. Problemy peredachi informatsii **15**(4), 73–87 (1979)
14. Krikidis, I., Devroye, N., Thompson, J.S.: Stability analysis for cognitive radio with multi-access primary transmission. IEEE Trans. Wirel. Commun. **9**(1), 72–77 (2010)
15. Loynes, R.M.: The stability of a queue with non-independent inter-arrival and service times. In: Mathematical Proceedings of the Cambridge Philosophical Society, vol. 58, pp. 497–520. Cambridge University Press (1962)
16. Jackson, J.R.: Jobshop-like queueing systems. Manag. Sci. **50**(Suppl. 12), 1796–1802 (2004)
17. Boyd, S., Vandenberghe, L.: Convex Optimization. Cambridge University Press, Cambridge (2004)
18. Mehanna, O., Huang, K., Gopalakrishnan, B., Konar, A., Sidiropoulos, N.D.: Feasible point pursuit and successive approximation of non-convex QCQPs. IEEE Sig. Process. Lett. **22**(7), 804–808 (2015)

Radio Environment Maps for Military Cognitive Networks: Density of Sensor Network vs. Map Quality

Marek Suchański, Paweł Kaniewski, Janusz Romanik, Edward Golan, and Krzysztof Zubel[✉]

Radiocommunications Department, Military Communications Institute, Warszawska 22A, 05-130 Zegrze Poludniowe, Poland
{m.suchanski, p.kaniewski, j.romanik, e.golan, k.zubel}@wil.waw.pl

Abstract. In this paper we present the dependency between density of sensor network and map quality in the Radio Environment Map (REM) concept. The architecture of REM supporting military communications systems is described. The map construction techniques based on spatial statistics and transmitter location determination are presented. The problem of REM quality and relevant metrics are discussed. The results of field tests for UHF range with different number of sensors are shown. Exemplary REM maps with different interpolation algorithms are presented. Finally, the problem of density of sensors network versus REM map quality is analyzed.

Keywords: Cognitive radio · Radio Environment Map · Spectrum monitoring · Density of sensor network · Deployment of sensors

1 Introduction

In recent years, in many fields of technology there has been a growing trend towards creating intelligent solutions that autonomously make decisions about their actions. This trend can also be noticed in wireless communications. It is worth mentioning here such solutions as self-organizing networks [1, 2], disruption-tolerant networks [3], dynamic spectrum management [4, 5] and cognitive radio [6]. In military communications new technical solutions are adopted with great caution as they are used in very specific conditions and must be extremely reliable. Military wireless networks must be immune to deliberate interference and operational even in the case of systematic destruction of telecommunications infrastructure.

The problem of efficient frequency management in common operations has been noticed by NATO and, as a consequence, the IST panel has established a working group whose tasks include among others checking potential benefits resulting from the implementation of the Radio Environment Map concept.

The aim of the IST-146 RTG-069 group is to work out a concept of REM enabling their users to obtain the spectrum operational picture and to minimize the level of interferences between wireless systems of coalition forces. One of the main goals of the

© ICST Institute for Computer Sciences, Social Informatics and Telecommunications Engineering 2019
Published by Springer Nature Switzerland AG 2019. All Rights Reserved
A. Kliks et al. (Eds.): CrownCom 2019, LNICST 291, pp. 195–207, 2019.
https://doi.org/10.1007/978-3-030-25748-4_15

research group is to define the architecture of the system and to specify interfaces to other systems in the area of frequency management.

According to the NATO IST RTG-050 plan, REM is also needed to make a significant step towards a coordinated spectrum management system in NATO [4].

In the paper we discuss the concept of REM and the problem of the number of sensors from the point of view of tactical operation. We also present exemplary maps created using different interpolation methods and analyze how the number of sensors affects the quality of the maps. Additionally we focused on the possibility of localization of the TX antenna in reference to selected interpolation techniques.

The rest of the paper is organized as follows: related works (Sect. 2), map construction techniques (Sect. 3), test scenario and exemplary maps (Sect. 4), analysis of the results (Sect. 5) and conclusions (Sect. 6).

2 Related Works

In general, REM is considered as a database which maintains comprehensive and up-to-date information on the radio spectrum. It is assumed that this information is composed of geographical features, available services, spectral regulations, positions and activities of radios, policies adopted by the user and/or service providers, and knowledge from the past [7].

The simplified architecture of REM excerpted from [8, 9] and adapted to military applications is presented in Fig. 1. REM architecture comprises the following modules: REM Manager, REM storage and data collection, REM Acquisition, sensors and GUI. REM Manager processes the data and controls the REM database in terms of measurement configuration, e.g. monitoring subranges, measurement mode (continuous or on request), active sensors. REM storage and collection module is an interface between the database, REM acquisition modules and REM Manager. REM acquisition modules are interfaces to various systems of sensors.

In the literature [11] sensors are generally named MCDs (Measurement Capable Devices). MCDs are controlled through REM Acquisition modules and monitor spectrum. In civilian applications the function of MCDs can be performed by various devices with measurement capability, such as simple mobile phones, smart phones, notebooks, etc.

When military systems are considered, spectrum measurements can be taken by dedicated receivers, cognitive radios, Electronic Warfare (EW) systems or Intelligence, Surveillance, Reconnaissance (ISR) systems [10, 18].

In the literature on the topic the spectrum sampling method for REM has not been thoroughly researched. Although the process of collecting the results of measurements to construct REM can be carried out by dedicated sensors with fixed positions and mobile devices (e.g. cognitive radios), the resources of mobile devices are more limited since they have to use their battery efficiently [17]. Therefore, the problem how the density of sensor network affects the quality of the REM must be addressed.

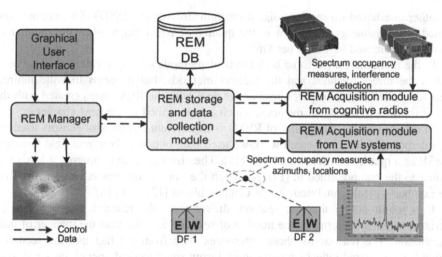

Fig. 1. REM architecture to support tactical operation [10]

In [13] the authors performed an experiment in real conditions whose aim was to determine the position of a transmitter operating at 800 MHz frequency with the application of the indirect method. The transmitter was placed inside a grid consisting of 49 nodes in a 7×7 arrangement, spaced 5 m apart. The results of measurements and calculations showed that at least 20 randomly selected sensors are necessary in order to determine the position of the transmitter with sufficient accuracy. In such a case the error of determining the position of the transmitter was about 1.5 m. When the results of measurements from 46 sensors were taken into account, the error of position determining decreased to about 1 m, which is 20% of the distance between the sensors in the grid.

In [14] the authors discussed a method of searching for White Spaces in UHF band (470–900 MHz) which could be used for Cognitive Radio (CR). Some field tests were performed with 100 measurement units deployed in the area of 5 km^2 and distributed in two ways: regular lattice (Cartesian) and pseudo-random. The authors noticed the relation between the number of measuring sensors and the required terrain resolution of the REM map being created and the number of CR users per square km.

In [15] the authors presented three methods of creating REM: path loss based method, Kriging based method and their own method. To compare the efficiency of the proposed methods a series of simulations were performed for scenario with: (a) one transmitting node, (b) 81 sensing nodes and (c) 8 validating nodes which do not overlap with the 81 sensors. All the nodes were deployed on the area 70 m by 70 m. To assess the quality of the created REMs the Root Mean Square Error (RMSE) was calculated for 8 validating nodes.

The accuracy of determining the location of the transmitter in meters was used as a measure of the quality of REM maps in [16]. The environment considered in the research work was a simulated urban macro-cell square area of 1 km^2. In this area one transmitter and up to 20 measuring sensors were placed randomly. REM maps were developed using two indirect methods: one based on received signal strength (RSS) and the other one based on received signal strength difference (RSSD). The authors confirmed a noticeable improvement in the quality of REM maps when the number of sensors is increased to 14–20 per km^2.

In the literature on the topic both kinds of methods of map creation are analyzed, that is the direct methods and the indirect methods, but it seems that the indirect methods prevail. In our paper, however, we deal with the REM maps created with the use of a few selected direct methods, which are described in the next chapter.

In order to assess the quality of REMs with different numbers of sensors used for the interpolation in our research work we used data obtained from real field tests and RMSE as a quality metric, similarly to [15]. The size of the area (approx. 4 km^2) was similar to the one presented in [14]. Although the number of sensors was smaller than the number typically analysed, it was comparable to [13] and [16].

It is worth noting that our research differs from the research described in the literature not only in terms of the number of sensors used but also the manner of their distribution. The reasons for these differences stem from the fact that the scenarios which we considered reflect networks used during small tactical operations, i.e. dozens of sensors operating in the area of several square kilometres. In military operations, the role of sensors is played by cognitive radio stations, and therefore the tactical situation determines their distribution. The scenarios presented in the literature usually assume that there are hundreds of sensors spaced quite regularly or arranged in controlled manner.

3   Map Construction Techniques

In the literature on the topic there is a description of three main categories of the REM construction techniques, namely *direct*, *indirect* and *hybrid* [11, 12]. *Direct* methods, also called *spatial statistics based methods*, are based on the interpolation of the measured data, while *indirect* methods, also known as *transmitter location based methods*, apply transmitter location and propagation model to obtain the estimated value. *Hybrid* methods combine both manners.

Spatial statistics based methods use measurement data taken at certain locations. In the case of REM the measurement is done at the location of the sensors. It is understandable that placing sensors in all required locations is impractical or simply impossible. For this reason samples from sensors are used as an input for the estimation process that can employ different kinds of techniques.

When REM is considered the most promising estimation techniques described in the literature are as follows: Nearest Neighbor (NN), Inverse Distance Weighting (IDW) and Kriging.

The Nearest Neighbor method is considered to be one of the simplest methods but it offers little accuracy. NN uses Thiessen (or Voronoi) polygons, which are defined by boundaries with equal distances from the points at which measurements were taken. A specific feature of these polygons is the fact that their boundaries are exactly in the middle of the distance between neighboring points.

IDW method is based on the assumption that the signal value P_1 at a given point (x_1, y_1) is much more dependent on the values in the nearest measurement points than on samples taken at distant points. To interpolate the signal value the IDW uses weighting factors w_i that are inversely proportional to the distance between the given point (x_l, y_l) and the sampling point (x_i, y_i) and raised to the power p. The power p determines how the weighting factors decrease with the distance. If the power p value is set high, the points which are close/nearby have stronger impact. When the power p value is set at zero, regardless of the distance, the weighting factors remain at the same level. In the rest of the paper we use the following notation for IDW method: IDW px where x is the power.

Kriging is one of the geostatic methods of interpolation. Like IDW, Kriging uses weighting factors but they are determined on the basis of the semivariogram. This semivariogram is based on the distance between measurement points and the variation between measurements of signal levels as a function of the distance. Kriging is considered to be the most accurate, though quite a complex method of interpolation.

In the literature of REM the use of Kriging in combination with another method of the signal level determining or the modification of Kriging is proposed [19, 20].

A more detailed description of the estimation techniques mentioned above is presented in [10].

4 Test Scenario and Exemplary Maps

In order to investigate the impact of the number of sensors on the REM quality several tests were conducted for UHF frequency band. First, measurements were taken in a real environment with 39 sensors to get input data and then, exemplary maps were created using different construction techniques, namely Nearest Neighbor, IDW and Kriging. After that, the analysis of calculated Root Mean Square Error (RMSE) for various numbers of sensors was made.

To assess the quality of the maps created with the selected interpolation techniques we analyzed the results for three scenarios with different number of sensors each, see Table 1. Each of the three scenarios consisted of 2 tests which were performed with different (random) deployment of sensors. It is worth noting that the sensors were arranged irregularly due to the fact that the measurements were taken in a real environment.

Table 1. Scenarios and tests for RMSE analysis

Number of sensors used for interpolation per number of control sensors	The name of scenario	The name of test
13/26	Scenario_13	Test_13a
		Test_13b
20/19	Scenario_20	Test_20a
		Test_20b
26/13	Scenario_26	Test_26a
		Test_26b

The initial distribution of 39 sensors is shown in Fig. 2. For the interpolation process the sensors selected in each test were chosen in a random process. For remaining sensors, in each test, the differences between the measured and the interpolated signal level were compared and used for calculating the RMSE. Finally, average values of the RMSE were calculated for each scenario.

In order to perform measurements in a real environment we established the test bed composed of a transmitting part and a receiving part.

The transmitting part of the system consisted of a signal generator connected to a controlling computer, an amplifier and an antenna mounted on the roof of a building at the height of 8 m.

The receiving part consisted of an antenna installed on a vehicle, a radio receiver and a computer controlling the receiving operation and recording the results of the measurements. The antenna was installed at the height of 3 m. The vehicle was moving within a preliminarily selected area, Fig. 2. The following configuration of the testbed was used: (1) UHF frequency: 1997 MHz, (2) modulation type: CW, (3) output power: 10 W, (4) measured parameter: avg. RSS, (5) number of averages: 10, (6) antenna type: omnidirectional.

The measurements were taken in the area of Zegrze lake in Central Poland (the area of approximately 4 km^2 presented in Fig. 2). The test area was diverse in terms of coverage (partly an open meadow neighboring a forest and partly an urbanized area with medium-sized and high buildings).

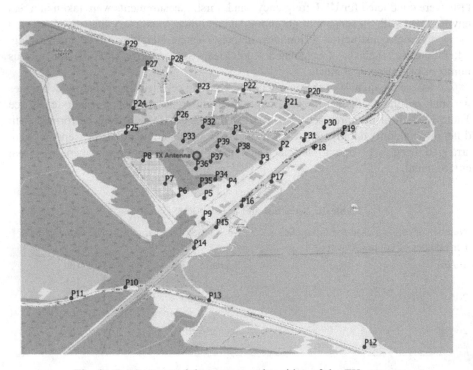

Fig. 2. Deployment of the sensors and position of the TX antenna

Some exemplary maps for scenario with 26 sensors constructed with four interpolation techniques are presented in Fig. 3.

The NN method (Fig. 3a) creates polygons around each sensor. The size and the shape of the polygons depends on the number and the arrangement of neighboring sensors. Within each polygon the signal strength takes the value measured by the sensor. For this reason the signal strength changes suddenly at the edges of polygons, e.g. between the orange polygon close to the center and the dark blue one to its right.

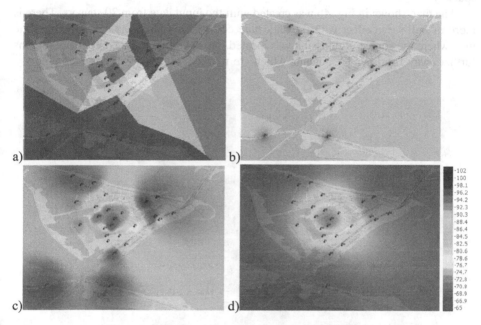

Fig. 3. Exemplary maps for scenario with 26 sensors constructed with different interpolation techniques (signal value in dBm): (a) NN, (b) IDW p1, (c) IDW p3, (d) Kriging (Color figure online)

The IDW method (Fig. 3b and c) generates smoother maps when compared to NN. However, the bull's-eye effect occurs and the size of eyes depends on the power p used in the interpolation process. The estimation of the signal strength is quite accurate if the power p is set at 3 or higher and the sensors are deployed densely.

When Kriging is applied (Fig. 3d), the signal value changes smoothly within the whole area. Kriging seems to be a method which is least sensitive to the deployment of the sensors. Neither bull's-eye effects nor rapid changes in the signal value are observed even if the sensors are deployed sparsely or irregularly.

In the presented scenario the position of the TX antenna can be determined with the accuracy of approximately:

- 350 m for IDW p1,
- 300 m for NN,
- 250 m for IDW p3,
- 150 m for Kriging.

Exemplary maps for IDW p3 interpolation technique for various numbers of sensors are shown in Fig. 4. The lowest signal level is represented by the dark blue color while the highest level - by the red color. The map presented in Fig. 4a (13 sensors) seems to be unnatural since there is quite an extensive yellow and green area representing the medium signal strength, even for those regions that are distant from the TX antenna. The bull's-eye effect with the dark blue color is present in a few places only. The general conclusion is that there are too few sensors and that they are deployed too sparsely.

The map shown in Fig. 4b was created with the input data from 20 sensors. There is more of bull's-eye effect with the dark blue color surrounding the central part of the map where the source of emission was located. However, there are quite many regions further away from the TX antenna which are marked with yellow and green color.

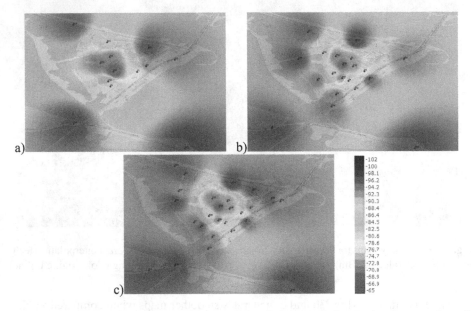

Fig. 4. Exemplary maps constructed for IDW p3 interpolation technique and various numbers of sensors (signal value in dBm): (a) 13 sensors, (b) 20 sensors, (c) 26 sensors (Color figure online)

The map presented in Fig. 4c (26 sensors) looks more natural when compared to the maps shown in Fig. 4a and b. Since the sensors are arranged much more densely the red-orange center of the map is quite regularly enclosed by the dark blue color of the bull's-eye effect. Moreover, the increased number of sensors caused better reflection of the signal level for these areas that are distant from the TX antenna (medium low signal level imitated by the blue color).

Exemplary maps for Kriging interpolation technique for various density of sensor network are shown in Fig. 5. The dark blue color represents the lowest signal level while the red color - the highest. As the number of sensors increases, the map seems to look more natural, that is the area where the signal level is high (the red-orange color) becomes smaller, whereas regions around the TX antenna where the signal level is low become more distinct (marked with the dark blue color). Moreover, if there are more sensors, the position of the TX antenna can be determined with better precision. This effect can be easily noticed when the sizes of the red-orange areas in Fig. 5c and a are compared.

A more detailed analysis of the impact of the density of sensor network on the quality of maps mentioned above is given in the next section.

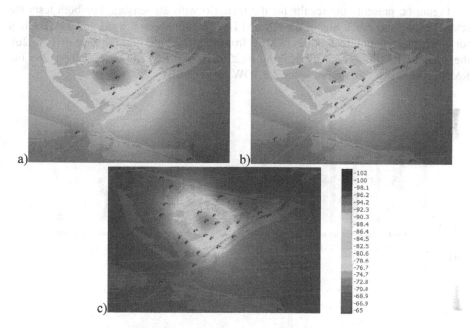

Fig. 5. Exemplary maps constructed for Kriging interpolation technique with various numbers of sensors (signal value in dBm): (a) 13 sensors, (b) 20 sensors, (c) 26 sensors (Color figure online)

5   Analysis of the Results

The RMSEs calculated for Nearest Neighbor, Kriging and IDW methods with power p from 1 to 6 are shown in Fig. 6.

Figure 6a presents the results for the scenarios with 13 sensors. The differences between the results for individual tests are quite significant. The comparison shows that, irrespectively of the interpolation technique, the RMSE values are smaller for Test_13b than for Test_13a. The RMSE for Test_13a reaches 9.1 dB for IDW p3 and 7.8 dB - for Kriging. The RMSE for Test_13b reaches 10.95 dB for IDW p3 and 9.6 - for Kriging. The results for NN method are comparable for both tests (RMSE oscillates around 11.85 dB). When Kriging was applied, the RMSE values were the smallest for both compared tests.

The results for the scenario with 20 sensors are shown in Fig. 6b. Independently of the applied interpolation technique, the RMSE values are smaller for Test_20a when compared to Test_20b, except the results for IDW p1, which are in fact the worst case (RMSE over 10 dB). The RMSE for Test_20a for IDW p3 reaches 8.5 dB and for Kriging - 6.7 dB, while for Test_20b the RMSE reaches 8.8 dB for IDW p3 and 8 dB for Kriging. For both compared tests in this scenario: (a) Kriging offers the best results, (b) RMSE drops as the power p increases for IDW method. The differences between the results for individual tests are within 1.3 dB.

Figure 6c presents the results for the scenario with 26 sensors. For both tests the RMSE values are much higher for NN and IDW p1 (between 8.8 dB and 11 dB) than for other interpolation techniques (RMSE from 6.2 to 7.5 dB). In the case of Test_26b the smallest RMSE occurs for Kriging (6.25 dB), while in the case of Test_26a the RMSE reaches the minimum value for IDW p4 (6.3 dB).

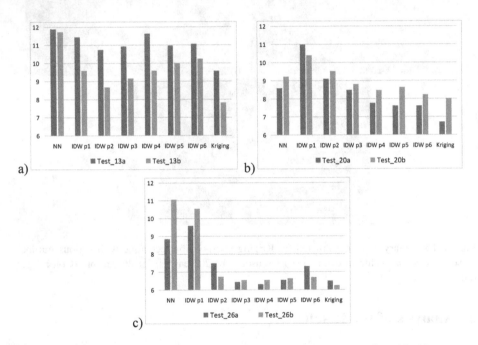

Fig. 6. RMSE (in dB) for selected interpolation techniques: (a) scenario with 13 sensors, (b) scenario with 20 sensors, (c) scenario with 26 sensors

The average values of RMSE for each scenario are shown in Fig. 7. The effect of the drop in the RMSE as the number of sensors increases is clearly visible for IDW with power p higher than 1 and for Kriging interpolation technique. When IDW p1 method was applied, the benefit of having more sensors in the network was inconsiderable. If NN method was applied, the smallest RMSE value occurred for the scenario with 20 sensors. In general, the trend in the changes of RMSE confirms that placing more sensors in the network makes the quality of REM higher.

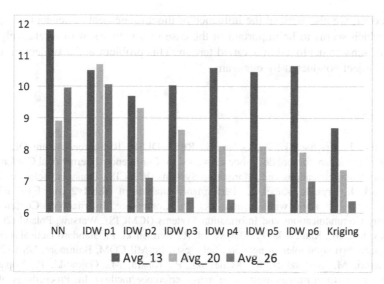

Fig. 7. The average RMSE (in dB) for selected interpolation techniques and scenarios with 13, 20 and 26 sensors

6 Conclusions

The quality of maps depends on several factors, among others the density and regularity of deployment of sensors, the distance between sensors, the propagation environment and the interpolation technique. In this paper we analyzed the impact of the number of sensors on the REM quality.

In the literature on the subject mainly scenarios with several hundred measurement points located in the area of around 5 km^2 are studied. In some real applications this number is much lower, e.g. reaching dozens of sensors in the area of approximately 4 km^2. That is why we focused on the scenarios with a small number of sensors that reflect, for example, a small-scale tactical operation or CR networks operating in suburban areas.

In our research work we used data from real field tests with 39 sensors deployed within the area of 4 km^2. We analyzed results of the tests with different numbers of sensors (13, 20 and 26) used for the interpolation process. For each scenario two tests with various arrangements of sensors were analyzed. To create REM maps the following interpolation techniques were applied: NN, IDW and Kriging. To assess the quality of maps the calculated RMSE values were compared. In general, the increase in the number of sensors from 13 to 26 caused a visible improvement in the quality of REM maps. The average RMSE values dropped from 8.7 dB to 6.3 dB for the Kriging method and from 10 dB to 6.5 dB for the IDW p3 method.

In the literature on the topic several methods of interpolation are analyzed. Analyzing our results the smallest RMSE values were noticed for Kriging and IDW with the power of 3 or 4. For this reason these interpolation techniques should be recommended for REM construction.

Moreover, we also noticed the influence of the arrangement of sensors on the map quality, which seems to be important in the case of a network with a relatively small number of sensors deployed in a varied terrain. This problem is the subject of another research project conducted by our team.

References

1. Romanik, J., Krasniewski, A., Golan, E.: RESA-OLSR: RESources-Aware OLSR-based routing mechanism for mobile ad-hoc networks. In: Conference: International Conference on Military Communications and Information Systems (ICMCIS), Brussels, Belgium (2016)
2. Romanik, J., Brys, R., Zubel, K.: Performance analysis of OLSRv2 with ETX, ETT and DAT metrics in static wireless networks. In: Conference: International Conference on Military Communications and Information Systems (ICMCIS), Warsaw, Poland (2018)
3. Malowidzki, M., Kaniewski, P., Matyszkiel, R., Berezinski, P.: Standard tactical services in a military disruption-tolerant network: field tests. In: MILCOM, Baltimore, USA (2017)
4. Suchanski, M., Matyszkiel, R., Kaniewski, P., Kustra, M., Gajewski, P., Lopatka, J.: Dynamic spectrum management as an anti-interference method. In: Proceedings of SPIE, SPIE, Bellingham, WA, vol. 10418, 20 April 2017 https://doi.org/10.1117/12.2269294. 2269288, ISSN 0277-786X, ISSN 1996-756X (electronic)
5. Suchański, M., Gajewski, P., Łopatka, J., Kaniewski, P., Matyszkiel, R., Kustra, M.: Coordinated dynamic spectrum management in legacy military communication systems. In: WinnComm-Europe 2016, Wireless Innovation Forum European Conference on Communications Technologies and Software Defined Radio, Paris, France (2016)
6. Matyszkiel, M., Kaniewski, P., Kustra, M., Jach, J.: The evolution of transmission security functions in modern military wideband radios. In: Book Series: Proceedings of SPIE, vol. 10418, Article no. UNSP 104180E (2017)
7. Pesko, M., Javornik, T., Košir, A., Štular, M., Mohorčič, M.: Radio environment maps: the survey of construction methods. KSII Trans. Internet Inf. Syst. 8(11) (2014). https://doi.org/10.3837/tiis.2014.11.008
8. Flexible and Spectrum Aware Radio Access through Measurements and Modeling in Cognitive Radio Systems, FARAMIR project, Document Number D2.4, Final System Architecture
9. Cai, T., et al.: Design of layered radio environment maps for RAN optimization in heterogenous LTE systems. In: IEEE 22nd International Symposium on Personal, Indoor and Mobile Radio Communications, pp. 172–176 (2011). https://doi.org/10.1109/pimrc.2011.6139803
10. Suchanski, M., Kaniewski, P., Romanik, J., Golan, E.: Radio environment maps for military cognitive networks: construction techniques vs. map quality. In: ICMCIS, Warsaw. IEEE Xplore (2018). https://doi.org/10.1109/icmcis.2018.8398723
11. Yilmaz, B.H., Tugcu, T.: Location estimation-based radio environment map constructing techniques in fading channels. Wirel. Commun. Mob. Comput. 15(3), 561–570 (2015). https://doi.org/10.1002/wcm.2367
12. Yilmaz, H.B., Tugcu, T., Alagöz, F., Bayhan, S.: Radio environment map as enabler for practical cognitive radio networks. IEEE Commun. Mag. 51(12), 162–169 (2013). https://doi.org/10.1109/mcom.2013.6685772
13. Ezzati, N., Taheri, H., Tugcu, T.: Optimised sensor network for transmitter localisation and radio environment mapping. Inst. Eng. Technol. IET Commun. 10(16), 2170–2178 (2016). https://doi.org/10.1049/iet-com.2016.0341

14. Patino, M., Vega, F.: Model for measurement of radio environment maps and location of white spaces for cognitive radio deployment. In: IEEE-APS Topical Conference on Antennas and Propagation in Wireless Communications (2018). https://doi.org/10.1109/apwc.2018.8503755
15. Mao, D., Shao, W., Qian, Z., Xue, H., Lu, X., Wu, H.: Constructing accurate radio environment maps with Kriging interpolation in cognitive radio networks. In: Cross Strait Quad-Regional Radio Science and Wireless Technology Conference (CSQRWC 2018) (2018). https://doi.org/10.1109/csqrwc.2018.8455448
16. Alfattani, S., Yongacoglu, A.: Indirect methods for constructing radio environment map. In: IEEE Canadian Conference on Electrical & Computer Engineering (CCECE) (2018). https://doi.org/10.1109/ccece.2018.8447654
17. Suchanski, M., Kaniewski, P., Romanik, J., Golan, E.: Radio environment map to support frequency allocation in military communications systems. In: Baltic URSI Symposium, Poznan, 15–17 May (2018)
18. Suchanski, M., Kaniewski, P., Romanik, J., Golan, E., Zubel, K.: Electronic warfare systems supporting the database of the Radio Environment Maps. In: Proceedings of SPIE, XII Conference on Reconnaissance and Electronic Warfare Systems, vol. 11055 (2018). https://doi.org/10.1117/12.2524594
19. Kliks, A., Kryszkiewicz, P., Kułacz, Ł.: Measurement-based coverage maps for indoor REMs operating in TV band. In: IEEE International Symposium on Broadband Multimedia Systems and Broadcasting (2017). https://doi.org/10.1109/bmsb.2017.7986162
20. Ojaniemi, J., Kalliovaara, J., Poikonen, J., Wichman, R.: A practical method for combining multivariate data in radio environment mapping. In: IEEE 24th Annual International Symposium on Personal, Indoor, and Mobile Radio Communications (2013). https://doi.org/10.1109/pimrc.2013.6666232

An Out-of-Sample Extension for Wireless Multipoint Channel Charting

Tushara Ponnada[1], Hanan Al-Tous[1(✉)], Olav Tirkkonen[1],
and Christoph Studer[2]

[1] Department of Communications and Networking, Aalto University, Espoo, Finland
{tushara.ponnada,hanan.al-tous,olav.tirkkonen}@aalto.fi
[2] School of Electrical and Computer Engineering, Cornell University,
Ithaca, NY, USA
studer@cornell.edu

Abstract. Channel-charting (CC) is a machine learning technique for learning a multi-cell radio map, which can be used for cognitive radio-resource-management (RRM) problems. Each base-station (BS) extracts features from the channel-state-information samples (CSI) from transmissions of user-equipment (UE) at different unknown locations. The multi-path channel components are estimated and used to construct a dissimilarity matrix between CSI samples at each BS. A fusion center combines the dissimilarity matrices of all base-stations, performs dimensional reduction based on manifold learning, constructing a Multipoint-CC (MPCC). The MPCC is a two dimension map, where the spatial difference between any pair of UEs closely approximates the distance between the clustered features. MPCC provides a mapping for any given trained UE location. To use MPCC for cognitive RRM tasks, CSI measurements for new UEs would be acquired, and these UEs would be placed on the radio map. Repeating the MPCC procedure for out-of-sample CSI measurements is computationally expensive. For this, extensions of MPCC to out-of-sample UE CSIs are investigated in this paper, when Laplacian-Eigenmaps (LE) is used for dimensional reduction. Simulation results are used to show the merits of the proposed approach.

Keywords: Massive MIMO · Channel charting ·
Laplacian eigenmaps · Out-of-sample mapping

1 Introduction

Massive-multiple-input-multiple-output (mMIMO) technology is a promising technology for fifth-generation (5G) cellular communications, with the potential to provide high spectral and power efficiency. In a mMIMO cell, each base-station (BS) has a large number of antennas, which can provide a simultaneous use of the resource (e.g., frequency and/or time slots) for multiple user equipment (UEs) in the cell [3,4,13]. Furthermore, the high spatial resolution exploited by

© ICST Institute for Computer Sciences, Social Informatics and Telecommunications Engineering 2019
Published by Springer Nature Switzerland AG 2019. All Rights Reserved
A. Kliks et al. (Eds.): CrownCom 2019, LNICST 291, pp. 208–217, 2019.
https://doi.org/10.1007/978-3-030-25748-4_16

the large-scale antenna arrays used at the mMIMO BSs can be used for many applications, such as UE positioning and environment mapping [7,8,11].

To efficiently mange a mMIMO network, and to perform cognitive networking tasks, the network states which include the spatial distribution and trajectories of the UEs, neighborhood relationships among the UEs, and handover boundaries among neighboring cells need to be estimated. A novel framework called channel charting (CC) based on the massive amounts of channel-state-information (CSI) available at the base-stations is proposed for a single cell MIMO system in [12]. CC is based on using unsupervised machine learning techniques to create a radio map of the cell served by the BS, which preserves the neighborhood relations of UEs, using features that characterize the large scale fading effects of the channel. The obtained CC can be used for local radio-resource-management (RRM) in the cell. However, cell edge UEs may not be accurately located in the chart due to their low signal-to-noise-ratio (SNR) at the cell edge.

In [5], a multi-point CC (MPCC) framework is proposed to support advanced multi-cell RRM and to accurately map cell edge UEs. For improved charting performance, features are extracted and clustered based on advanced signal processing and machine learning techniques. Each BS generates it own dissimilarity matrix between the users it can decode, then the dissimilarity matrices are fused at a fusion center and then used to construct the MPCC. The trustworthiness and continuity measures show that the proposed MPCC is capable to preserve the neighborhood structure between UEs in the network.

To use the MPCC framework for different RRM functionalities, it is important to generalize its capability, allowing to incorporate new data to an existing MPCC and/or to estimate the features related to a location in the chart. As the CSI of a UE can change rapidly in a small distance, it is important to accurately estimate the location of an out-of-sample UE location. In this paper, the extension of MPCC to out-of-sample data points based on Laplacian Eigenmaps (LE) is considered.

The remainder of this paper is organized as follows. In Sects. 2 and 3, the system model and the MPCC are introduced, respectively. In Sect. 4, the problem formulation is presented. Numerical results are presented and discussed in Sect. 5. Finally, conclusions are drawn in Sect. 6.

2 System Model

The system under consideration is schematically shown in Fig. 1. Each BS b for $b = 1, \cdots, B$ has M antenna elements and each UE k for $k = 1, \cdots, K$ has a single antenna element. For a mMIMO system, the channel vector of UE k using a uniform-linear-array (ULA) at BS b for a coherence bandwidth can be modeled as [5]:

$$h_b^{(k)} = \sum_{l=1}^{L_k} \beta_b^{(k)}(l) a_b(\phi_b^{(k)}(l)), \qquad (1)$$

where L_k is the number of multi-path components for the wireless channel between UE k and BS b, $\phi_b^{(k)}(l)$ is the direction of arrival of the lth path and

$\beta_b^{(k)}(l)$ is the gain of the lth path, and $a_b(\cdot)$ is BS b steering vector. For ULA, $a(\phi)$ can be modeled as:

$$a(\phi) = [1, e^{i\frac{2\pi}{\lambda}s\sin(\phi)}, \cdots, e^{i\frac{2\pi}{\lambda}(M-1)s\sin(\phi)}]^T, \tag{2}$$

where λ is the carrier wave-length and s is antenna spacing. The covariance $Y_b^{(k)} \in \mathbb{C}^{M \times M}$ of the CSI $h_b^{(k)}$ used to extract the features becomes

$$Y_b^{(k)} = \mathbb{E}[h_b^{(k)}h_b^{(k)H}] = A_b^{(k)}S_b^{(k)}A_b^{(k)H}, \tag{3}$$

where $\mathbb{E}$ is the expectation operator, $A_b^{(k)} = [a(\phi_b^{(k)}(1)), \cdots, a(\phi_b^{(k)}(L_k))]$ is a matrix of array steering vectors, and $S_b^{(k)} = \mathrm{diag}(\mathbb{E}[|\beta_b^{(k)}(1)|^2], \cdots,$ $\mathbb{E}[|\beta_b^{(k)}(L_k)|^2])$ is a diagonal matrix of multi-path power components. Channel charting is based on the assumption that there is a continuous mapping from the spatial location $p^{(k)}$ of UE k to the covariance CSI $Y_b^{(k)}$ given as [5,12]:

$$\mathcal{H}_b : \mathbb{R}^d \to \mathbb{C}^{M \times M}; \mathcal{H}_b(p^{(k)}) = Y_b^{(k)}, \tag{4}$$

where d is the spatial dimension.

3   MPCC

Multi-point channel charting extends CC to multiple BSs. A block diagram representing MPCC is shown in Fig. 2. Using the estimated covariance CSI $\{\{Y_b^{(k)}\}_{k=1}^K\}_{b=1}^B$ collected at B BSs form K unknown UE spatial locations $P = \{p_k\}_{k=1}^K$, the MPCC finds a low dimension channel chart $Z = \{z_k\}_{k=1}^K$, such that:

$$\|z_k - z_m\| \approx \alpha \|p_k - p_m\|, \text{ for } k, m \in \{1, \cdots, K\}, \tag{5}$$

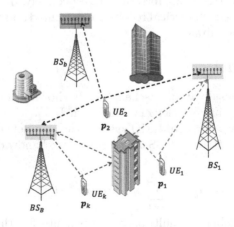

Fig. 1. Multipoint mMIMO system.

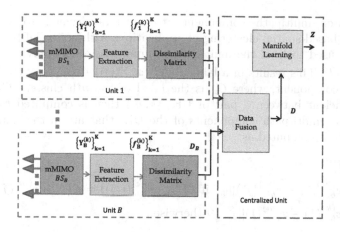

Fig. 2. MPCC block diagram.

where α is a scaling factor. Note that neither the UEs spatial locations $\boldsymbol{P}$ nor the locations of the BSs are known; the MPCC is computed solely based on the covariance CSI $\{\{\boldsymbol{Y}_b^{(k)}\}_{k=1}^K\}_{b=1}^B$.

3.1 Feature Extraction and Dissimilarity Matrix

The feature vector $\boldsymbol{f}_b^{(k)}$ for UE k at BS b is selected based on the multi-path components as [5]:

$$\boldsymbol{f}_b^{(k)} = [\lambda_b^{(k)}(1), \cdots, \lambda_b^{(k)}(L_k), \phi_b^{(k)}(1), \cdots, \phi_b^{(k)}(L_k)], \qquad (6)$$

where $\lambda_b^{(k)}(l) = \mathbb{E}[|\beta_b^{(k)}(l)|^2]$. The multi-path components (power and phase) $\{\lambda_b^{(k)}(l)\}_{l=1}^{L_k}$ and $\{\phi_b^{(k)}(l)\}_{l=1}^{L_k}$ of UE k at BS b are estimated from the covariance matrix $\boldsymbol{Y}_b^{(k)}$ using the multiple-signal-classification (MUSIC) algorithm [10]. The dissimilarity between two UEs (k, m) is based on identifying multi-path components in their feature vectors that are similar. For this, the components of feature vectors are transformed to Cartesian coordinates as [5]:

$$\mathcal{F}\{\boldsymbol{f}_b^{(k)}\} = [\boldsymbol{x}_b^{(k)}(1), \cdots, \boldsymbol{x}_b^{(k)}(L_k)], \qquad (7)$$

where $\boldsymbol{x}_b^{(k)}(l) = [\lambda_b^{(k)}(l) \cos(\phi_b^{(k)}(l)), \lambda_b^{(k)}(l) \sin(\phi_b^{(k)}(l))]^T$.

A 2D non-linear transformation $\mathcal{N}_b^\nu : \mathbb{R}^2 \to \mathbb{R}^2$ with a set of parameters $\boldsymbol{\nu}$ is hand-crafted based on the statistical and geometrical characteristics of the multi-path components of all K UEs that a BS is seeing. This function is applied to the geometrical representations of the multi-path components $\boldsymbol{x}_b^{(k)}(l)$. The transformation $\mathcal{N}_b^\nu$ is used in order to make clusters of multipath components separable.

To cluster multipath components to clusters deemed to be similar, the density-based-spatial-clustering-of-applications-with-noise (DBSCAN) algorithm [6] is used to label the multi-path components after applying the transformation $\mathcal{N}_b^\nu$. This results in a label $\mathcal{L}(\mathcal{N}_b^\nu(\boldsymbol{x}_b^{(k)}(i))) \in \{C_1, \cdots, C_N\}$ for each multi-path component, where C_n is the label of the nth cluster. The dissimilarity coefficient between a pair of UEs (k, m) then is computed taking into consideration multi-path components of the UEs that are in the same cluster. Dissimilarity is computed as:

$$
d_b(\boldsymbol{f}_b^{(k)}, \boldsymbol{f}_b^{(m)})
$$
$$
= \begin{cases} ||\boldsymbol{x}_b^{(k)}(i^*) - \boldsymbol{x}_b^{(m)}(j^*)||_2 & \text{if } \mathcal{L}(\mathcal{N}_b^\nu(\boldsymbol{x}_b^{(k)}(i^*))) = \mathcal{L}(\mathcal{N}_b^\nu(\boldsymbol{x}_b^{(m)}(j^*))), \\ ||\boldsymbol{x}_b^{(k)}(1) - \boldsymbol{x}_b^{(m)}(1)||_2, & \text{otherwise,} \end{cases} \tag{8}
$$

where $[i^*, j^*] = \underset{i,j}{\arg\min}(\lambda_b^{(k)}(i), \lambda_b^{(m)}(j))$. The dissimilarity matrix $\boldsymbol{D}_b \in \mathbb{R}^{K \times K}$ then has the elements $D_b(k, m) = d_b(\boldsymbol{f}_b^{(k)}, \boldsymbol{f}_b^{(m)})$ for $k, m = 1, \cdots, K$.

The benefits of having multiple spatially distributed BSs can be utilized by merging the BS-specific dissimilarity matrices $\{\boldsymbol{D}_b\}_{b=1}^B$ into a global dissimilarity matrix $\boldsymbol{D}$, where the (k, m)th element $D(k, m)$ can be computed as:

$$
D(k, m) = \frac{1}{\sum_{b=1}^B \omega_b(k, m)} \sum_{b=1}^B \omega_b(k, m) D_b(k, m), \tag{9}
$$

where $\omega_b(k, m)$ is a weighting factor computed as $\omega_b(k, m) = \min(\gamma_b^{(k)}, \gamma_b^{(m)})^2$ and $\gamma_b^{(k)}$ is the SNR of the wireless link between UE k and BS b.

4   Out-of-Sample Extension

For a given dissimilarity matrix, different dimension reduction techniques (i.e., linear, non-linear, convex and non-convex optimization approaches) have been proposed in the literature. The performance of a given technique is problem dependent, as discussed in [9]. The single cell CC problem has been solved using the principle-component-analysis (PCA), Sammon's-mapping (SM) and Autoencoder reduction techniques in [12], whereas the MPCC is solved using SM, Laplacian Eigenmaps (LE) and t-distributed-stochastic-neighbor-embedding (t-SNE) in [5]. In this paper, LE is considered, and extended for the out-of-sample MPCC problem.

LE is a computationally efficient non-linear dimensionality reduction algorithm based on the graph Laplacian, that preserves neighborhood properties and clustering connections [1]. LE constructs a graph from neighborhood information of the dissimilarity matrix. The LE problem can be formulated as [1]:

$$
\min_X \text{trace}(\boldsymbol{X}^T \boldsymbol{L} \boldsymbol{X}), \tag{10a}
$$
$$
\text{s.t. } \boldsymbol{X}^T \boldsymbol{S} \boldsymbol{X} = \boldsymbol{I}_K, \tag{10b}
$$

Algorithm 1. The LE for MPCC.

1: **Given**: the dissimilarity matrix D.
2: **Construct**: the adjacency matrix, two approaches can be considered:
   - The ϵ-neighborhood, nodes k and m are connected by an edge if $D(k, m) \leq \epsilon$.
   - Nodes k and m are connected by an edge if m is among the N nearest neighbors of k or k is among the N nearest neighbors of m.
3: **Choosing**: the weight matrix W with $[W]_{k,m} = W(k, m)$ with two approaches can be considered:
   - Using the heat kernel with temperature T; if nodes k and m are connected, $W(k, m) = e^{-\frac{D^2(n,m)}{T}}$, otherwise $W(k, m) = 0$.
   - Simple approach, if nodes k and m are connected, $W(k, m) = 1$, otherwise $W(k, m) = 0$.
4: **Compute**: the Laplacian Matrix $L = S - W$, where S is the degree matrix (diagonal matrix) with $S(k, k) = \sum_{i=1}^{K} W(k, i)$,
5: **Compute**: the eigenvalues λ_i for $i = 0, \cdots, K - 1$ and eigenvectors v_i for $i = 0, \cdots, K - 1$ for the generalized eigenvector problem: $Lv = \lambda Sv$,
6: **Order**: the eigenvectors $v_0, v_1, \cdots, v_{K-1}$ according to their eigenvalues, with $0 = \lambda_0 < \lambda_1 \leq \lambda_2 \leq \cdots \leq \lambda_{K-1}$.
7: **Return**: the position of the kth UE on the MPCC as: $z(k) = [v_1(k), v_2(k)]$ for $d = 2$.

---

where trace is the trace function, I_K is the identity matrix of order K, $X = [x(1)^T, \cdots, x(K)^T]^T$ represents the optimization variables in a matrix form, L is the Laplacian matrix and S is the degree matrix as detailed below. The solution of (10) can be obtained in a closed form as the solution of a generalized eigenvector problem [1].

The MPCC is obtained by computing the eigenvectors of the LE as described in Algorithm 1. Since the MPCC is constructed by processing the data of all UEs from all BSs, it is computationally expensive to repeat the MPCC process if an out-of-sample data item is available, and needs to be inserted into the chart. If the original MPCC is based on a sufficient number of samples, it is expected that the out-of-sample data will not change the MPCC positions.

Here, we address out-of-sample extension of MPCC in this sense, aiming to estimate the location of the new sample on the MPCC, to be used for RRM functions, such as hand-over prediction. It is worth mentioning that an out-of-sample data item needs to be processed using the same non-linear transformation $\mathcal{N}_b^\nu$ at each BS b for $b = 1, \cdots, B$ and then, the cluster labeling based on the original data has to be applied for each multi-path component.

In [2], a generalized framework for out-of-sample extension is proposed for several algorithms, providing that these algorithms can be seen as learning eigenfunctions of a data dependent kernel. The out-of-sample mapping can be formulated as an optimization problem, where the objective is to find a normalized kernel function that minimizes the mean squared-error. The normalized kernel vector is used as a weight vector to find the out-of-sample mapping. Using this on MPCC is called E-MPCC. For LE, the normalized kernel functions (weights)

Algorithm 2. The E-MPCC for UE j, $j \notin \{1, \cdots, K\}$.

1: **Given:** ν of $\mathcal{N}_b^\nu$, $\hat{\boldsymbol{W}} \in \mathbb{R}^{K \times K}$, and the corresponding eigenvectors $\hat{\boldsymbol{v}}_1$, and $\hat{\boldsymbol{v}}_2$ for $d = 2$.

2: **Estimate:** the multi-path components $\{\boldsymbol{f}_b^{(j)}\}_{b=1}^B$ for $j \notin \{1, \cdots, K\}$.

3: **Compute** the dissimilarity coefficient $d_b(\boldsymbol{f}_b^{(j)}, \boldsymbol{f}_b^{(m)})$, $m = 1, \cdots, K$ of the out-of-sample UE j for $j \notin \{1, \cdots, K\}$ at B base-stations.

4: **Compute** the dissimilarity fusion vector for the out-of-sample UE j.

5: **Compute** the weight vector $\hat{\boldsymbol{W}}_j = [\hat{W}(j, 1), \cdots, \hat{W}(j, K)]$.

6: **Map** the position $\boldsymbol{z}(j)$ on the MPPC using (12).

---

are computed as [2]:

$$\hat{W}(k, i) = \frac{1}{K} \frac{W(k, i)}{\sqrt{\mathbb{E}_x[W(k, x)]\mathbb{E}_y[W(i, y)]}}, \ k, i \in \{1, \cdots, K\}, \qquad (11)$$

where the expectation is taking with respect to the original data set. The E-MPCC position of an out-of-sample data $\boldsymbol{z}(j)$ for $j \notin \{1, \cdots, K\}$ can be computed as:

$$\boldsymbol{z}(j) = \left[\sum_{k=1}^K \hat{W}(j, k)\hat{\boldsymbol{v}}_1(k), \sum_{k=1}^K \hat{W}(j, k)\hat{\boldsymbol{v}}_2(k)\right], \qquad (12)$$

where the weight $\hat{W}(j, i)$ for $j \notin \{1, \cdots, K\}$ is computed based on the dissimilarity of the location with respect to the points in the original set, and the eigenvectors $\hat{\boldsymbol{v}}_1$ and $\hat{\boldsymbol{v}}_2$ are computed based on the normalized weighting matrix $\hat{\boldsymbol{W}}$.

5 Simulation Results

An urban outdoor multi-cell mmWave scenario is considered as discussed in [5]. The system parameters are shown in Table 1. A ray tracing channel model is used to generate multi-path channels. We generate K UE locations on the streets of a Manhattan grid. The CSI of the UEs are estimated at multiple BSs. The number of nearest neighbors N are selected as 5% of UEs. The number of new samples J for which out-of-sample extension is applied is 10% of the total number of UEs.

Two scenarios are considered. In Scenario I, the MPCC is generated based on the channel features of K UE locations, and then J UE locations are removed randomly. The proposed E-MPCC is used for mapping the J locations to the chart. In scenario II, J UE locations are selected randomly and the MPCC is generated based on the channel features of $K - J$ UE locations. The proposed E-MPCC is used for mapping the J locations to the chart.

An example instance for MPCC/-EMPCC of Scenarios I&II for different settings is shown in Fig. 3. Clearly, the depicted chart shows that the J out-of-sample locations are accurately mapped by E-MPCC. For settings 1 (Set. 1), the parameters are $K = 500$, $J = 100$ and $B = 4$, and for settings 2 (Set. 2),

Table 1. Simulation parameters [5].

Parameter	Value	Parameter	Value
Carrier frequency	28 GHz	Bandwidth	256 MHz
UE Tx power	23 dBm	BS noise power	−86 dBm
OFDM subcarriers	256		

Table 2. CT and TW performance measures.

	MPCC		E-MPCC	
	CT	TW	CT	TW
Settings 1	0.725	0.682	0.7203	0.677
Settings 2	0.755	0.701	0.758	0.701

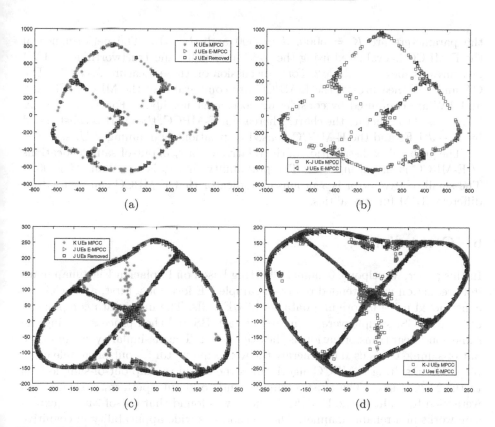

Fig. 3. E-MPCC and MPCC for: (a) Scenario I and Set. 1; (b) Scenario II and Set. 1; (c) Scenario I and Set. 2 (d) Scenario II and Set. 2.

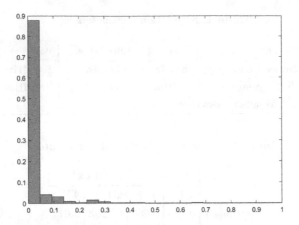

Fig. 4. The probability of relative-error ϵ_r.

the parameters are $K = 5000$, $J = 500$ and $B = 10$. The performance of the E-MPCC is evaluated using the continuity (CT) and trustworthiness (TW) measures as shown in Table 2. For a discussion on these measures, see [12]. The CT and TW measures of the E-MPCC are comparable of the MPCC. The CT and TW are computed by considering 20 nearest neighbors. For MPCC, all K UE are used to generate the chart, whereas for E-MPCC, the chart is constructed by $K - J$ UEs and the E-MPCC is used to position the reaming J UEs.

The probability distribution of the relative-error ϵ_r of out-of-sample locations of E-MPCC is shown in Fig. 4. The probability that ϵ_r is less than 7% is 90%. The small relative error is a promising indicator that E-MPCC can be used for different RRM functionalities.

6   Conclusion

In this paper, Multipoint Channel Charting based on Laplacian Eigenmap manifold reduction was extended to out-of-sample UE locations. First, a MPCC was constructed using an original data set of UE CSIs. The multi-path components of the new CSI sample were estimated at each BS and then processed using the same non-linear transformation as the original set. The dissimilarity vector of the out-of-sample UE is used to generate the weighting vector for out-of-sample mapping. The resulting E-MPCC algorithm is then used to map out-of-sample UEs to the MPCC map. The trustworthiness and continuity performance measures were used to evaluate the E-MPCC, and it was found that out-of-sample extension works in a reliable manner. The method has wide applicability in cognitive radio resource management, where predictions of UE connectivity parameters would be used.

In future work, the out-of-sample extension of MPCC for different dimension reduction techniques are going to be addressed. Machine learning techniques based on neural networks can be used to parametrize the MPCC and then used for out-of-sample mapping. The efficiency and accuracy of out-of-sample extensions need to be evaluated for different RRM functions such as handover and identifying cell boundaries.

Acknowledgement. This work was funded in part by the Academy of Finland (grant 319484).

References

1. Belkin, M., Niyogi, P.: Laplacian eigenmaps for dimensionality reduction and data representation. Neural Comput. **15**(6), 1373–1396 (2003)
2. Bengio, Y., Paiemcnt, J.F., Vincent, P., Delalleau, O., Roux, N.L., Ouimet, M.: Out-of-sample extensions for LLE, Isomap, MDS, eigenmaps, and spectral clustering. In: Advances in Neural Information Processing Systems, pp. 177–184. MIT Press (2004)
3. Bjornson, E., Larsson, E.G., Marzetta, T.L.: Massive MIMO: ten myths and one critical question. IEEE Trans. Commun. **54**(2), 114–123 (2016)
4. Busari, S.A., Huq, K.M.S., Mumtaz, S., Dai, L., Rodriguez, J.: Millimeter-wave massive MIMO communication for future wireless systems: a survey. IEEE Commun. Surv. Tutor. **20**(2), 836–869 (2018)
5. Deng, J., Medjkouh, S., Malm, N., Tirkkonen, O., Studer, C.: Multipoint channel charting for wireless networks. In: Proceedings of 52nd Asilomar Conference on Signals, Systems, and Computers, pp. 286–290, October 2018
6. Ester, M., Kriegel, H.P., Sander, J., Xu, X.: A density-based algorithm for discovering clusters in large spatial databases with noise, pp. 226–231. AAAI Press (1996)
7. Garcia, N., Wymeersch, H., Larsson, E.G., Haimovich, A.M., Coulon, M.: Direct localization for massive MIMO. IEEE Trans. Sig. Process. **65**(10), 2475–2487 (2017)
8. Guidi, F., Guerra, A., Dardari, D., Clemente, A., D'Errico, R.: Environment mapping with millimeter-wave massive arrays: System design and performance. In: Proceedings of IEEE Globecom Workshops (GC Wkshps), pp. 1–6, December 2016
9. van der Maaten, L., Postma, E.O., van den Herik, H.J.: Dimensionality reduction: a comparative review (2008)
10. Schmidt, R.: Multiple emitter location and signal parameter estimation. IEEE Trans. Antennas Propag. **34**(3), 276–280 (1986)
11. Shahmansoori, A., Garcia, G.E., Destino, G., Seco-Granados, G., Wymeersch, H.: Position and orientation estimation through millimeter-wave MIMO in 5G systems. IEEE Trans. Wireless Commun. **17**(3), 1822–1835 (2018)
12. Studer, C., Medjkouh, S., Gönültaş, E., Goldstein, T., Tirkkonen, O.: Channel charting: locating users within the radio environment using channel state information. IEEE Access **6**, 47682–47698 (2018)
13. Yang, S., Hanzo, L.: Fifty years of MIMO detection: the road to large-scale MIMOs. IEEE Commun. Surv. Tutor. **17**(4), 1941–1988 (2015)

Spectrum-Agile Cognitive Interference Avoidance Through Deep Reinforcement Learning

Mohamed A. Aref and Sudharman K. Jayaweera$^{(\boxtimes)}$

Communications and Information Sciences Laboratory (CISL), ECE Department, University of New Mexico, Albuquerque, NM, USA
{maref,jayaweera}@unm.edu

Abstract. This work introduces a spectrum-agile wideband autonomous cognitive radio (WACR) that is capable of avoiding interference and jamming signals. Proposed cognitive technique is based on deep reinforcement learning (DRL) that uses a double deep Q-network (DDQN). Moreover, it introduces new definitions for the state and the operation parameters that enable the WACR to collect information about the RF spectrum of interest in both time and frequency domains. The simulation results show that the proposed technique can efficiently learn an effective strategy to avoid harmful signals in a wideband partially observable environment. Furthermore, the experiments on an over-the-air channel inside a laboratory show that the proposed algorithm can rapidly adapt to sudden changes in the surrounding RF environment making it suitable for real-time applications.

Keywords: Deep Q-network · Deep reinforcement learning · Interference avoidance · Wideband autonomous cognitive radios

1 Introduction

With its ability to automatically extract important features from data, deep learning (DL) has made major breakthroughs in many applications such as computer vision, natural language processing, medical diagnosis, image and speech recognition [1–3]. In recent years, this has prompted researchers to investigate application of DL techniques in the wireless communications domain. The RF spectrum domain, however, has different characteristics compared with other domains including high data rates, representation of RF waveforms as complex numbers and time-varying multipath wireless channels. These all make the task of applying DL in the RF spectrum domain challenging because it requires modifications to existing DL algorithms or develop new ones. In the coming years, the DL is expected to play an important role in future wireless communications networks design including Internet of things (IoT), Unmanned Aerial Vehicles (UAVs) and the 6th generation (6G) cellular communication systems.

© ICST Institute for Computer Sciences, Social Informatics and Telecommunications Engineering 2019
Published by Springer Nature Switzerland AG 2019. All Rights Reserved
A. Kliks et al. (Eds.): CrownCom 2019, LNICST 291, pp. 218–231, 2019.
https://doi.org/10.1007/978-3-030-25748-4_17

Recently, the wideband autonomous cognitive radios (WACRs) have been proposed as an emerging technology to achieve spectrum situational awareness and signal intelligence [4–6]. With its ability to sense, learn and take decisions, a WACR may be a good candidate to apply DL techniques and especially deep reinforcement learning (DRL) to effectively address challenges that may be difficult to solve with the traditional machine learning techniques. The DRL is one of the widely used DL techniques in applications that require autonomous decision-making [7–9]. The DRL explores the advantage of deep neural networks to improve the training and the learning process of the traditional reinforcement learning making it suitable for systems with a large state-action space [7,9,10]. Most existing DRL techniques are based on deep Q-network (DQN) algorithm that extends the Q-leaning by using a convolutional neural network (CNN) instead of the Q-table to learn an approximate Q-function [7,10].

The DRL has previously been proposed for several applications in cognitive radio networks (CRNs) including power control, network access and connectivity preservation [9,11–16]. Another important application is the network security in which the CR adopts DRL to avoid jamming and other malicious attacks. One of the first works that uses DQN for the anti-jamming in CRN can be found in [14]. The system model in [14] assumes one secondary user (SU), one primary user (PU) and two jammers. The SU adopts a DQN with CNN to learn an efficient frequency hopping policy and decide whether to leave the area of heavy jamming and connect to another base station. One of the drawbacks of the proposed approach in [14] is that the state definition is based on the signal-to-interference-plus-noise ratio (SINR) estimates of the signals. In practice, SINR may take arbitrary value and the SINR estimates may not be perfect.

The authors in [15] extend the model in [14] by adding mobility features to the receiver allowing it to change its location. Using the same state and utility definitions in [14], the receiver is considered an agent that needs to learn an optimal policy using the DQN. However, the mobility capabilities may not be available for the SU and its corresponding receiver in many real-time applications. In [16], the authors considered the same problem formulation as in [14] in which the SU attempts to learn an optimal frequency hopping strategy. The authors in [16] used the spectrum vector as their system state that contains the received power spectral density (PSD) function at different time instants. This framework, however, is not applicable for wideband applications where the agent cannot sense all frequency channels simultaneously.

The goal of this paper is to design a spectrum-agile WACR that is capable of finding spectrum opportunities in a heterogeneous RF environment contested by jamming and crowded with interference signals. We propose a cognitive interference and jamming resilience technique that is suitable for real-time applications and mitigates limitations in the above mentioned previous work. Our proposed technique is based on double deep Q-network (DDQN) algorithm [17]. The advantages of the proposed approach can be summarized as follows:

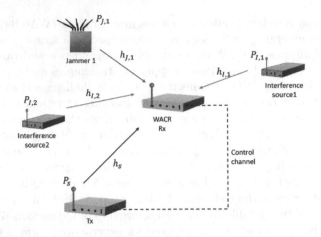

Fig. 1. System model.

- Ability to work in a partially observable wideband spectrum environment making it suitable for existing hardware, including the ones with limited instantaneous bandwidth.
- New simple definitions for the state and operation parameters that can represent more information about the surrounding RF environment in both time and frequency domains.
- A fast learning algorithm that can rapidly reconfigure to tackle sudden changes in the RF environment making it suitable for real-time applications in heterogeneous environments.

The rest of this paper is organized as follows: the system model is introduced in Sect. 2. Next, the proposed DDQN algorithm is discussed in details in Sect. 3. The performance evaluation is shown in Sect. 4 including both simulation and experimental results in an over-the-air channel inside a laboratory. Finally, Sect. 5 contains the concluding remarks.

2   System Model

Let us consider a WACR that is operating in a heterogeneous RF environment that includes multiple interference and jamming signals as shown in Fig. 1. The WACR is considered as the receiver in the communications link of interest, while the transmitter device may or may not have cognitive capabilities. The objective of the WACR is to choose a frequency channel with highest SINR for communications at every time instant. It is assumed that the frequency synchronization between the receiver and the transmitter is done through a secured common control channel as shown in Fig. 1. A centralized controller (e.g. a base station) or frequency rendezvous algorithms could be used as alternatives for the common control channel to maintain the frequency synchronization between the two nodes [18,19].

The RF spectrum of interest is assumed to have N possible channels. At time t, the WACR chooses an action, denoted by $a^t \in \{1, \cdots, N\}$, that represents the index of the channel for communications at time $t + 1$. The transmitter sends the signal of interest with a given power P_s. The channel power gain from the transmitter to the WACR is given by h_s. The interference source i and the jammer j send their signals with given powers $P_{I,i}$ and $P_{J,j}$, while their channel power gains to the WACR are $h_{I,i}$ and $h_{J,j}$, respectively. The received SINR of the WACR at channel n and time t can be expressed as

$$\mu_n^t = \frac{h_s P_s}{\sigma^2 + \sum_i h_{I,i} P_{I,i} + \sum_j h_{J,j} P_{J,j}} , \tag{1}$$

where σ^2 is the receiver noise power, assuming additive white Gaussian noise.

Due to hardware constraints, the WACR may not be able to sense all the N channels simultaneously. Assume that at any time instant the WACR can sense only N_s channels, with $N_s \leq N$. At time t, the WACR can estimate the power spectral density c_n^t for the sensed channel n. The WACR can then identify the availability of channel by comparing c_n^t with an appropriate threshold c_{th} that is designed based on noise floor estimation [4]. Let $f(c_n^t) = 1$ denotes the unavailability of the channel for $c_n^t > c_{th}$, otherwise $f(c_n^t) = 0$. At any given time, sensing is assumed to be performed on a different channel than the one used for communications. Thus, the WACR can sense the surrounding RF spectrum while maintaining the communications link. The sensing process can adopt any strategy (e.g. sweeping or random selections) based on the application of interest. In the following, we will assume that the WACR adheres to sweeping sensing strategy that sweeps sequentially over the spectrum of interest. Then, at time t, a sensing matrix W^t that stores the sensing results of all channels for T successive time instants up to time t is defined as follows:

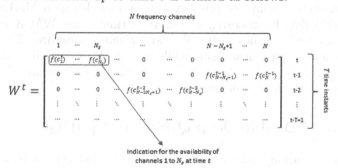

Indication for the availability of channels 1 to N_s at time t

The columns of W^t represent the different channels, while the rows represent the temporal memory depth. For each row, there are N_s values that indicate the availability of the sensed channels at the corresponding time instant. If $N_s \neq N$, remaining entries in each row are filled with zeros. Since the sensing matrix contains rich information about the RF environment in both frequency and time domains, it is used as a part of the state. In addition, the state definition also includes the index of the current channel used for communications in addition to an indication whether the communications over this channel is successful or not.

Let μ_{th} denote the required SINR threshold for successful communications. Then, an indicator function for the communications over channel n at time t can be defined as follows:

$$g(\mu_n^t) = \begin{cases} 1 & \text{if } \mu_n^t > \mu_{th} \text{ (success)} \\ 0 & \text{if } \mu_n^t \leq \mu_{th} \text{ (failure)} \end{cases} \tag{2}$$

The state at time t is then represented by a $(T+1) \times N$ matrix as shown below:

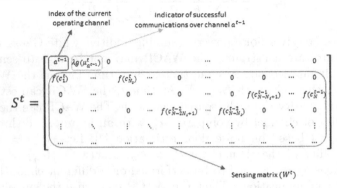

where $\lambda > 1$ is a weighting factor that may be optimized to achieve efficient learning. For sufficiently large T, the state may include information about all the channels of interest, ordered in time. Since there is only two possible values: 0 and 1 (denoting availability and unavailability, respectively), for each channel, the proposed state definition is less complicated compared with previous definitions that include SINR estimates as in [14] or received PSD as in [16].

The interference avoidance problem can be modeled as a Markov decision process (MDP) [20]. By choosing action a^t at time t, the WACR moves from its current state S^t to a new state S^{t+1} and receives a reward. The reward of choosing channel a^t for transmission while in state S^t is defined as the received SINR value $r(S^t, a^t) = \mu_{a^t}^{t+1}$. Note that, the reward value of state S^t and action a^t is obtained in the next time instant $t + 1$.

3   Proposed Double Deep Q-Network (DDQN) Algorithm

Reinforcement learning (RL) has shown to be a good candidate for learning in MDP environments [10]. It is based on delayed-reward principle in which the agent receives a reward from the environment after executing each action [4]. The value of the reward indicates how good or bad the action is. The objective of the agent is then to choose actions that maximize the rewards. In our scenario, the WACR attempts to learn a channel selection policy that maximizes the received SINR at each time instant.

The traditional RL approaches such as Q-learning, however, may not be the best technique in our scenario for several reasons. First, we are dealing

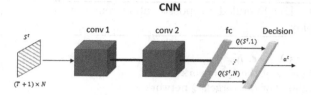

Fig. 2. CNN network structure of the proposed DDQN-based interference avoidance technique.

with a two-dimensional state. Second, the number of possible states can become extremely large even with few channels and a short memory depth. Furthermore, the rate of convergence of Q-learning may not be sufficient for real-time applications because it needs long time to explore and gain knowledge of the entire system. Hence, in this paper we propose using DDQN algorithm, an extension of the DQN that is developed by Google DeepMind team [17].

The basic idea of the DQN is to combine reinforcement learning with deep neural networks, more specifically, a CNN [7]. For each time t, the previously defined state S^t is used as an input to the proposed CNN. Then, the CNN attempts to estimate the Q-value $Q(S^t, a^t)$ for each possible action $a^t \in \{1, \cdots, N\}$. Several tests were performed to determine the best CNN design and the configuration of each layer to achieve consistently high performance while keeping the structure as simple as possible. Figure 2 shows the network structure of the proposed CNN which consists of 2 convolutional layers and 1 fully connected layer. The first convolutional layer (conv1) includes 10 filters with size 1×1 and stride 1. The second convolutional layer (conv2) has 20 filters of size 2×2 and stride 1. Both convolutional layers use rectified linear unit (ReLU) as the activation function. The fully connected layer (fc), on the other hand, has N rectified linear units that are used to output the Q-value estimates for each possible action. Finally, the WACR decides the action a^t corresponding to the maximum Q-value estimate.

For training, the DQN uses experience replay in which we store WACR's experiences $x^t = (S^t, a^t, \mu_{a^t}^{t+1}, S^{t+1})$ at each time t in a data set $\mathcal{D}^t = \{x^1, \cdots, x^t\}$. Let θ^t represents the weights of the proposed Q-network (CNN) at time t. During learning at time t, we draw an experience $x^k \sim U(\mathcal{D}^t)$, where U denotes the uniform distribution on $\mathcal{D}^t$ with $1 \leq k \leq t$, from the set of the stored experiences. The network parameters θ^t are then updated according to a stochastic gradient descent algorithm using the following loss function [7]:

$$L(\theta^t) = \mathbb{E}_{(S^t, a^t, \mu_{a^t}^{t+1}, S^{t+1}) \sim U(D)}[(\eta - Q(S^t, a^t; \theta^t)^2] \tag{3}$$

where η is the target optimal Q-value given by

$$\eta = \mu_{a^t}^{t+1} + \gamma \max_{a'} Q(S^{t+1}, a'; \hat{\theta}^t) \tag{4}$$

Algorithm 1. DDQN-aided proposed interference avoidance algorithm with experience replay

---

1: **Initialize:**
    Parameters λ, γ, ϵ, K
    The weights θ of the Q-network
    The weights $\hat{\theta}$ of the target Q-network
2: **for** each time t **do**
3:    Observe $\mu^t_{a^{t-1}}$, c^t_i, $\forall i \in \mathcal{C}^t$
4:    Obtain W^t and S^t
5:    With probability ϵ:
        Choose $a^t \in \{1, \cdots, N\}$ at random
6:    Otherwise:
        Obtain $Q(S^t, a')$ from the proposed CNN $\forall a'$
        Select $a^t = \arg\max_{a'} Q(S^t, a'; \theta^t)$
7:    Use channel a^t for communications at time $t+1$
8:    Store new experience $x^{t-1} = (S^{t-1}, a^{t-1}, \mu^t_{a^{t-1}}, S^t)$ in data set $\mathcal{D}$
9:    **for** k$=1, \cdots$, K **do**
10:       Select $x^k = (S^k, a^k, \mu^{k+1}_{a^k}, S^{k+1}) \sim U(\mathcal{D})$
11:       Compute η from (5)
12:       Compute the gradient of the loss function (3)
13:       Update θ^t
14:    **end for**
15:    Reset $\hat{\theta}^t = \theta^t$ for every fixed number of iterations.
16: **end for**

---

with $\hat{\theta}^t$ representing the weights of the target Q-network. This process can be repeated for K times at each time t in which θ^t is updated according to K randomly selected experiences.

The max operator in (4) uses the same value $Q(S^{t+1}, a'; \hat{\theta}^t)$ to decide which action is the best and to evaluate the optimal Q-value which might produce overestimated values degrading the learning process and the convergence rate [17, 21]. In order to overcome this problem, we use DDQN to decouple the selection and the evaluation operations. In this case, the original Q-network (with weights θ^t) is used for action selection and the target Q-network (with weights $\hat{\theta}^t$) is used to estimate the Q-value associated with the selected action. Thus, the target value η of (4) can be rewritten as follows:

$$\eta = \mu^{t+1}_{a^t} + \gamma\, Q(S^{t+1}, \arg\max_{a'} Q(S^{t+1}, a'; \theta^t); \hat{\theta}^t) \tag{5}$$

Algorithm 1 summarizes the proposed DDQN-based interference avoidance approach. For each time t, the WACR computes the received SINR $\mu^t_{a^{t-1}}$ on the current channel a^{t-1}. Let $\mathcal{C}^t$ represent the set of N_s channel indices that the WACR is sensing at time t. The WACR identifies the power spectral density c^t_i at each channel $i \in \mathcal{C}^t$ and updates the sensing matrix W^t. With the knowledge of a^{t-1}, $\mu^t_{a^{t-1}}$ and W^t, the WACR can obtain the current state S^t. The DDQN algorithm takes the state S^t as an input and estimates the Q-values for all possible actions. The optimal action $a^t = \arg\max_{a'} Q(S^t, a'; \theta^t)$ is chosen with a high probability $1 - \epsilon$, and a random action $a^t \in \{1, \cdots, N\}$ is selected uniformly with low probability ϵ to avoid staying in a local optima.

4   Performance Evaluation and Experimental Results

4.1   Simulation Results

Simulations have been performed to evaluate the performance of our proposed interference avoidance technique. The following parameters are used: $N = 6$, $T = 5$, $N_s = 2$, $K = 5$, $\epsilon = 0.1$, $\gamma = 0.4$, $\lambda = 10$, $\sigma^2 = 1$, $c_{th} = 2$, $\mu_{th} = 2$ and learning rate of 0.1. With these parameter values, state S^t at any time t is a 6×6 matrix which is the input to the CNN. Jamming signal j is transmitted with power $P_{J,j} = 8\,\text{mW}$ with a channel power gain to the WACR $h_{J,j} = 0.7$. On the other hand, any interference signal i has a transmit power of $P_{I,i}$ that can take any value between $3\,\text{mW}$ and $6\,\text{mW}$, while the channel power gain to the WACR $h_{I,i}$ is ranging from 0.4 to 0.9. For each interference source i, the values of $P_{I,i}$ and $h_{I,i}$ are chosen randomly from the predefined sets. Our signal of interest is transmitted with power $P_s = 5\,\text{mW}$ and the channel power gain to the WACR is $h_s = 0.8$. Hence, the optimal SINR value at any channel is 4 which corresponds to WACR selecting a channel free of interference and jamming.

As a benchmark, we used DQN, Q-learning and random channel selection techniques to evaluate our proposed DDQN technique [5]. Similar to the DDQN, the action and the reward of the DQN and Q-learning at time t are the index of the channel $a^t \in \{1, \cdots, N\}$ and the received SINR value μ^t, respectively. The DQN uses the same state definition S^t as in the proposed algorithm. The Q-learning, however, uses a simplified version of the original proposed state that does not include the sensing matrix W^t. Instead, the state of the Q-learning algorithm at time t is represented by $S_Q^t = [a_Q^{t-1}, \lambda g(\mu_{a_Q^{t-1}}^t)]$ so that the number of possible states is $2N$. On the other hand, in the random technique, the WACR randomly chooses a channel for communications.

Three test cases are considered with different interference and jamming signal scenarios. Table 1 shows the performance comparison with a scenario description for each test case. Test case 1 represents a simplified scenario in which there are

Table 1. Performance comparison: normalized accumulated reward values after 10,000 iterations.

Test case	Scenario	Proposed	DQN	Q-learning	Random	Optimal
1	2 interference signals	3.73	3.68	3.62	3.02	4
2	3 interference signals	3.65	3.56	3.52	2.57	4
3	3 interference signals and Markov jammer	3.12	3.07	2.84	2.14	4

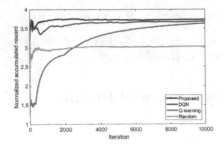
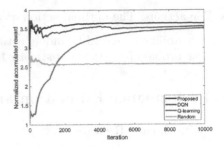

Fig. 3. $T = 5$: normalized accumulated reward (SINR) for test case 1.

Fig. 4. $T = 5$: normalized accumulated reward (SINR) for test case 2.

only two interference sources that transmit continuously their signals over two dedicated channels. Figure 3 shows the normalized accumulated reward for this scenario. Two main observations can be obtained from Fig. 3: (1) The proposed DDQN technique achieves a higher SINR than DQN, Q-learning and random techniques. (2) The proposed DDQN technique has a faster convergence than both the DQN and Q-learning.

In test case 2, an extra interference source is added on a third dedicated channel besides the two interference sources described above. This source, however, does not operate continuously. Instead, it switches between ON and OFF in a random manner. From Fig. 4, we may observe that the proposed DDQN technique outperforms both Q-learning and random techniques while having a similar performance to the DQN.

In test case 3, there is a Markov jammer operating besides the 3 interference signals described in test case 2. The Markov jammer selects a channel to jam based on a Markov chain as shown in Fig. 5 where $p_h = 0.8$ and $p_l = 0.2$. Figure 6 shows the normalized accumulated reward for this scenario: (a) for 10,000 iterations (b) for 2,000 iterations to have a closer look on the convergence rate. Again, from Fig. 6, the proposed DDQN technique shows better performance in terms of SINR and convergence rate compared to those achieved with the DQN and Q-learning.

Figure 7 shows the normalized accumulated reward for test case 3 for $T = 1$, $T = 5$ and $T = 10$. Part (a) of the figure shows the full iterations while part (b) only focuses on the beginning of the iterations to analysis the convergence rate. Note that, the state matrix dimensions at any time t in the case of $T = 1$ and $T = 10$ are 2×6 and 11×6, respectively. Figure 7 shows that reducing the temporal memory depth to $T = 1$ has a negative impact on the performance especially if the number of sensing channels is less than the total number of channels ($N_s = 2$ and $N = 6$).

On the other hand, both cases of $T = 10$ and $T = 5$ converge to the same accumulated reward value after 10,000 iterations as shown in Fig. 7 (a). This is because when $T = 5$, the state includes information about all the channels arranged in time from the newest to the oldest. Increasing the memory depth to

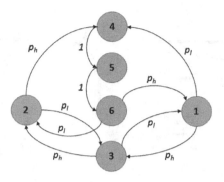

Fig. 5. Markov jammer selection strategy for test case 3 with 6 channels.

$T = 10$, will only add outdated information about the same channels. It does not seem to provide any significant new information since the most updated information about all the channels are already included with $T = 5$. However, this outdated information increases the state size that makes the computations more complex. These results show that choosing a suitable T value can be essential depending on the values of N and N_s.

4.2 Experimental Results

The experiments are performed inside the Communications and Information Sciences Laboratory (CISL) in the ECE Department at the University of New Mexico. The experiment setup consists of a USRP 2943R from National Instruments that is used as the WACR. The proposed cognitive interference-avoidance technique is implemented in LabVIEW on a DELL PRECISION TOWER 5810 PC with a built-in MATLAB interface to run the deep learning algorithm. The USRP interacts with the LabVIEW through a high speed PCIe connection.

The spectrum of interest is 240 MHz from 1.92 GHz to 2.16 GHz which is divided into 10 channels with 24 MHz each. The parameters used in the proposed DDQN are as follows: $N = 10$, $N_s = 1$, $T = 7$, $K = 5$, $\epsilon = 0.1$, $\gamma = 0.4$ and $\lambda = 10$. From spectrum observation, the noise floor threshold is set to -95 dBm. Any channel other than the one used by the WACR with received power above this threshold is considered unavailable. The USRP uses an IQ rate of $24 M samples/sec$, acquisition time of 0.16ms and RX gain $= 20$ dB. Figure 8 shows the whole spectrum of interest as observed on the KEYSIGHT N9952A spectrum analyzer. It is clear from Fig. 8 that all but channel 5, 6 and 7 are occupied with different signals. Hence, if the proposed cognitive interference-avoidance algorithm works properly, the WACR has to choose a channel from these three channels.

The experiment consists of two stages. In the first stage we evaluate our proposed algorithm in the spectrum described above. We ran this stage for 300 iterations, in which each iteration represents a single sensing duration. The total time for this stage is about 489 s. The WACR adopts a random sensing strategy

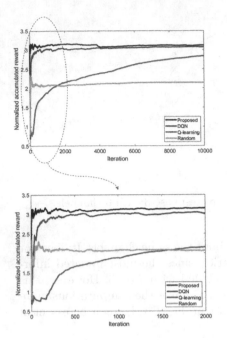

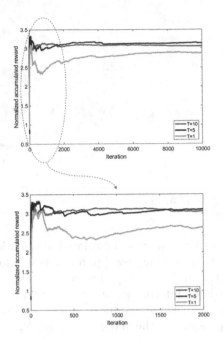

Fig. 6. $T = 5$: normalized accumulated reward (SINR) for test case 3.

Fig. 7. Normalized accumulated reward (SINR) for test case 3 for different temporal memory depth values using the proposed algorithm.

in which it randomly selects a channel to sense for each iteration. Figure 9 shows the number of times that the WACR was able to avoid channels with interference as a percentage of the total number of iterations. Figure 10 shows whether the actions selected by the WACR correspond to a channel free of interference or not. From the figures, we can notice that the proposed DDQN algorithm was able to learn an optimal policy after a few number of iterations (approx. 40 iterations). In this experiment the WACR learned to operate in channel 6 which is free of interference.

An interesting question is how the WACR will react to sudden changes in the RF environment. A good learning algorithm should make the WACR adjust to this new condition rapidly. Thus, in the second stage of our experiment we generated an interference signal in channel 6 starting at the 301st iteration. It can be observed from Fig. 10 that proposed DDQN algorithm reacts very fast and switch to a new interference-free location (channel 5).

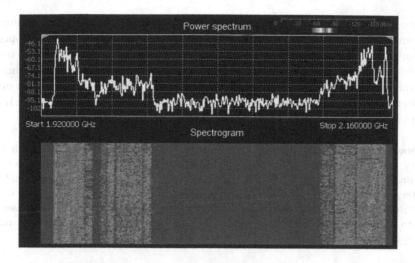

Fig. 8. Power spectrum and its corresponding spectrogram for start freq. = 1.92 GHz and stop freq. = 2.16 GHz.

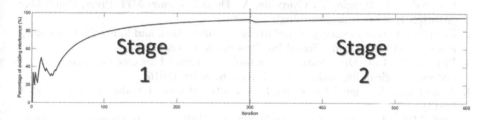

Fig. 9. The percentage of selecting interference-free channels.

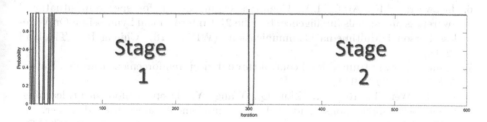

Fig. 10. Probability of selecting an interference-free channel for each iteration

5   Conclusion

In this paper, we have studied cognitive interference avoidance through spectrum agility. The proposed technique is based on DDQN algorithm with CNN. The WACR uses two separate channels for sensing and communications. The sensing operation is used to create the sensing matrix that includes information about the availability of different channels of interest. The sensing matrix along with the chosen communications channel and an indication of the success/failure of the communications over this channel form the state of the DDQN. The proposed technique was evaluated through various test cases that include multiple interference and jamming signals. Both simulation and experimental results showed that the proposed algorithm is suitable for real-time applications and can operate over wideband spectrum. Furthermore, the proposed technique was shown to rapidly adapt to sudden changes in the surrounding RF environment.

Acknowledgment. This research was sponsored in part by the Army Research Laboratory and was accomplished under Grant Number W911NF-17-1-0035. The views and conclusions contained in this document are those of the authors and should not be interpreted as representing the official policies, either expressed or implied, of the Army Research Laboratory or the U.S. Government. The U.S. Government is authorized to reproduce and distribute reprints for Government purposes notwithstanding any copyright notation herein.

References

1. Goodfellow, I., Bengio, Y., Courville, A.: Deep Learning. MIT Press, Cambridge (2016). http://www.deeplearningbook.org
2. Deng, L.: A tutorial survey of architectures, algorithms, and applications for deep learning. APSIPA Trans. Signal Inf. Process. **3**, e2 (2014)
3. Litjens, G., et al.: Deep learning as a tool for increased accuracy and efficiency of histopathological diagnosis. Nat. Sci. Rep. **6**, 26286 (2016)
4. Jayaweera, S.K.: Signal Processing for Cognitive Radios, 1st edn. Wiley, New York (2014)
5. Aref, M.A., Jayaweera, S.K., Machuzak, S.: Multi-agent reinforcement learning based cognitive anti-jamming. In: IEEE Wireless Communications and Networking Conference (WCNC 17) , San Francisco, CA, March 2017
6. Jayaweera, S.K., Aref, M.A.: Cognitive engine design for spectrum situational awareness and signals intelligence. In: The 21st International Symposium On Wireless Personal Multimedia Communications (WPMC 18), Chiang Rai, Thailand (2018)
7. Mnih, V., et al.: Human-level control through deep reinforcement learning. Nature **518**(7540), 529 (2015)
8. Li, H., Wei, T., Ren, A., Zhu, Q., Wang, Y.: Deep reinforcement learning: framework, applications, and embedded implementations: invited paper. In: 2017 IEEE/ACM International Conference on Computer-Aided Design (ICCAD), Irvine, CA, USA, November 2017
9. Luong, N.C., et al.: Applications of deep reinforcement learning in communications and networking: a survey. arXiv, eprint arXiv:1810.07862 [cs.NI] (2018)
10. Sutton, R.S., Barto, A.G.: Reinforcement Learning: An Introduction. MIT Press, Cambridge (1998)
11. Huang, W., Wang, Y., Yi, X.: Deep q-learning to preserve connectivity in multi-robot systems. In: Proceedings of the 9th International Conference on Signal Processing Systems (ICSPS 2017) (2017)
12. Li, X., Fang, J., Cheng, W., Duan, H., Chen, Z., Li, H.: Intelligent power control for spectrum sharing in cognitive radios: a deep reinforcement learning approach. IEEE Access **6**, 25463–25473 (2018)

13. Liu, S., Hu, X., Wang, W.: Deep reinforcement learning based dynamic channel allocation algorithm in multibeam satellite systems. IEEE Access **6**, 15733–15742 (2018)
14. Han, G., Xiao, L., Poor, H.V.: Two-dimensional anti-jamming communication based on deep reinforcement learning. In: IEEE International Conference on Acoustics, Speech and Signal Processing, New Orleans, LA, USA, March 2017
15. Xiao, L., Jiang, D., Wan, X., Su, W., Tang, Y.: Anti-jamming underwater transmission with mobility and learning. IEEE Commun. Lett. **22**(3), 542–545 (2018)
16. Liu, X., Xu, Y., Jia, L., Wu, Q., Anpalagan, A.: Anti-jamming communications using spectrum waterfall: a deep reinforcement learning approach. IEEE Commun. Lett. **22**(5), 998–1001 (2018)
17. Van Hasselt, H., Guez, A., Silver, D.: Deep reinforcement learning with double q-learning. In: The Thirtieth AAAI Conference on Artificial Intelligence (AAAI-16), AZ, USA, Phoenix, February 2016
18. Theis, N.C., Thomas, R.W., DaSilva, L.A.: Rendezvous for cognitive radios. IEEE Trans. Mobile Comput. **10**(2), 216–227 (2011)
19. Pu, D., Wyglinski, A.M., McLernon, M.: An analysis of frequency rendezvous for decentralized dynamic spectrum access. IEEE Trans. Veh. Technol. **59**(4), 1652–1658 (2010)
20. Puterman, M.L.: Markov Decision Processes: Discrete Stochastic Dynamic Programming. Wiley, Hoboken (2014)
21. Van Hasselt, H.: Double q-learning. In: Advances in Neural Information Processing systems 23 (NIPS 2010), pp. 2613–2621 (2010)

Localization Techniques for 5G Radio Environment Maps

Marcin Hoffmann[✉] and Hanna Bogucka

Chair of Wireless Communications, Poznan University of Technology,
60-965 Poznan, Poland
marcin.r.hoffmann@student.put.poznan.pl

Abstract. Localization techniques are going to be a significant part of 5G networks, not only for user-plane services (e.g., navigation) but they can also be used to improve network performance using Radio Environment Maps (REMs). The REM's operation requires an accurate localization technique that can work in adequate (radio) environments with high reliability. This paper firstly, provides comprehensive overview of the existing localization techniques that can be used in 5G systems focusing on cellular network-based solutions and advancement in satellite-based localization. Secondly, these techniques are analysed and assessed against the requirements stated for 5G REM systems.

Keywords: 5G · Localization · Radio Environment Maps

1 Introduction

The 5G networks will have to support many new, challenging use-cases [12]. These can vary from high throughput Enhanced Mobile Broadband (eMBB) service, through Ultra Reliable Low Latency Communications (URLLC) up to Massive Machine Type Communications (mMTC). The network has to be fully adaptive to support either high spectral efficiency (for eMBB), high reliability (for URLLC) or massive connectivity and high energy efficiency (for mMTC). In order to meet these requirements various radio access technologies are to be utilized, e.g., Non-Orthogonal Multiple Access (NOMA), Massive MIMO (M-MIMO) or transmission in mm-waves. However, as in all previous generation, 5G requires significant amount of spectrum to be used for transmission. The electromagnetic spectrum is scarce, therefore, the promising solution is to increase spectrum utilization by Dynamic Spectrum Access (DSA) [7]. An efficient architecture for DSA utilizes location-based database, e.g., Radio Environment Map (REM) [16]. Such a database can be used not only to support dynamic spectrum licenses assignment in various scenarios (e.g., Vehicle-to-Vehicle communications [18]) but also other advanced control mechanisms in a network, e.g., interference management [8]. However, efficient operation of REM requires the obtained/controlled data to be tagged with accurate localization.

© ICST Institute for Computer Sciences, Social Informatics and Telecommunications Engineering 2019
Published by Springer Nature Switzerland AG 2019. All Rights Reserved
A. Kliks et al. (Eds.): CrownCom 2019, LNICST 291, pp. 232–246, 2019.
https://doi.org/10.1007/978-3-030-25748-4_18

The demand on localization of a user terminal in cellular networks began with definition of enhanced 911 (E911) location requirements for emergency calls by Federal Communications Commission (FCC) of United States (U.S.) in 1990's. In the cellular networks, from 2G to 4G, localization was used at the beginning to fulfill these FCC emergency calls requirements, and later also in some location-based services like e.g. navigation, mapping and geo-marketing [15]. In 5G, utilizing DSA and REM, the reliability, accuracy and low energy consumption of localization techniques plays a key role.

The aim of this paper is to overview localization techniques currently available for user devices and those to be available in the near future, for the REM-based DSA applications. The presented methods are compared taking their accuracy into account, their implementation requirements as well as the robustness in both indoor and outdoor conditions, to choose proper positioning technique to specified environment conditions.

The paper is organized as follows: in Sect. 2, the basics of the localization techniques are introduced. In Sect. 3, their utilization in the contemporary systems is presented, with the main focus on Global Navigation Satellite System (GNSS) and LTE networks. In Sect. 4, the presented methods are compared, and their usefulness for 5G REM is discussed. The conclusions are formulated in Sect. 5.

2   Localization Techniques Overview

This section describes the main positioning approaches, that are utilized to implement various localization systems, i.e., cellular, satellite and the other ones described in Sect. 3 in detail. The first reasonable classification can be done as proposed in [15] by splitting positioning techniques into two groups:

- *Mobile-based*, where the responsibility of calculating position lays on the side of a user device, with the use of its own measurements and network (which may, but does not have to be cellular) assistance data.
- *Network-based*, where, in opposite to the *Mobile-based* approach, a user device sends its measurements to the network location server which computes the final position.

Regarding specific techniques used for the localization purpose there are four basic ones. They are presented below, as concepts; their implementations and accuracy in existing systems are discussed in the following section.

2.1   Trilateration

The user position is obtained on the basis of measured distances to the reference stations of known locations [10]. Due to fact that radio wave has the known value of the propagation speed, the simplest idea is to measure the time of the radio wave travel from the reference station to the user receiver, and to compute the distance (this concept is system based, and it is called the Time of Arrival

(TOA) method [10]). The user position is where the circles, indicating distances of a user from the reference (base) stations, intersect (see Fig. 1). To obtain the 2D position, at least three user-to-reference station distances are required. In the mathematical form, the trilateration may be expressed as solving the set of Eq. (1) for $k = 1, \ldots, K$ (where K is the reference-stations number):

$$r_k = \sqrt{(x - x_k)^2 + (y - y_k)^2}, \tag{1}$$

where r_k is the distance between the user and the k-th reference station, x, y are the 2-D user position coordinates and x_k, y_k are the k-th reference station coordinates. In [10], the authors also mention the position calculation based on the difference in TOA (TDOA). There, two pairs of measurements are necessary for obtaining the 2-D position. In [15], the received signal strength (RSS) is considered as the third option that requires at least three measurements for 2-D positioning as in TOA.

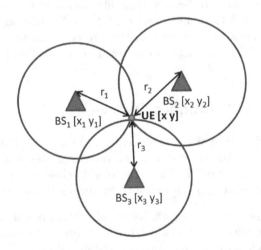

Fig. 1. The concept of the trilateration positioning technique.

The time-based positioning requires precise time interval measurements, i.e., to achieve centimeter- accuracy level, the time-errors must be of the order of 1ns [5]. The receiver and the reference stations clocks synchronization is crucial to achieve this requirement. In [5], it is noticed that a user receiver has an internal clock with much worse accuracy than that of a reference station. One additional measurement is necessary to provide the estimation of the receiver clock offset, however the reference station transmitters still must be synchronized [10].

2.2   Triangulation

In the triangulation method, the user position is obtained on the basis of angles measured between the user device and at least two reference stations of known position (for 2-D positioning), however, performing more then two measurements can improve accuracy. This type of measurement is called the Angle of Arrival (AoA). The user position is defined where the directions of the received signals (AoAs) intersect, as it is shown in Fig. 2.

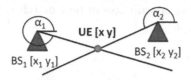

Fig. 2. The concept of triangulation positioning technique.

In the mathematical form, the triangulation method for obtaining the 2D user's position may be expressed as solving set of linear equations:

$$y - y_k = \tan \alpha_k \cdot (x - x_k), \quad \text{for} \quad k = 1, \dots, K \tag{2}$$

where x and y are the user coordinates, x_i and y_i are the ith reference station coordinates, and α_i is AoA measured between the user position and the ith reference station.

The triangulation technique does not require time synchronization, but it depends on the quality of the AoA measurements. The authors of [20] claim that the AoA measurement error is proportional to the inverse number of the antenna array elements.

2.3 Proximity

Proximity method is probably the simplest positioning method, where the user position is approximated by the closest reference station position (see Fig. 3). For example, the cell-ID of a connected cell can be used to approximately locate the user-equipment (UE) [15].

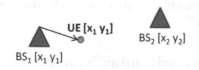

Fig. 3. The concept of the proximity positioning technique.

2.4 Radio Frequency Pattern Matching

Radio Frequency Pattern Matching (RFPM), also called fingerprinting, is based on the comparison between the measured value (for example RSS or the propagation time delay [15]) with the value stored in a database. The fingerprint is typically a geographic position associated with the signal measurement. The user position is obtained as coordinates of best matched fingerprint in database [15], as illustrated in Fig. 4.

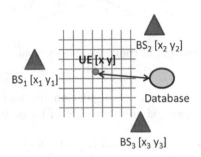

Fig. 4. The concept of RFPM positioning technique.

The fingerprinting position estimation process is divided into three steps [13]:

– *Database Building Phase*; In this step the database of fingerprints is created on the basis of performed measurements in the reference points of known locations.
– *Acquisition Phase*; In this phase, the user device is performing the same measurements as in the previous step, but in unknown location.
– *Matching Phase*; In this final step, the values measured by a user are compared (with the use of predefined matching rule) with the ones stored in database. The user position is claimed to be the position of the best matching fingerprint from database.

The position accuracy in fingerprinting is highly related to the quality of location-depend measurements, calibration of the database and the density of the measurement grid [15].

3 Contemporary Positioning Systems

In this section, utilization of the presented positioning methods in the existing systems is discussed. Implementation of REM databases requires robust positioning techniques with predictable and well tested accuracy. Thus, the main focus is on the practical positioning systems, i.e., the Global Navigation Satellite System and on the solutions described in [2] for LTE networks. Then, some other independent positioning systems and expected 5G improvements are mentioned.

3.1 Global Navigation Satellite System (GNSS)

GNSS allows users to estimate their position on the basis of signals received from satellites orbiting the Earth. Position calculations are based on TOA trilateration method [10]. Each satellite broadcasts the information about ephemerids (the detailed orbit parameters used to compute its position), almanac (basic ephemerids for the whole constellation) and clock corrections [10]. Signal transmitted by satellites is code-modulated, and can be expressed as [10]:

$$s(t) = \sqrt{2P}D(t)c(t)cos(2\pi ft + \phi), \tag{3}$$

where $\sqrt{2P}$ is the signal amplitude, $D(t)$ is the navigation data signal, $c(t)$ is the spread spectrum code, f is the carrier frequency, and ϕ is the phase [10]. All satellites internal clocks are synchronized thanks to the use of the atomic clocks. The receiver, on the basis of the received ephemerids and the TOA measurements, estimates its position. Due to the receiver's clock uncertainty, additional measurement is necessary to estimate the clock offset. To obtain the 3D position, at least four satellites must be visible, more satellites can improve accuracy [10].

Currently available GNSS systems are, e.g., GPS (USA), Galileo (EU), GLONASS (Russia), BeiDou(China). However, GPS is the most popular.

GNSS Range Measurements. There are two types of measurements, that allow to estimate the receiver-to-satellite distance. The first approach is to track the pseudo-random code transmitted by the desired satellite, the second one to obtain the range on the base of the carrier phase measurements.

Code Phase Measurements. In this approach, the distance between the user and the ith satellite is computed on the basis of aligning the spreading code replica with the one received from satellite. The distance estimate suffers from some errors, and that is why it is called pseudorange. The error sources with the biggest impact on measurements are: both satellite and receiver clock biases, the ionosphere and troposphere propagation delays and the multipath fading [10].

Code phase measurements are common in "mass market" receivers, e.g., in smart-phones. The accuracy achieved with this method in stand-alone GPS receivers is about 10 m for 95% of cases [10].

Carrier Phase Measurements. In this approach, the pseudorange is obtained as the number of carrier signal wavelengths between the satellite and the receiver [10]. The measured value is the phase offset between the carrier replica and the one received from the satellite. Under the assumption of ideally synchronized clocks and no other errors, this phase offset is expressed (in cycles) as [10]

$$: \phi(t) = \phi_u(t) - \phi^s(t - \tau) + N, \tag{4}$$

where $\phi_u(t)$ is the carrier-phase replica, $\phi^s(t - \tau)$ is the phase of the signal from the satellite received at time t, τ is the propagation time, and N is called *integer ambiguity*. (It's estimation, namely, the integer ambiguity resolution, is the most complex part of the carrier phase measurements, but when done it allows to achieve cm-level accuracy positioning [10]).

Apart from the integer ambiguity carrier, the phase measurements are affected by the same errors as the code phase measurements (clock biases, propagation delays and multipath fading), although they can be eliminated by double difference measurements discussed later.

The carrier Phase measurement is much more accurate, but it also requires more complex computations than the code phase estimation method. Centimeter-accurate GNSS positioning with phase carrier measurements is common in geodesy, agriculture and surveying. In mass market, the receivers' main problem is in low-quality antennas causing not high enough multipath-fading effect suppression and long time correlations in phase errors, resulting in long (tens of seconds) initialization time (time to ambiguity resolution- TAR) [17].

GNSS Improvements. As the basic GNSS positioning technique for stand-alone receiver is utilizing code phase measurements, let us discuss how to improve GNSS accuracy, and how to take advantage of the carrier phase measurements with Real Time Kinematic approach. In Table 1, requirements of the presented GNSS improvements are compared.

The Dual-Frequency GNSS comes as the improvement for a stand-alone GPS receiver. All GPS satellites broadcast L1 civil signal at frequency 1575,42 MHz with the chip rate of 1.023 Mcps. The newer ones provide also the L5 signal [10] broadcast at 1176.45 MHz frequency and higher chipping rate of 10.23 Mcps [11]. Thanks to that L5 suffer less from the multipath reflections than L1. When combining these two signals in the receiver, it is possible to eliminate the ionospheric error [10]. With dual frequency receiver, about 30 cm accuracy is expected to be provided [11]. Moreover, the Broadcom company released BCM47755 chip, as the first mass-market chip which uses both L1 and L5 signals. The first smart phones equipped with that chip have been expected to appear in the end of 2018 [11]. Currently, the only phone with GPS dual frequency receiver identified by us is Xiaomi Mi 8, however no reliable localization tests results could be found.

Differential-GNSS. In Differential-GNSS (DGNSS), measurements are performed using code phase approach, but the idea of reducing errors is based on the fact that ionosphere, troposphere, ephemerids and satellite clock errors are correlated in time and space [10]. A reference receiver with known location is used which compares the position obtained on the basis of satellites' signals measurements with the known true location, and computes pseudorange corrections that are then send to a user (called rover) by the side link. The differential approach provides accuracy up to 1 m [10].

Real-Time Kinematics. Real Time Kinematics (RTK) is based on relative positioning [10]. Similarly to DGNSS, the reference station with known coordinates is required, but it sends (to a rover) its raw measurements instead of the computed corrections. RTK also takes advantage of the carrier-phase measurements instead of the code-phase. To eliminate clock biases and atmospheric delays, position (relative to that of the reference station) estimation is performed on the basis of the double difference.

To obtain one double difference measurement, carrier-phase in both the reference station and in the rover must be measured using the same pair of satellites. The measurement result of the rover is subtracted from the reference station measurement result using the same satellite signals (resulting in a single difference). The same happens with measurements based on another satellite signal. Finally, the mentioned single differences are subtracted resulting in the double difference [10]. As the satellites' and the receivers' clocks biases are the same, they cancel out in the equation. The same happens to the atmospheric delays, when the distance between rover and reference station is small enough (lower then 5 km [17]). The main component of the remaining error is multipath fading.

Table 1. Measurements requirements for GNSS variants.

	Code phase	Carrier phase	Reference station	Dual frequency chip
stand alone single frequency	✓	✗	✗	✗
stand alone dual frequency	✓	✗	✗	✓
DGNSS	✓	✗	✓	✗
RTK	✗	✓	✓	✗

3.2 LTE Positioning Techniques

The LTE standard utilizes some of the above mentioned methods [2]. An overview of the most important ones is presented below [3].

Observed Time Difference of Arrival (OTDoA). OTDoA uses trilateration. Measurements are performed with the use of special positioning reference signal (PRS), defined by a pseudo-random QPSK sequence mapped to resource elements and bandwidth (the number of resource blocks), time offset, duration in subframes and its periodicity [3]. PRS is transmitted by each eNB. To obtain the position, at least four pairs of time differences must be measured. In this method, the obtained OTDoA values are send to the location server (E-SMLC) which computes the final position and sends it back to UE [10]. It may be reasonable to use OTDoA in dense urban and indoor environments where not enough GNSS satellites are visible. In [15], the accuracy of this method is claimed to be < 50 m for 67% of cases.

Enhancement Cell-ID (E-CID). E-CID is based on the proximity method and network-based approach [3], where initial UE coordinates are assumed to be coordinates of its serving base station. To improve the accuracy, several parameters specified in [2] may be requested from UE or eNB by E-SMLC. They are presented in Table 2.

Table 2. Information that may be transferred from eNB/UE to the E-SMLC. [2]

Information
Timing Advance (T_{ADV})
Angle of Arrival (AoA)
E-UTRA Measurement Results List: - Evolved Cell Global Identifier (ECGI)/Physical Cell ID   - Reference signal received power (RSRP)   - Reference Signal Received Quality (RSRQ)
GERAN Measurement Results List: - Base Station Identity Code (BSIC)   - ARFCN of Base Station Control Channel (BCCH)   - Received Signal Strength Indicator (RSSI)
UTRA Measurement Results List: - UTRAN Physical ID   - Common Pilot Channel Received Signal Code Power (RSCP)   - Common Pilot Channel Ec/Io
WLAN Measurement Results List: - WLAN Received Signal Strength Indicator (RSSI)   - SSID   - BSSID   - HESSID   - Operating Class   - Country Code   - WLAN Channel(s)   - WLAN Band

In [3], three scenarios of additional measurements are mentioned: measuring the distance to serving base station (on the basis of timing advance), measuring the distance to three base stations (trilateration) and measuring AoA (triangulation) from at least two base stations (requires large antenna arrays to be accurate [3]). In [15], the E-CID method achieves horizontal accuracy of 50 m.

Radio Frequency Pattern Matching. Although RFPM is not explicitly defined in [2], it is proposed in [1] to be implemented in the existing E-CID infrastructure with the use of timing advance and RSRP measurements already available in E-SMLC. The results provided in [1] show the accuracy up to 85 m in 67 % cases, in urban conditions using the reference signal of 10 MHz bandwidth. In [19], the use of RSRP combined with WLAN RSSI measurements for creating the location database gave the accuracy of 13 m in 67% of cases in the outdoor environment and 10-by-10 m measurement grid, however, it decreased to 22 m with the use of larger 40-by-40 m grid. In [15] WLAN supported fingerprinting methods are found to be promising for indoor localization.

Assisted GNSS. The idea of supporting UE embedded GNSS receiver by distributing the assistance data through cellular network, is implemented in systems from GSM through UMTS and LTE. The main target of this method is to improve robustness of UE receiver positioning in urban conditions and shorten the start-up time [10]. It is reasonable to split the assistance data into the data assisting measurements e.g., visible satellite list or code phase search window and the data assisting position calculation like: reference time, satellite ephemeris [3]. According to [10], in a stand alone GPS receiver, it takes 30 s to receive satellite ephemeris and up to 12.5 min. to receive almanac containing information about the whole constellation. The accuracy of AGNSS localization equals about 10 m [15]. However, its quality depends on the implementation used by a given network operator.

RTK Support. As can be seen in [2], currently (up to LTE rel. 15) DGNSS corrections are already available. Interestingly, rel. 15 introduced new types of the assistance data in LTE Positioning Protocol (LPP) with the main purpose to support high accuracy (HA) GNSS, i.e., mentioned RTK [17]. As RTK requires continuous carrier phase tracking, in [2], it is proposed to apply LPP Periodic Assistance Data Transfer procedure for measurements transmission between UE and E-SMLC. UE GNSS receivers do not support carrier phase measurements yet, but [17] shows that it is possible, and including RTK assistance in LTE rel 15. should be commonly available in a near future.

3.3   Other Positioning Systems

Apart from mentioned cellular and satellite systems, there are several other methods, suitable mostly for indoor environment. One of them is trilateration used in IEEE 802.15.4a ultra wide band (UWB) system [9]. Large bandwidth ($> 500\,$MHz) results in accurate TOA measurements, and easy multipath components estimation. In [6], UWB is expected to provide the accuracy below 1 m for at least 90% of cases. So far, it is common in public or enterprise domains using WLAN 802.11 standard, that the use of RSS measurements give positioning accuracy from 3 to 30 m [9]. WLAN and UWB are only mentioned examples, other ideas are based on utilizing Bluetooth, barometer or augmenting existing methods with various sensors' or cameras' data [20].

3.4   5G Expected Improvements

Although 5G positioning protocol is not fully defined yet, some of the expected improvements of the basic concepts can be discussed. Requirements for positioning in 5G systems expect accuracy up to 10 cm [15]. The way to fulfil them is to take advantage of 5G features like: high carrier frequencies, larger bandwidth, MIMO and high density networks [20]. Higher carrier frequencies (near 30 GHz and more) result in the possibility of utilizing larger bandwidth (hundreds of megahertz), what causes improvement in trilateration based methods (TOA, TDOA, OTDoA). In [14], below 20 cm accuracy was achieved for 100 MHz bandwidth. By using MIMO systems AoA estimation may be much improved. In [5], a single-anchor (with only one reference station) method is mentioned as the benefit from 5G system features. Apart from cellular methods, high accuracy GNSS (e.g. RTK) is also taken into account in [15].

4   Localization Techniques for 5G REM

The existing positioning techniques have been presented above with the main focus on LTE and GNSS systems, as they are well described, and researched solutions available. Below, in this section, their utilization for radio-network efficiency-improving REMs in various environment conditions and applications is considered by analysis of their accuracy (in Table 3) and other features (in Table 4).

The REM may be utilized on various network levels for different purposes, e.g., global REM for Quality of Service (QoS) or Quality of Experience (QoE) management or local REM for distribution of available radio resources. Data stored in REM may be used for several purposes: DSA, traffic steering or massive MIMO (M-MIMO) transmission optimization. In all mentioned cases UE is the unit obtaining it's position. There are different requirements on accuracy and update rate of localization service for different REM levels (associated with the OSI layers) and applications. The higher the level of REM, the lower the update rate and required accuracy [4]. The key feature of localization for REM in all cases is robustness. Without accurate enough localization data, the network performance will be decreased, and in the worst case, may lead to disconnection of UE from the network.

For global-level REM with QoS and QoE data, focused on large scale traffic analysis, high accuracy is not necessary. Stand alone GNSS should be enough. When not enough satellites are visible (e.g., in an indoor environment) any of the presented cellular-system-based positioning may be utilized. Even E-CID without additional measurements could be enough for high level traffic analysis.

Local REMs responsible for radio resource management requires higher accuracy (which is further dependent on utilized radio frequency band) and high updates rate. For outdoor scenarios utilizing LTE frequency band, 5 m spatial resolution should be enough for REM [4]. Such a localization accuracy can be provided by a dual frequency or differential GNSS. For outdoor, mm-waves M-MIMO transmission optimization, spatial resolution of 0.5 m is required (as mm-waves wavelength is order of magnitude smaller then cm waves utilized in LTE scenario). This accuracy can be supported by RTK. However, it is a very energy consuming solution, resulting in fast UE battery discharge. The way out of this problem may be to use RTK for REM data acquisition, and less accurate and energy consuming dual frequency GNSS (UE must have dual frequency

Table 3. Mobile positioning methods accuracy.

Method	System	Horizontal accuracy	Type	Reference
OTDoA	LTE	63 m (95% cases)	Trilateration	[15]
E-CID	LTE	~50 m (67% cases)	Proximity	[15]
RFPM	LTE	85 m (67% cases)	Fingerprinting	[1]
RFPM	LTE+WLAN	13 m (67% cases)	Fingerprinting	[19]
GNSS	GPS	10 m (95% cases)	Trilateration	[10]
DGNSS	GPS	$\geq$1 m (95% cases)	Trilateration	[10]
RTK	GPS	~1 cm	Trilateration	[10]
DF-GNSS	GPS	~30 cm	Trilateration	[11]
A-GNSS	GPS/LTE	~10 m (67 % cases)	Trilateration	[15]

Table 4. Mobile positioning methods comparison.

Method	Advantages	Disadvantages
OTDoA	- No satellites - Operates indoor	- Reference stations clocks must be synchronized - Accuracy depends on bandwidth, - Dense BS network is necessary
E-CID	- Simplicity - Base infrastructure for RFPM - No satellites - Operates indoor	- Low accuracy
RFPM	- No satellites - Implementation without Infrastructure modification - Operates indoor	- Accuracy depends on matching algorithm, database design and measurements grid - Requires huge number of measurements
GNSS	- Good accuracy, - Almost in all smartphones	- At least 4 satellites visible - Clocks, atmospheric and multipath errors,
DGNSS	- Reduces clocks and Atmospheric errors	- Requires reference station
RTK	- Centimeter level accuracy	- Requires reference station - Computationally complex and energy consuming carrier phase measurements
DF-GNSS	- Decimeter level accuracy	- Requires dual-frequency chip
A-GNSS	- UE GNSS receiver cellular Assistance data, - Improves GNSS robustness - DGNSS or RTK corrections Possible	- The same errors like previous GNSS methods

chip) to send current location to the REM manager to obtain the transmission parameters. Discussed RTK and DGNSS corrections are claimed to be provided by A-GNSS.

In the indoor or urban-canyon conditions, GNSS methods couldn't be utilized due to the lack of enough satellites visible. RFPM utilizing LTE combined with WLAN signals measurements provides the best accuracy. However, it is much worse in terms of accuracy then the GNSS methods, and sure not good enough to support mm-waves transmission optimization. Fingerprinting is also highly implementation dependent (various matching algorithms, database design and measurement grid). It should be taken into account that measured values chosen for RFPM, should be stable in time (not dependent on the network optimization protocols). However, OTDoA method does not seem to be accurate enough for local REM. With the introduction of larger bandwidths and mm-waves it may be much improved and serve as indoor localization method. Also E-CID could be improved with the use of M-MIMO providing accurate AoA measurements.

5   Conclusions

The GNSS-based methods seem to be best solutions for the REM and should be used whenever is possible. Highly accurate RTK, highly energy-consuming though, could be helpful for REM data acquisition to provide extremely accurate localization tags for measurements. Due to lower accuracy of cellular positioning techniques, their utilization is reasonable under urban canyons or indoor conditions (where not enough satellites are visible), however it is expected to be improved in 5G networks with the use of larger bandwidths, M-MIMO and mm-waves. As the REM contains a lot of measurements-data related to position, an interesting field of future research is to check if REM could serve as fingerprints database, or even if it could perform fingerprinting by it's own to determine UE position. Another direction of future research is to obtain trustful position error models for mentioned localization methods to be used in REM operation modeling.

Acknowledgement. The presented work has been funded by the Polish Ministry of Science and Higher Education within the status activity task "Cognitive and sustainable communication systems" in 2018-19.

References

1. 3GPP: Radio Frequency (RF) pattern matching location method in LTE. Technical Report (TR) 36.809, 3rd Generation Partnership Project (3GPP), version 12.0.0, September 2013
2. 3GPP: Evolved Universal Terrestrial Radio Access Network (E-UTRAN); Stage 2 functional specification of User Equipment (UE) positioning in E-UTRAN. Technical Specification (TS) 36.305, 3rd Generation Partnership Project (3GPP), version 14.1.0, March 2017

3. Ahmadi, S.: Chapter 15 - Positioning and multimedia broadcast/multicast services. In: Ahmadi, S. (ed.) LTE-Advanced, pp. 1069–1105. Academic Press (2014). https://doi.org/10.1016/B978-0-12-405162-1.00015-0

4. Beek, J.V.D., et al.: How a layered rem architecture brings cognition to today's mobile networks. IEEE Wirel. Commun. **19**(4), 17–24 (2012). https://doi.org/10.1109/MWC.2012.6272419

5. COST Action CA15104, IRACON, Pedersen, T., Fleury, B.: Whitepaper on new localization methods for 5G wireless systems and the internet-of-things (2018)

6. Dardari, D., Conti, A., Ferner, U., Giorgetti, A., Win, M.Z.: Ranging with ultra-wide bandwidth signals in multipath environments. Proc. IEEE **97**(2), 404–426 (2009). https://doi.org/10.1109/JPROC.2008.2008846

7. Kliks, A., Musznicki, B., Kowalik, K., Kryszkiewicz, P.: Perspectives for resource sharing in 5G networks. Telecommun. Syst. **68**(4), 605–619 (2018). https://doi.org/10.1007/s11235-017-0411-3

8. Kryszkiewicz, P., Kliks, A., Kułacz, Ł., Bogucka, H., Koudouridis, G.P., Dryjański, M.: Context-based spectrum sharing in 5G wireless networks based on radio environment maps. Wirel. Commun. Mob. Comput. **2018**, 1–5 (2018)

9. Liu, H., Darabi, H., Banerjee, P., Liu, J.: Survey of wireless indoor positioning techniques and systems. IEEE Transact. Syst. Man Cybern. Part C (Appl. Rev.) **37**(6), 1067–1080 (2007). https://doi.org/10.1109/TSMCC.2007.905750

10. Misra, P., Enge, P.: Global Positioning System: Signals, Measurements, and Performance. Ganga-Jamuna Press, Kathmandu (2006)

11. Moore, S.K.: Superaccurate gps coming to smartphones in 2018 (2017). https://spectrum.ieee.org/semiconductors/design/superaccurate-gps-coming-to-smartphones-in-2018

12. Osseiran, A., et al.: Scenarios for 5G mobile and wireless communications: the vision of the metis project. IEEE Commun. Mag. **52**(5), 26–35 (2014). https://doi.org/10.1109/MCOM.2014.6815890

13. Pecoraro, G., Domenico, S.D., Cianca, E., Sanctis, M.D.: LTE signal fingerprinting localization based on CSI. In: 2017 IEEE 13th International Conference on Wireless and Mobile Computing, Networking and Communications (WiMob), pp. 1–8, October 2017. https://doi.org/10.1109/WiMOB.2017.8115803

14. del Peral-Rosado, J.A., López-Salcedo, J.A., Kim, S., Seco-Granados, G.: Feasibility study of 5G-based localization for assisted driving. In: 2016 International Conference on Localization and GNSS (ICL-GNSS), pp. 1–6, June 2016. https://doi.org/10.1109/ICL-GNSS.2016.7533837

15. del Peral-Rosado, J.A., Raulefs, R., López-Salcedo, J.A., Seco-Granados, G.: Survey of cellular mobile radio localization methods: From 1G to 5G. IEEE Commun. Surv. Tutor. **20**(2), 1124–1148 (2018). https://doi.org/10.1109/COMST.2017.2785181

16. Perez-Romero, J., et al.: On the use of radio environment maps for interference management in heterogeneous networks. IEEE Commun. Mag. **53**(8), 184–191 (2015). https://doi.org/10.1109/MCOM.2015.7180526

17. Pesyna, K.M.: Advanced techniques for centimeter-accurate GNSS positioning on low-cost mobile platforms. Ph.D. thesis, The University of Texas at Austin (2015)

18. Sybis, M., Kryszkiewicz, P., Sroka, P.: On the context-aware, dynamic spectrum access for robust intraplatoon communications. Mob. Inf. Syst. **2018**, 1–2 (2018)

19. Turkka, J., Hiltunen, T., Mondal, R.U., Ristaniemi, T.: Performance evaluation of LTE radio fingerprinting using field measurements. In: 2015 International Symposium on Wireless Communication Systems (ISWCS), pp. 466–470, August 2015

20. Wymeersch, H., Seco-Granados, G., Destino, G., Dardari, D., Tufvesson, F.: 5G
    mmwave positioning for vehicular networks. IEEE Wirel. Commun. **24**, 80–86
    (2017)

A Hybrid Chain Based Incentive Mechanism for Resource Leasing in NDN

Xin Wei[1] [ID], Zhuo Yu[2], Shaoyong Guo[1]([✉]), Jing Shen[3], Feng Qi[1], and Xuesong Qiu[1]

[1] State Key Laboratory of Networking and Switching Technology of Beijing University of Posts and Telecommunications, Beijing 100876, China
vaisy@bupt.edu.cn
[2] Beijing China-Power Information Technology Co. Ltd, Beijing 100192, China
[3] State Grid Henan Electric Power Company Information and Telecommunication Company, Zhengzhou 450000, China

Abstract. Since the main feature of Named Data Network (NDN) is in-net caching, it is crucial to motivate users to offer resource such as bandwidth and storage. However, few research works on incentive mechanism design for NDN. This paper proposes a market for NDN to lease bandwidth and storage from Access Points (APs). Since blockchain can supply a traceable and credible environment while public chain has long latency and low throughput, the paper combines permissioned chain with public chain, constructs a hybrid chain based environment without hurting its truthfulness. Furthermore, the paper formulates the market as a reverse auction running by a Content Provider (CP) who aims to serve more users for profit by leasing resource from APs, and investigates incentive mechanism for motivating APs. Especially, the paper designs an optimal mechanism, which could overcome defects of traditional mechanism, get the most profit for CP with guaranteeing interest of AP. Evaluation results compare effectiveness of mechanism proposed with traditional incentive mechanism, and prove that the mechanism we designed could get better results.

Keywords: Blockchain · Incentive mechanism · Named Data Network · Reverse auction

1 Introduction

Named Data Networking (NDN) is a promising framework fetching contents by name instead of IP address, which liberates content from host location, then users could fetch contents by cache instead of origin server. The design could ease storage and access pressure of Content Producer (CP), help users to fetch contents faster, alleviate the transmission pressure to network. Up to now, most of research on NDN focus on fetch content efficiently by cache and routing strategy design, while how can content be accessible is rarely researched. According to [1], the most potential scenario for deploying NDN would be that CPs lease resource from access network. CP needs to extend its users by providing access to its content and offloading its content to network, while it is not obligated to construct an access network to support this. Therefore,

© ICST Institute for Computer Sciences, Social Informatics and Telecommunications Engineering 2019
Published by Springer Nature Switzerland AG 2019. All Rights Reserved
A. Kliks et al. (Eds.): CrownCom 2019, LNICST 291, pp. 247–261, 2019.
https://doi.org/10.1007/978-3-030-25748-4_19

motivating nodes in network to contribute their own available unexploited resource for access is crucial. The paper introduces incentive mechanism into NDN networks for resource leasing.

To motivate nodes in a truthful way, the paper adopts blockchain to construct a marketplace for resource leasing. Blockchain is a technology which could record transaction in a transparent, tamper-resistant, secure way. Lots of works [2–4] have been done on blockchain based name resolution for NDN, which help users to retrieve contents and verify them in a credible way. On base of these works, the paper takes blockchain to deploy incentive mechanism for resource leasing. To improve efficiency and throughput of system, the paper combines public chain and permissioned chain to construct market.

Since monetary mechanism is the most flexible way to incentive users, it is widely adopted. Auction policy provides solution for CP to select APs and remunerate them. However, traditional auction policy used in resource leasing always ignore the requirement of Budget Balance (BB). Considering BB and Individual Rational (IR), the paper proposes an optimal mechanism for CP to get the largest profit with guaranteeing interest of AP.

Contributions of the paper can be summarized as follows.

(1) The paper designs a blockchain based marketplace for resource leasing in NDN, which supplies a credible environment for NDN. Contents can be fetched from AP without often using backhaul bandwidth, and users can verify contents easily.
(2) The paper divides the process of motivation into AP selection and remuneration. After proving the AP selection problem is NP-hard and designing greedy algorithm to work out it, the paper investigates several incentive mechanisms to remunerate APs, and designs an optimal mechanism.
(3) The paper provides performance comparisons and analyzes impact of different factors such as number of attending APs, files distribution, AP distribution, and profit per bandwidth.

This paper is structured as follows: Sect. 2 briefly introduces access NDN, incentive mechanism and blockchain. Section 3 describes the network architecture and construct a model of allocation problem for remuneration, whereas Sect. 4 proposes a mechanism to motivate APs. Section 5 describes experiments and analyzes of the proposed mechanisms. Finally, concluding remarks are presented in Sect. 6.

2 Preliminaries

2.1 Named Data Network

NDN is a promising architecture which makes caching be integrated into network layer. To motivate nodes to contribute their own resources, game theory is widely adopted. In [5], interactions between CP and Internet Service Providers (ISPs) are modeled as a Stackerlberg game, CP decides price for contents at first, Access ISPs act as followers to make caching decisions. In [6], the problem is modeled as a reverse auction, CPs claim bid for content while Access ISP choose content to cache. In researches above, motivation of Access ISP is profiting from selling contents to users.

There may be another economic relationship between CPs and nodes maintaining access resources. Different with CP sells content to Access ISP, CP can also lease access resource from APs. In [7], Access Points claim bid for access resource including storage and bandwidth while CP choose which AP to serve users. However, incentive mechanism it adopted may cause CP out of budget balance under extreme conditions. The paper follows the scenario in [7] and designs an optimal mechanism for resource leasing in NDN.

Besides resource leasing, content verification is another crucial problem for NDN. As users must verify content regardless of where it comes from, they must have public key of CP before they request it. Reference [8] pointed that NDN need a rootless scheme to supply name resolution. As a tamper-resistant distributed ledger, blockchain become a solution to this. Reference [4] designs a blockchain based identifier management system for NDN.

Since CPs need to construct blockchain for name resolution, and motivation should be executed truthfully, the paper designs incentive mechanism for resource leasing and deploys incentive mechanism into the blockchain.

2.2 Incentive Mechanism

Incentive mechanism is a field in game theory which aims to encourage players to reveal true information. Up to now, incentive mechanisms are always classified as three types: reputation based mechanisms, Tit-for-tat, and monetary mechanisms [9]. Reputation based mechanisms aim to identify selfish nodes and punish them, and Tit-for-tat takes services as incentives to discourage free-ridding behaviors. These two types of mechanisms are always limited to apply to long-term users and lack formal specification. Monetary mechanisms encourage players by designing payoff structure, which is more flexible. For these reasons, monetary mechanism is widely adopted in resource leasing especially in spectrum leasing and data offloading.

Common monetary mechanisms including auction mechanisms and pricing mechanisms are based on the Stackerlberg game. Reference [5, 10, 11] design pricing strategy with Stackerlberg game to motivate access network to distribute contents. Reference [7, 12] use Vickrey–Clarke–Groves (VCG) auction policy to motivate APs to contribute their access resources. Reference [13] proposes a VCG based multi object auction for access resource leasing. However, VCG auction policy has its defect: it cannot ensure BB. While Arrow-d' Aspremont-Gerard-Varet (AGV) auction policy could ensure BB at the cost of IR. Reference [14] proposes an AGV auction policy to motivate nodes for content delivery. According to [15], an auction policy which could meet IR and minimize the cost of acquirer is optimal. In fact, some optimal auction policies have been designed for task assignment in crowdsensing [16, 17], while rarely used in resource leasing.

The paper creates a reverse auction model for CP and AP, designs an optimal mechanism for it, and proves it is Incentive Compatible (IC), IR and BB. Finally, the paper provides performance comparisons of these proposed mechanisms.

2.3 Blockchain Based Resource Management

Blockchain is a technology meant to store, read and validate transactions in a distributed data-base system [18]. In blockchain, a group of anonymous strangers can work together to share and secure a perpetually growing set of data without anyone having to trust anyone else [19]. Smart contracts are self-executing scripts that reside on the blockchain, which allow for proper, distributed, heavily automated workflows operated in blockchain [20].

Some attempts on resource allocation with blockchain have been tried in these years. Public blockchain for Internet source transactions is proposed in [21], which could record Internet core transactions like IP address assignments, domain name assignments and AS-Path advertisements, thus allowing Internet peers to verify core Interment resource usage and assignment authorizations.

Specifically, blockchain also be used in spectrum access in [22], which provided a verification and validation scheme to ensure the security of the network by introducing public chain as a CA without central node. In this way, the robustness and security of the system can be strengthened, however, there are several problems:

1. The prospect of public chain is unpredictable. Public chains rely on a whole community of developers contributes to the open-source code, the process lacks formal governance [23]. To motivate other nodes to record the transaction into blockchain, who invoke smart contract need to pay a few fee. As the fee is also defined by the public chain, value trend of digital currency is unpredictable too.
2. As public blockchains have to coordinate the resources of multiple unaffiliated participants, they are slower and less private than traditional databases [23].

To solve reliability and efficiency problem of public chain, there are lots of solution been proposed, such as lightning network [24], and hybrid chain which combining permissioned chain [25]. Lightning network solve it by complete transactions off the public chain, lots of operation can be executed without consensus. While permissioned chain reduces the scale of consensus by limiting the node who can attend the blockchain.

The paper adopts permissioned chain to build a marketplace for resource leasing. To execute leasing truthfully, CP could add smart contract to its origin system, while all bidding process be recorded in blockchain. Besides, the paper presents several possible incentive mechanisms for the scenario.

3   System Model and Problem Formulation

As auction should be operated truthfully, while there is no need for all operation to be credible in global, the paper introduces permissioned chain to make up the deficiency of public chain in performance. After creating a credible market, motivation of CP and APs to attend the market is discussed. Then, we construct the auction as a problem model.

3.1   Hybrid Chain Based Auction Market

The whole architecture can be demonstrated as Fig. 1. Within CP, authorization management module distribute certificates to nodes inside CP, consensus module package transactions into block, these blocks construct a permissioned chain. And, authorization management module set rules into smart contract and publish it into permissioned chain. By exposing itself in public chain, CP can be accessed and verified by AP and Mobile Clients (MCs). APs attend bidding by invoking smart contract in permissioned chain, selected APs would become NDN router and help MCs to fetch contents from CP.

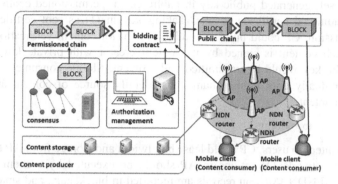

Fig. 1. System architecture

The process for building a credible marketplace can be divided into three steps as following. Firstly, ensure the credibility of CP, then AP and MC could verify information published by CP. Secondly, construct NDN network, MC could get data via AP credibly. Thirdly, build permissioned chain and develop smart contract for APs.

(1) Ensure the credibility of CP

As central authority may introduce risks and bottlenecks into the whole system, the paper adopts a decentral way to ensure the credibility for CP.

By generating a pair of asymmetric keys according to rules of public chain, CP could get an identity. Taking its own critical message as a transaction, CP need to signature the transaction with private key and broadcast it to the whole public chain. Then the transaction will be packaged into a block. Constructed by blocks, public chain is shared among each participant by consensus. As records in blockchain are tamper-resistant, message published by CP is credible. In other word, everyone can confirm the message is published by the CP. They can communicate with the CP via the access address claimed in the message, and verify contents published by CP with its public key, which is the source of transaction including the message.

Taking location of the message in public chain as name, CP can get a unique name, and others can find its access address, public key according it in a credible way. For instance, if the message can be found in 108th record in 10th block, it can be named as 10.108. As NDN is name based, creating credible name mapping service is the basis of NDN. After this, CP could supply credible service.

(2) Ensure content credible for NDN

At the beginning, it is necessary to make content credible. Users need retrieve desired content by name and authenticate the result regardless of where it comes from, which means the network should supply a framework for content verification.

Since there may be lots of content creators inside CP, the paper adopts permissioned chain for their cooperation. Credibility in permissioned chain relies on relationships in reality instead of proof algorithm, so consensus among permissioned chain is more efficient.

Not like self-generated public key in public chain, permissioned chain is compatible with traditional certificate system. Taking key in (1) as root certificate for the whole permission, CP could build a credible system for content verification.

Since each content is signed by its content creator with private key, certificate include public key would be transmitted with content. Mobile consumer could verify certificate gradually until public chain, thus identify source of content and confirm integrity of content.

(3) Attract AP to bidding

To deliver content to users, CP could lease bandwidth and storage from AP to construct NDN access network. Operations in AP should be executed by a client combining blockchain and NDN. Auction records are recorded in blockchain, and smart contracts supply faces for APs to attend bidding.

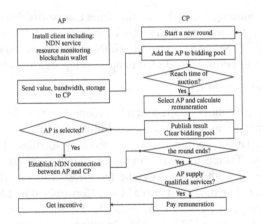

Fig. 2. Bidding process

The core logic should be like Fig. 2. At the beginning, AP installs a client which could monitor resource occupation, convert NDN packet, generate blockchain based identity. APs submit information to CP to attend current bidding, supply service when it is selected, if it supplies services as it promised, it will get incentive as bidding result. When an auction come to AP selection, next round would start for coming APs, and when APs in current round end services, new round would come to AP selection.

By the design, content can be verified in a credible way, bidding and remuneration can be executed automatically and traceably, a market for resource leasing in NDN is formed. What drives different roles to attend the market is discussed in next part.

3.2 Motivation of Different Roles

Motivation of different roles including CP and AP to attend the market is discussed in this part.

Firstly, we discuss motivation for CP. According to model defined in Sect. 3.1, the major business of CP is distributing contents. More accurately, CP profit from MCs, and the more users it serves, the more income it can earn. Hence, CP want to supply more connections to cover more MCs. However, building access networks is beyond service scope of CP, there is no need for CP to complete that. Hence it is necessary for CP to lease bandwidth to distribute contents. In addition, by introducing in-network caching ability into network, MCs may get contents from APs instead of CP, access pressure to CP can be significantly reduced. In summary, CP has ample motivation to construct a market to get resources from AP.

Secondly, we discuss the motivation of AP. To attend the market, AP submits to CP the bid consists of $[\hat{v}_j, \hat{b}_j, \hat{s}_j]$, v_j represents the cost, b_j represents the backhaul bandwidth that AP j obtains, while s_j represent the storage that AP j can share. Let p_j be price paid by the CP to lease AP j. The utility function can be expressed as (1):

$$u_j = \begin{cases} p_j - v_j, & \text{if AP } j \text{ is selected} \\ 0, & \text{otherwise} \end{cases} \tag{1}$$

According to IR, the utility of each player is always non-negative. Only if u_j is non-negative, AP would attend to bid. It is noteworthy that the bidding of AP may not be truth. When CP has collected the information of APs, it will decide to lease which AP and how much should pay to AP. It is obvious the process can be divided into two parts: AP selection and p_j computation.

3.3 Problem Formulation

This part would construct problem model for CP to design appropriate mechanism for remunerating. The notation used in this paper is summarized in Table 1.

Table 1. Symbol of parameters

Symbol	Quantity
P	Average profit per unit of bandwidth
C	Cache miss lost per unit of bandwidth
A	Set of AP
M_j	Set of MCs covered by AP j
h_j	Hit ratio of AP j
t_i	Traffic demand of MC i
$r_{i,j}$	Maximum WiFi access rate for AP j and MC i
$d_{i,j}$	Distance between AP j and MC i
$x_{i,j}$	Binary variable that indicates whether MC i is connected to AP j
y_j	Binary variable that indicates whether AP j is selected
F	The objective function of the problem

Since all variables are integers, this problem is an Integer Linear Programming (ILP) problem. And, the problem can be formulated as follows:

Maximize:

$$P \sum_{i \in M} \sum_{j \in A} x_{i,j} d_i - \sum_{j \in A} y_j v_j - \sum_{i \in M} \sum_{j \in A} x_{i,j} d_i (1 - h_j) C \qquad (2)$$

Subject to:

$$\sum_{i \in M} \frac{x_{i,j} t_i}{r_{i,j}} \leq 1, \forall j \in A \qquad (3)$$

$$\sum_{i \in M} x_{i,j} t_i (1 - h_j) \leq b_j, \forall i \in M_j, \forall j \in A \qquad (4)$$

$$x_{i,j} \leq y_j, \forall i \in M, \forall j \in A \qquad (5)$$

$$x_{i,j} \in \{0, 1\}, \forall i \in M, \forall j \in A \qquad (6)$$

$$y_j \in \{0, 1\}, \forall j \in A \qquad (7)$$

The objective function (2) maximizes the total revenue of the system, which is given by (1) the traffic-proportional profit the CP is going to obtain for serving the MCs' demand: $P \sum_{i \in M} \sum_{j \in A} x_{i,j} d_i$ (2) the cost for procuring the selected APs: $\sum_{j \in A} y_j v_j$, (3) the cost for cache misses: $\sum_{i \in M} \sum_{j \in A} x_{i,j} d_i (1 - h_j) C$, while h_j is proportional to s_j. In the paper, (4) represents social welfare.

Constraints (3) and (4) limit the number of MCs assigned to each AP, give radio access network and backhaul Internet connection a bounded capacity. Constraint (3) imposes that the total demand served by an AP does not exceed the capacity of radio access network. Constraint (4) considers the fact that backhaul Internet connection serves only the aggregate demand that generates a cache miss. Constraint (5) ensures that MCs can associate only to the APs selected by mechanism. Finally, the sets of constraints (3) and (7) express the integrality conditions on decision variables.

Taking the selection of AP as a knapsack problem, CP would choose some AP to lease access resources. Similarly, Choosing MCs to serve is another knapsack problem for AP. Since knapsack problem is NP-hard, AP selection is NP-hard. The paper chooses greedy algorithm to solve it as Algorithm1.

Algorithm1	The greedy algorithm to solve the model
Input: M,A,b,s,d,r,h,R,C,P	
Output : y,p,x	
1	$L <= \text{sort}(\, j \in A \,, \dfrac{b_i}{(1-h_i)}(P - C(1-h_i)) - v_i \,,\text{descend})$
2	**while** demand of CP has not been met
3	$j=\text{next}(L)$, $y_j = 1$; //next AP is selected
4	$V_j <= \text{sort}(\, i \in M_j, d_{i,j} \,,\text{ascend})$
5	**while** AP j satisfy constraints (3-4)
6	$i=\text{next}(V_j)$p; $x_{i,j}=1$;
7	**if** constraints (3-4) are not satisfied
8	$x_{k,j} = 0$;
9	**end**
10	**end**
11	**end**

In the algorithm, CP always select the AP which helps it profit most from all APs until all its demand is met, the demand maybe offloading data enough or covering users enough. To cover users, CP aim to get enough bandwidth, while offloading data means get enough storage. When an AP is selected, it would serve the nearest MC until its access ability is run out according to constraints (3-4).

4   Proposed Mechanisms

In VCG based mechanism, payment for an AP is the extra profit to system because its participation, while may cause CP out of BB. In AGV based mechanism, each AP contributes a participation fee for paying other APs. In this way, CP need not supply remuneration, CP could gain most by this way while it cannot satisfy IR for APs. Reference [26] pointed that, if a mechanism is Bayesian incentive compatible and individually rational, then the mechanism is optimal. According to [27], an optimal leasing mechanism should satisfy following:

1: The offered surplus is of the form:

$$p_i(\hat{v}_i, \hat{q}_i) = p_i(\bar{v}_i, \hat{q}_i) + \int_{\hat{v}_i}^{\bar{v}} X_i(v, \hat{q}_i)dv \tag{8}$$

2: Expected allocation $X_i(v_i, \hat{q}_i)$ is nonincreasing in the cost parameter $\hat{v}_i$ suppliers.

In (8), p_i refers to payment for AP i, $\hat{v}_i$ is bid of AP i, while $\hat{q}_i$ is quality of resource supplied by AP i. $\bar{v}_i$ is the mean bid for selected $\hat{q}_i$. (8) pointed that the payment for AP i should be the highest bid of selected AP which have the same quality with AP i, and incentive for its low price. Instead of taking quantity as $\hat{q}_i$ in [26], the paper take maximum profit may get from AP i as $\hat{q}_i$, and sort AP with (10).

$$q_i = \frac{b_i}{(1 - h_i)}(P - C(1 - h_i)) - v_i \tag{9}$$

$$H_i(v_i, q_i) = v_i + \frac{F_i(\frac{v_i}{q_i})}{f_i(\frac{v_i}{q_i})} \tag{10}$$

Aim to meet requirements as above, we set X as follows:

$$X_i(\hat{v}_i, \hat{q}_i) = \int_{\underline{v}}^{\hat{v}_i} x_i(v, \hat{q}_i)dv \tag{11}$$

$x(\hat{v}, \hat{q})$ is a binary value represents that whether select AP when it bidding as value is $\hat{v}$ and quality is $\hat{q}$. It is obvious that (11) can satisfies the above properties and hence is optimal. Pseudocode is shown in the following:

Algorithm2	payment function for optimal mechanism
Input: M,A,b,s,d,r,h,R,C,P	
Output : y,p,x	
1	$(y, x) <=$Solve the model while order by H
2	Calculate distribution of $x(\hat{v}, \hat{q})$ and get $X(\hat{v}, \hat{q})$
3	**foreach** $j \in A$ **do**
	$p_j = \hat{v}_j y_j + \int_{\hat{v}_i}^{\bar{v}} X_i(v, \hat{q}_i)dv$
4	**end**

The algorithm proceeds in three steps. Step 1 computes the maximum revenue allocation. Different with VCG or AGV, when optimal mechanism solve model with Algorithm 1, it needs to sort APs with H. Step 2 calculates $X(\hat{v}, \hat{q})$ for step 3, while step3 computes incentives for APs.

Theorem 1: The payment rule satisfies trustfulness property (IC).

Proof 1: Since AP always wants to get more pay with less storage and less bandwidth. If AP j wants to bid $(\hat{v}_j, \hat{q}_j)$ untruthfully, there would be $\hat{v}_j > v_j, \hat{q}_j < q_j$.

$$u_j(\hat{v}_j, \hat{q}_j) = p_j(\hat{v}_j, \hat{q}_j) + (\hat{v}_j - v_j)X_j(\bar{v}, q_j)$$

$$= p_j(\bar{v}_j, \hat{q}_j) + \int_{\hat{v}_j}^{\bar{v}} X_j(y, \hat{q}_j)dy + (\hat{v}_j - v_j)X_j(\bar{v}, q_j)$$

$$= p_j(\bar{v}_j, \hat{q}_j) + \int_{\hat{v}_j}^{v_j} X_j(y, \hat{q}_j)dy$$

$$+ \int_{v_j}^{\bar{v}} X_j(y, \hat{q}_j)dy + (\hat{v}_j - v_j)X_j(\bar{v}, q_j) \tag{12}$$

$$\leq p_j(\bar{v}_j, \hat{q}_j) + \int_{v_j}^{\bar{v}} X_j(y, \hat{q}_j)dy$$

$$\leq p_j(\bar{v}_j, q_j) + \int_{v_j}^{\bar{v}} X_j(y, \hat{q}_j)dy$$

$$= u_j(v_j, \hat{q}_j)$$

From (12), it can be seen that AP would get most when it bid truthfully.

Theorem 2: The payment rule maximizes revenue of CP.

Proof 2: As $p(v, q)$ is increasing in q and non-increasing in v, we can ensure CP pay the lowest price to AP. Since the profit get from MCs is stable when it selects APs from certain set, CP maximizes revenue in optimal mechanism.

Theorem 3: The payment rule satisfies IR property.

Proof 3: According to (8) (11), $u(v, q)$ is non-negative, thus mobile users always could get profit.

5   Evaluation Results

Reference [27] has evaluated that time consumption for reaching consensus in permissioned chain would be almost 3 ms. Reference [28] evaluated throughput and verification speed. Without considering performance of blockchain, the section calculates social welfare gains by each proposed algorithm, and discusses impact of different factors. To be concise, this part use VCG, AGV, OPT to represent VCG, AGV based mechanism and optimal mechanism respectively. To make results more accurate, each scenario has been performed for 20 times and adopts data in the narrow 95% confidence intervals.

According to [7], bids of APs are uniformly selected in the [7, 15], the traffic-proportional cache miss cost is set to 5 Mb/s, while CP profit 15 from per Mb/s. Considering simulation settings in [7] [29], we choose the backhaul Internet bandwidth uniformly in the set $\{6; 8; 10; 15; 20\}$ Mb/s, whereas the size of the leased caching storage is uniformly selected in the range [10, 100] GB. Since the paper does not consider network selection, geographical conflictions between APs is neglected. The paper assumes the radio access network is composed of 802.11n wireless APs, and adopts datasheet of Atheros AR9342 chipset [30].

According to [31], the density of mobile user population in dense city is 12,000 mobile users per km². The paper assume MCs is uniformly distributed around AP as the density. Demands of every MC are generated in the range [0.5, 3] Mb/s.

Considering two different scenarios: getting access bandwidth or offloading data, we perform two groups of simulation. In the first group, CP aims to lease 600 Mbps bandwidth for access, while lease 3000 GB for data offloading in second. In the following simulation, CP has 10000 files being requested with zipf parameter is 0.8.

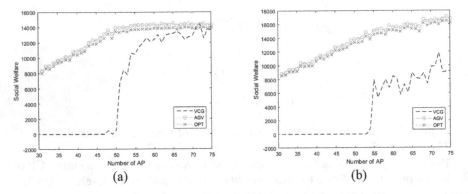

Fig. 3. Social welfare (a) getting bandwidth (b) offloading data

From Fig. 3, it can be observed that:

1. In each case, AGV based mechanism can get the best revenue, because APs pay others, while CP need pay APs in OPT and VCG. Since OPT mechanism pay APs as little as possible, social welfare of OPT is close to AGV. VCG gains least, which is consistent to our expectation.
2. The two scenarios demonstrate a similar trend. When attending AP is less, VCG may loss budget balance because it has to choose all AP with no choice, while profit of AGV and OPT are growing linearly. When CP has enough choices, VCG profit rises dramatically and tend to be stable quickly while AGV and OPT profit turns to be slow gradually.
3. Comparing (a) and (b), the fluctuation of VCG in (b) is more extreme. Since bandwidth affects social welfare by network traffic while storage affect it by hit ratio, effect of bandwidth would be more directly. To AGV and OPT, when bandwidth is stable, social welfare would get stable.

To find impact of file distribution on performance of incentive mechanism. We first adjust the parameter of Zipf. According to [32], the zipf parameter α always be approximately around 0.5–0.9. The larger α is, the more concentrated the requests for files are, and hit ratio would be higher. Besides, the less files CP has, the higher hit ratio would be reached by each AP, so we adjust number of files (Fig. 4).

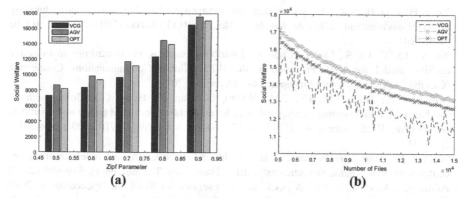

Fig. 4. Social welfare (a) Zipf parameter (b) number of files

As shown above, the more hit ratio would be reached by each AP, the more profit CP could gain. The phenomenon is consisted with our common sense, as the more hit ratio means less cache miss cost and less backhaul bandwidth occupied, the more users can get access to CP efficiently.

6 Conclusion

To motivate users to contribute their available resources to NDN, the paper proposes a hybrid chain based auction market for resource leasing. By combining public chain with permissioned chain, a transparent and credible marketplace is constructed. To overcome defects in traditional auction mechanism, the paper designs an optimal mechanism, which could obtain the most profit for CP with guaranteeing interests of AP. Proof and simulation results show the mechanism we proposed is efficient.

Acknowledgement. This work was supported in part by National Natural Science Foundation of China (Grant: 61702048) and industrial Internet platform standard management service public support platform.

References

1. Agyapong, P., Sirbu, M.: Economic incentives in information - centric networking: implications for protocol design and public policy. IEEE Commun. Mag. **50**(12), 18–26 (2012)
2. Fotiou, N., Polyzos, G.C.: Decentralized name-based security for content distribution using blockchains. In: Proceedings of the IEEE Conference on Computer Communications Workshops (INFOCOM WKSHPS), San Francisco, CA, USA, pp. 415–420 (2016)
3. Jin, T., Zhang, X., Liu, Y., Kai, L.: BlockNDN: a bitcoin blockchain decentralized system over named data networking. In: Presented at the 2017 Ninth International Conference on Ubiquitous & Future Networks (ICUFN), Milan, Italy, 4–7 July 2017

4. Yang, H., Cha, H., Song, Y.: Secure identifier management based on blockchain technology in NDN environment. IEEE Access. https://doi.org/10.1109/access.2018.2885037. (to be published)
5. Xu, Y., Li, Y., Ci, S., Lin, T., Chen, F.: Distributed caching via rewarding: an incentive caching model for ICN. Presented at the IEEE Global Communications Conference (GLOBECOM), Singapore, Singapore, 4–8 December 2017
6. Ndikuma, A., Tran, N.H., Ho, T.M., Niyato, D., Han, Z., Hong, C.S.: Joint incentive mechanism for paid content caching and price based cache replacement policy in named data networking. IEEE Access **6**, 33702–33717 (2018). https://doi.org/10.1109/access.2018.2848231
7. Mangili, M., Martignon, F., Paris, S.: Bandwidth and cache leasing in wireless information-centric networks: a game-theoretic study. IEEE Trans. Veh. Technol. **66**(1), 679–695 (2017)
8. Afanasyev, A., et al.: NDNS: A DNS-like name service for NDN. In: Proceedings of 2017 26th International Conference on Computer Communication and Networks (ICCCN), Vancouver, Canada, pp. 1–9 (2017)
9. Zhang, X., Bai, X., Liu, Q.: A research of vehicle ad hoc network incentive mechanism. Presented at the 2018 8th International Conference on Electronics Information and Emergency Communication (ICEIEC), Beijing, China, 15–17 June 2018
10. Wu, D., Yan, J., Wang, H., Wu, D., Wang, R.: Social attribute aware incentive mechanism for device-to-device video distribution. IEEE Trans. Multimedia **19**(8), 1908–1920 (2017)
11. Shang, B., Zhao, L., Chen, K.: Operator's economy of device-to-device offloading in underlaying cellular networks. IEEE Commun. Lett. **21**(4), 865–868 (2017)
12. Paris, S., Martignon, F., Filippini, I., Chen, L.: An efficient auction-based mechanism for mobile data offloading. IEEE Trans. Mob. Comput. **14**(8), 1573–1586 (2014)
13. Bousia, A., Kartsakli, E., Antonopoulos, A., Alonso, L., Verikoukis, C.: Multiobjective auction-based switching-off scheme in heterogeneous networks: to bid or not to bid? IEEE Trans. Veh. Technol. **65**(11), 9168–9180 (2016)
14. Deng, J., Zhang, R., Song, L., Han, Z., Jiao, B.: Truthful mechanisms for secure communication in wireless cooperative system. IEEE Trans. Wirel. Commun. **12**(9), 4236–4245 (2013)
15. Myerson, R.B.: Optimal auction design. In: Mathematics of Operations Research, vol. 6, no. 1, pp. 58–73, February 1981
16. Chatzopoulos, D., Gujar, S., Faltings, B., Hui, P.: Privacy preserving and cost optimal mobile crowdsensing using smart contracts on blockchain. In: 2018 IEEE 15th International Conference on Mobile Ad Hoc and Sensor Systems (MASS), Chengdu, China, 9–12 October 2018
17. Qin, H., Zhang, Y., Li, B.: Truthful mechanism for crowdsourcing task assignment. In: 2017 IEEE 10th International Conference on Cloud Computing (CLOUD), Honolulu, CA, USA, 25–30 June 2017
18. Bozic, N., Pujolle, G., Secci, S.: A tutorial on blockchain and applications to secure network control-planes. In: 2016 3rd Smart Cloud Networks & Systems (SCNS), Dubai, UAE, pp. 1–8 (2016)
19. Hari, A., Lakshman, T.V.: The internet blockchain: a distributed, tamper-resistant transaction framework for the internet. In: ACM Workshop on Hot Topics in Networks, pp. 204–210. ACM, Atlanta (2016)
20. Christidis, K., Devetsikiotis, M.: Blockchains and smart contracts for the internet of things. IEEE Access **4**(4), 2292–2303 (2016)
21. Ali, M., Nelson, J., Shea, R., Freedman, M.J.: Blockstack: a global naming and storage system secured by blockchains. In: 2016 USENIX Annual Technical Conference (USENIX ATC 16), Denver, CO, USA, pp. 181–194 (2016)

22. Kotobi, K., Bilen, S.G.: Secure blockchains for dynamic spectrum access: a decentralized database in moving cognitive radio networks enhances security and user access. IEEE Veh. Technol. Mag. **13**(1), 32–39 (2018)

23. Peck, M.E.: Blockchain world - do you need a blockchain? This chart will tell you if the technology can solve your problem. IEEE Spectr. **54**(10), 38–60 (2017)

24. The Bitcoin Lightning Network: Scalable off-chain instant payments. http://lightning. network/lightning-network-paper.pdf. Accessed Mar 2017

25. Wu, L., Meng, K., Xu, S., Li, S., Ding, M., Suo, Y.: Democratic centralism: a hybrid blockchain architecture and its applications in energy internet. In: IEEE International Conference on Energy Internet, pp. 176–181. IEEE, Beijing (2017)

26. Iyengar, G., Kumar, A.: Optimal procurement mechanisms for divisible goods with capacitated suppliers. Rev. Econ. Design **12**(2), 129–154 (2008)

27. Sukhwani, H., Martínez, J.M., Chang, X., Trivedi, K.S., Rindos, A.: Performance modeling of PBFT consensus process for permissioned blockchain network (Hyperledger Fabric). In: 2017 IEEE 36th Symposium on Reliable Distributed Systems. Philadelphia, HongKong, China, 26–29 September 2017

28. Du, M., Ma, X., Zhang, Z., Wang, X., Chen, Q.: A review on consensus algorithm of blockchain. In: 2017 IEEE International Conference on Systems, Man, and Cybernetics (SMC). Banff, AB, Canada, 5–8 October 2017

29. Poularakis, K., Iosifidis, G., Pefkianakis, I., Tassiulas, L., May, M.: Mobile Data offloading through caching in residential 802.11 wireless networks. IEEE Trans. Netw. Serv. Manag. **13**(1), 71–84 (2016)

30. Atheros AR9342 Data Sheet. https://docs.wixstatic.com/ugd/8e9475_182546e1cd744158 8622012e50974ab3.pdf. Accessed Dec 2019

31. Joe-Wong, C., Seny, S., Ha, S.: Offering supplementary wireless technologies: adoption behavior and offloading benefits. In: Proceedings of IEEE Conference on Computer Communications (INFOCOM), pp. 1061–1069, April 2013

32. Li, Y., Xie, H., Wen, Y., et al.: Coordinating in-network caching in content-centric networks: model and analysis. In: IEEE International Conference on Distributed Computing Systems, Philadelphia, PA, USA. IEEE, 8–11 July 2013

Reinforcement Learning-Based Radio Access Network Slicing for a 5G System with Support for Cellular V2X

Haider Daami R. Albonda$^{(\boxtimes)}$ and J. Pérez-Romero

Universitat Politècnica de Catalunya (UPC), Barcelona, Spain
{haider.albonda, jorperez}@tsc.upc.edu

Abstract. 5G mobile systems are expected to host a variety of services and applications such as enhanced mobile broadband (eMBB), massive machine-type communications (mMTC), and ultra-reliable low-latency communications (URLLC). Therefore, the major challenge in designing the 5G networks is how to support different types of users and applications with different quality-of-service requirements under a single physical network infrastructure. Recently, Radio Access Network (RAN) slicing has been introduced as a promising solution to address these challenges. In this direction, our paper investigates the RAN slicing problem when providing two generic services of 5G, namely eMBB and Cellular Vehicle-to-everything (V2X). We propose an efficient RAN slicing scheme based on offline reinforcement learning that allocates radio resources to different slices while accounting for their utility requirements and the dynamic changes in the traffic load in order to maximize efficiency of the resource utilization. A simulation-based analysis is presented to assess the performance of the proposed solution.

Keywords: Vehicle-to-everything (V2X) · Network slicing · Reinforcement learning

1 Introduction

The 5G system has the ambition to meet the widest range of service and applications in the history of mobile and wireless communications. Supported services are classified as (a) enhanced Mobile Broad Band (eMBB) that include services that require high bandwidth requirements, such as high definition (HD) video and Virtual Reality (VR); (b) Ultra Reliable and Low Latency Communications (URLLC) that aim to support low-latency transmissions of small payloads with extremely high reliability for a range of active terminals and (c) massive Machine Type Communications (mMTC) that aim to meet the demands of a large number of Internet Things (IoT). In responding to the very different requirements of these services and applications, the 5G system aims to provide a flexible platform to enable new business cases and models to integrate vertical industries, such as, automotive, manufacturing, and entertainment [1, 2].

In order to realize the above vision, network slicing is one of the key capabilities that will provide the required flexibility, as it allows multiple logical networks to be created on top of a common shared physical infrastructure. Each one of these logical

© ICST Institute for Computer Sciences, Social Informatics and Telecommunications Engineering 2019
Published by Springer Nature Switzerland AG 2019. All Rights Reserved
A. Kliks et al. (Eds.): CrownCom 2019, LNICST 291, pp. 262–276, 2019.
https://doi.org/10.1007/978-3-030-25748-4_20

networks is referred to as network slice and can be used to serve a particular service category (e.g. applications with different functional requirements) through the use of specific control plane (CP) and/or user plane (UP) functions [3]. Network slicing will help new services and new requirements to be quickly addressed, according to the needs of the industries [4].

Different works in the literature have investigated different aspects of network slicing, addressing both the slicing of the core network and the slicing of the Radio Access Network (RAN). For example, a low complexity heuristic algorithm and slicing for joint admission control in virtual wireless networks is proposed in [5]. In turn, the deployment of function decomposition and network slicing as a tool to improve the Evolved Packet Core (EPC) is presented in [6]. In [7], a model for orchestrating network slices based on the service requirements and available resources is introduced. They proposed a Markov decision process framework to formulate and determine the optimal policy that manages cross-slice admission control and resource allocation for the 5G networks.

Focusing on the RAN, some research studies have dealt with managing the split of the available radio resources among different slices to support different services (e.g. eMBB, mMTC, and URLLC) with main focus on the Packet Scheduling (PS) problem through different approaches. For example, a novel radio resource slicing framework for 5G networks with haptic communications is proposed in [8] based on virtualization of radio resources. The author adopted a reinforcement learning (RL) approach for dynamic radio resource slicing in a flexible way, while accounting for the utility requirements of different vertical applications. Similarly, a network slicing strategy based on an auction mechanism is introduced in [9] to decide the selling price of different types of network segments in order to maximize the network revenue and to optimally satisfy the resource requirements. A network slicing scheme based on game theory for managing the split of the available radio resources in a RAN among different slice types is proposed in [10] to maximize utility of radio resources. Similarly, an adaptive algorithm for virtual resource allocation based on Constrained Markov Decision Process is proposed in [11]. An online network slicing solution based on multi-armed bandit mathematical model to maximize network slicing multiplexing gains and achieving the accommodation of network slice requests in the system with an aggregated level of demands above the available capacity is proposed in [12].

Although the above works have proposed different approaches for RAN slicing, none of them has dealt with scenarios including slices for supporting Vehicle-to-Vehicle (V2V) communications, which constitute the focus of this paper. V2V communications are a particular type of the so-called Vehicle-to-everything (V2X) services in which vehicles can communicate between them through two operational modes, namely sidelink (i.e. direct communication between vehicles via PC5 interface) and cellular mode (i.e. communication between vehicles in two hops with the support of the base station via the Uu interface). These different options have impact on the resource consumption in the different links of the radio interface and thus they have to be taken into account when devising a RAN slicing strategy that distributes the radio resources among different slices if one of them supports V2X services. It is worth mentioning that, although the support for V2X sidelink communications was already standardized in 3GPP in the context of LTE [13], V2X sidelink is not yet included in the current release 15 of 5G New Radio (NR) specifications, but it is subject to study for future release 16 [14]. Based on all the above considerations, the key contributions of this paper can be summarized as follows. Firstly, the paper formulates the RAN slicing problem to support one slice for eMBB and another one for cellular V2X services on the same RAN infrastructure. The problem considers the split of radio resources assigned to each slice considering the characteristics of the different involved links, i.e. uplink and downlink for the eMBB services and uplink, downlink and sidelink for V2V. Secondly, the paper proposes a novel strategy based on offline Q-learning and softmax decision-making to determine the amount of radio resources assigned to each slice. The proposed solution is evaluated through extensive simulations to demonstrate its capability to perform efficient resource allocation in terms of network utilization, latency, data rate and congestion probability.

The rest of the paper is organized as follows. Section 2 presents the system model assumptions and the RAN slicing problem formulation. Section 3 presents the proposed RL approach for splitting the radio resources among the involved RAN slices. This approach is evaluated through simulations in Sect. 4 and compared against a reference scheme. Finally, conclusions and future work are summarized in Sect. 5.

2   System Model and Problem Formulation

2.1   System Model

The considered scenario assumes a cellular Next Generation Radio Access Network (NG-RAN) with a gNodeB (gNB) [15] composed by a single cell. A roadside unit (RSU) supporting V2X communications is attached to the gNB. A set of eMBB cellular users (CUs) numbered as $m = 1,..., M$ are distributed randomly around the gNB and a flow of several independent vehicles move along a straight highway, as illustrated in the right part of Fig. 1. The highway segment is divided into sub-segments (clusters) by sectioning the road into smaller zones according to the length of the road. It is assumed that each vehicle includes a User Equipment (UE) that enables communication with the UEs in the rest of vehicles in the same cluster. Clusters are numbered as $j = 1,..., C$, and the vehicles in the j-th cluster are numbered as $i = 1,..., V(j)$.

The vehicles in the highway are assumed to enter the cell coverage following a Poisson process with arrival rate λ_a. The association between clusters and vehicles is managed and maintained by the RSU based on different metrics (e.g. position, direction, speed and link quality) through a periodic exchange of status information.

Regarding the V2X services, this paper assumes V2V communication between vehicles. They can be performed either in cellular or in sidelink mode. In cellular mode each UE communicates with each other through the Uu interface in a two-hops transmission via the gNB while in sidelink mode, direct V2V communications can be established over the PC5 interface. We assume that, when sidelink transmissions are utilized, every member vehicle can multicast the V2V messages directly to multiple member vehicles of the same cluster $1 \leq i \leq V(j)$ using one-to-many technology. The decision on when to use cellular or sidelink mode is done based on [16].

To simultaneously support the eMBB and the V2X services, the network is logically divided into two network slices, namely RAN_slice_ID = 1 for V2X and RAN_slice_ID = 2 for eMBB. The whole cell bandwidth is organized in Resource Blocks (RBs) of bandwidth B. Let denote as N_{UL} the number of RBs in the UpLink (UL) and N_{DL} the number of RBs in the DownLink (DL). The RAN slicing process should distribute the UL and DL RBs among the two slices. For this purpose, let denote $\alpha_{s,UL}$ and $\alpha_{s,DL}$ as the fraction of UL and DL resources, respectively, for the RAN_slice_ID = s with $s = 1, 2$. Regarding sidelink communications, and since the support for sidelink has not been yet specified for 5G in current 3GPP release 15, this paper assumes the same approach as in current LTE-V2X system, in which the SL RBs are part of the total RBs of the UL. For this reason, the slice ratio $\alpha_{s,UL}$ is divided into two slice ratios, namely $\bar{a}_{s,UL}$, which corresponds to the fraction of UL RBs that are used for uplink transmissions, and, $\alpha_{s,SL}$, which corresponds to the fraction of UL RBs used to support sidelink transmissions.

Each vehicle is assumed to generate packets randomly with rate λ_v packets/s according to Poisson arrival model. The length of the messages is S_m. When the vehicles operate in sidelink mode, the messages are transmitted using the SL resources allocated to the slice. Instead, when the vehicles operate in cellular mode, the messages are transmitted using the UL and DL resources. The average number of required RBs from V2X users of RAN_slice_ID = 1 per Transmission Time Interval (TTI) in UL, DL and SL, denoted respectively as $\Gamma_{1,UL}$, $\Gamma_{1,DL}$, $\Gamma_{1,SL}$ can be estimated as follows:

$$\Gamma_{1,x} = \frac{\sum_{t=1}^{T}\sum_{j=1}^{C}\sum_{i=1}^{V(j)} m(j,i,t) \cdot S_m}{T \cdot SP_{eff,x} \cdot B \cdot F_d} \tag{1}$$

where x denotes the type of link, i.e. $x \in \{UL, DL, SL\}$, $m(j, i, t)$ is the number of transmitted messages by the vehicles of the j-th cluster in the t-th TTI and $SP_{eff,x}$ is the spectral efficiency in the x link, F_d is the TTI duration, which is 0.1 ms and T is the number of TTIs that defines the time window used to compute the average.

Regarding the eMBB service, the average number of required RBs for eMBB users of RAN_slice_ID = 2 in UL and DL in order to support a certain bit rate R_b is denoted as $\Gamma_{2,UL}$, $\Gamma_{2,DL}$, respectively, and can be statistically estimated as follows:

$$\Gamma_{2,x} = \frac{\sum_{t=1}^{T}\sum_{m=1}^{M} \rho_x(m,t)}{T} \tag{2}$$

where x denotes the type of link, and $\rho_x(m, t)$ is the number of required RBs by the m-th user in the link x and in the t-th TTI in order to get the required bit rate R_b. It is given by $\rho_x(m, t) = R_b/(SP_{eff,x} \cdot B)$. The values $\Gamma_{2,UL}$, $\Gamma_{2,DL}$ are computed within a time window T TTIs. Note also that $\Gamma_{2,SL} = 0$, since the eMBB slice does not generate sidelink traffic.

2.2   Problem Formulation for RAN Slicing

The focus of this paper is to determine the optimum slicing ratios $\alpha_{s,UL}$, $\alpha_{s,DL}$ in order to maximize the overall resource utilization under the constraints of satisfying the resource requirements for the users of the two considered slices.

The total utilization of UL resources U_{UL} is given by the aggregate of the required RBs in the UL and SL for each slice, provided that the aggregate of a given slice s does not exceed the total amount of resources allocated by the RAN slicing to this slice, i.e. $\alpha_{s,UL} \cdot N_{UL}$. Otherwise, the utilization of slice s will be limited to $\alpha_{s,UL} \cdot N_{UL}$ and the slice will experience outage. Correspondingly, the optimization problem for the uplink is defined as the maximization of the UL resource utilization subject to ensuring an outage probability lower than a maximum tolerable limit p_{out}. This is formally expressed as:

$$\max_{\alpha_{s,UL}} U_{UL} = \max_{\alpha_{s,UL}} \sum_s \min\left(\Gamma_{s,SL} + \Gamma_{s,UL}, \alpha_{s,UL} \cdot N_{UL}\right) \tag{3}$$

$$\text{s.t.} \quad \Pr\left[\Gamma_{s,SL} + \Gamma_{s,UL} \geq \alpha_{s,UL} \cdot N_{UL}\right] < p_{out} \quad s = 1,2 \tag{3a}$$

$$\sum_s \alpha_{s,UL} = 1 \tag{3b}$$

Following similar considerations, the optimization problem to maximize the resource utilization U_{DL} in the DL subject to ensuring a maximum outage probability is given by:

$$\max_{\alpha_{s,DL}} U_{DL} = \max_{\alpha_{s,DL}} \sum_s \min\left(\Gamma_{s,DL}, \alpha_{s,DL} \cdot N_{DL}\right) \tag{4}$$

$$\text{s.t.} \quad \Pr\left[\Gamma_{s,DL} \geq \alpha_{s,DL} \cdot N_{DL}\right] < p_{out} \quad s = 1,2 \tag{4a}$$

$$\sum_s \alpha_{s,DL} = 1 \tag{4b}$$

3   Reinforcement Learning-Based RAN Slicing Solution

The problems in (3) and (4) with their constraints are nonlinear optimization problems. Such an optimization problem is generally hard to solve. The complexity of solving this problem is high for a network of realistic size with fast varying traffic conditions. For this reason, we propose the use of an offline reinforcement learning approach to solve the problem in a more practical way.

The general approach is depicted in Fig. 1. Specifically, a slicing controller is responsible for determining the slicing ratios $\alpha_{s,UL}$, $\alpha_{s,DL}$ for each slice by executing the RL algorithm. It is assumed that two separate RL algorithms are executed for the UL and the DL to determine respectively $\alpha_{s,UL}$ and $\alpha_{s,DL}$. In the general operation of RL, the optimum solutions are found based on dynamically interacting with the environment based on trying different actions $a_{k,x}$ (i.e. different slicing ratios) selected from a set of possible actions numbered as $k = 1,..., A_x$, where $x \in \{UL, DL\}$. As a result of the selected action, the RL process gets a reward $R_{TOT,x}(a_{k,x})$ that measures how good or bad the result of the action has been in terms of the desired optimization target. Based on this reward, the RL algorithm adjusts the decision making process to progressively learn the actions that lead to highest reward. The action selection is done by balancing the trade-off between exploitation (i.e. try actions with high reward) and exploration (i.e. try actions that have not been used before in order to learn from them). In case this inter-action with the environment was done in an on-line way, i.e. by configuring the slicing ratios on the real network and then measuring the obtained performance, this could lead to serious performance degradation since, during the exploration process, wrong or unevaluated decisions could be made at certain points of time due to the exploration, and affecting all the UEs of a given slice. To avoid this problem, this paper considers an off-line RL, in which the slicing controller interacts with a network model that simulates the behavior of the network and allows testing the performance of the different actions in order to learn the optimum one prior to configuring it in the real network. The network model is based on a characterization of the network in terms of traffic generation, propagation modelling, etc.

The specific RL algorithm considered in this paper is the Q-learning based on soft-max decision making [17], which enables an exploration-exploitation traversing all possible actions in long-term. In turn, the reward should be defined in accordance with the optimization problem, which in this paper intends to maximize the resource uti-lization subject to the outage probability constraint. The details about the reward function and the detailed operation of the Q-learning algorithm are presented in the following.

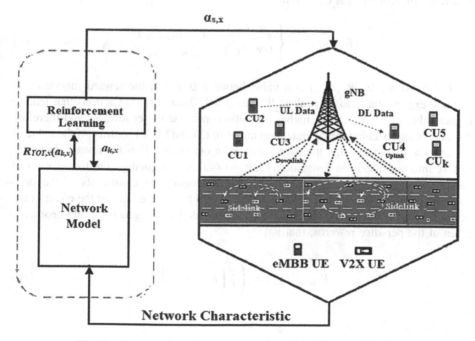

Fig. 1. General approach for the proposed RAN slicing solution.

3.1   Reward Computation

The reward function should reflect the ability of the taken action to fulfill the targets of the optimization problems (3) and (4). Based on this, and for a given action $a_{k,x}$ with associated slicing ratios $\alpha_{s,x}(k)$ the reward is computed as function of the normalized resource utilization $\Psi_{s,x}(a_{k,x})$ of slice s in link $x \in \{UL, DL\}$ defined as the ratio of used resources to the total allocated resources by the corresponding action. For the case of the V2X slice ($s = 1$), it is defined as:

$$\Psi_{1,UL}\left(a_{k,UL}\right) = \frac{\Gamma_{1,UL} + \Gamma_{1,SL}}{\alpha_{1,UL}(k) \cdot N_{UL}} \tag{5}$$

$$\Psi_{1,DL}\left(a_{k,DL}\right) = \frac{\Gamma_{1,DL}}{\alpha_{1,DL}(k) \cdot N_{DL}} \tag{6}$$

In turn, for the case of eMBB slice ($s = 2$), it is defined as:

$$\Psi_{2,UL}\left(a_{k,UL}\right) = \frac{\Gamma_{2,UL}}{\alpha_{2,UL}(k) \cdot N_{UL}} \tag{7}$$

$$\Psi_{2,DL}\left(a_{k,DL}\right) = \frac{\Gamma_{2,DL}}{\alpha_{2,DL}(k) N_{DL}} \tag{8}$$

Based on these expressions, the reward $R_{s,x}(a_{k,x})$ for the slice s in link $x \in \{UL, DL\}$ as a result of action $a_{k,x}$ is defined as

$$R_{s,x}\left(a_{k,x}\right) = \begin{cases} e^{\Psi_{s,x}(a_{k,x})} & \Psi_{s,x}\left(a_{k,x}\right) \leq 1 \\ 1/\Psi_{s,x}\left(a_{k,x}\right) & otherwise \end{cases} \tag{9}$$

In (9), whenever $\Psi_{s,x}(a_{k,x})$ is a value between 0 and 1, the reward function will increase exponentially to its peak at $\Psi_{s,x}(a_{k,x}) = 1$. Therefore, the actions that lead to higher value of $\Psi_{s,x}(a_{k,x})$ (i.e. higher utilization) provide larger rewards and therefore this allows approaching the optimization target of (3) and (4). In contrast, if the value of $\Psi_{s,x}(a_{k,x}) > 1$, it means that the slice s will be in outage and thus the reward decreases to take into consideration constraints (3a) and (4a). Consequently, the formulation of the reward function per slice in (9) takes into account the constraints of the optimization problem. In addition, since the total reward has to account for the effect of the action on all the considered slices $s = 1,\ldots, S$, it is defined in general as the geometric mean of the per-slice rewards, that is:

$$R_{TOT,x}\left(a_{k,x}\right) = \left(\prod_{s=1}^{S} R_{s,x}\left(a_{k,x}\right)\right)^{\frac{1}{s}} \tag{10}$$

3.2   Q-Learning Algorithm

The ultimate target of the Q-learning scheme at the slicing controller is to find the optimal action (i.e. the optimal slicing ratios for a given link $x \in \{UL, DL\}$) that maximizes the expected long-term reward to each slice. To achieve this, the Q-learning interacts with the network model over discrete time-steps of fixed duration and estimates the reward of the chosen action. Based on the reward, the slice controller keeps a record of its experience when taking an action $a_{k,x}$ and stores the action-value function (also referred to as the Q-value) in $Q_x(a_{k,x})$. Every time step, the $Q_{UL}(a_{k,UL})$ and $Q_{DL}(a_{k,DL})$ values are updated following a single-state Q-learning approach with a null discount rate [17] as follows:

$$Q_x(a_{k,x}) \leftarrow (1 - \alpha) \, Q_x(a_{k,x}) + \alpha \cdot R_{TOT,x}(a_{k,x}) \tag{11}$$

where $\alpha \in (0, 1)$ is the learning rate, and $R_{TOT,x}(a_{k,x})$ is the total reward accounting for both V2X and eMBB slices after executing an action $a_{k,x}$. At initialization, i.e. when action $a_{k,x}$ has never been used in the past, $Q_x(a_{k,x})$ is initialized to an arbitrary value.

The selection of the different actions based on the $Q_x(a_{k,x})$ is made based on the softmax policy [17], in which the different actions are chosen probabilistically. Specifically, the probability $P_x(a_{k,x})$ of selecting action $a_{k,x}$, $k = 1,..., A_x$, is defined as

$$P_x(a_{k,x}) = \frac{e^{Q_x(a_{k,x})/\tau}}{\sum\limits_{j=1}^{A_x} e^{Q_x(a_{j,x})/\tau}} \tag{12}$$

where τ is a positive integer called temperature parameter that controls the selection probability. With high value of τ, the action probabilities become nearly equal. However, low value of τ causes a greater difference in selection probabilities for actions with different Q-values. Softmax decision making allows an efficient trade-off between exploration and exploitation, i.e. selecting with high probability those actions that have yield high reward, but also keeping a certain probability of exploring new actions, which can yield better decisions in the future. The pseudo-code of the proposed RL-based RAN slicing algorithm is summarized in Algorithm 1. Once the offline RL algorithm has converge, i.e. the selection probability of one of the actions is higher than 99.99%, the selection ratios $\alpha_{s,x}$ associated to this action are configured on the network, as illustrated in Fig. 1.

Algorithm 1: RAN slicing algorithm based on RL
1. **Inputs:** N_{UL}, N_{DL}: Number of RBs in UL and DL. S: number of slices, Set of actions $a_{k,x}$ for link $x \in \{UL,DL\}$
  2.**Initialization of Learning**: $t \leftarrow 0$, $Q_x(a_{k,x}) = 0$, $k=1,...,A_x$, $x \in \{UL,DL\}$
3. **Iteration**
4. **While** learning period is active do
5.   **for** each link $x \in \{UL,DL\}$
6. | Apply softmax and compute $P_x(a_{k,x})$ for each action $a_{k,x}$ according to (12);
7. | Generate an uniformly distributed random number $u \in \{0,1\}$
8. | Select an action $a_{k,x}$ based on u and probabilities $P_x(a_{k,x})$
9. | Apply the selected action to the network and evaluate $\Psi_{s,x}(a_{k,x})$ based on (5)-(8).
10. | **If** $\Psi_{s,x}(a_{k,x}) \leq 1$ then
11. | $$R_{s,x}\left(a_{k,x}\right) = e^{\Psi_{s,x}(a_{k,x})}$$
12 | **else**
13. | $$R_{s,x}\left(a_{k,x}\right) = 1/\Psi_{s,x}\left(a_{k,x}\right)$$
14. | **End**
15. | **Compute** $R_{TOT,x}(a_{k,x})$ based on equation (10)
16. | **Update** $Q_x(a_{k,x})$ based on equation (11)
17. **End**
18. **End**

---

4 Performance Analysis

In this section, we evaluate the performance of the proposed RAN slicing solution through system level simulation performed in MATLAB. Our simulation model is based on a single-cell hexagonal layout configured with a gNB. The model considers vehicular UEs communicating through cellular mode (uplink/downlink) and via side-link (direct V2V) and use slice (RAN_slice_ID = 1) and eMBB UEs operating in cellular mode (uplink/downlink) and using slice (RAN_slice_ID = 2) based on the assumptions described in Sect. 2. Note that the slice ratio $\alpha_{1,UL} \cdot N_{UL}$ is divided into two ratios ($\bar{\alpha}_{1,UL} = 65\%$ of $\alpha_{1,UL} \cdot N_{UL}$ PRBs for V2X users in sidelink and $\alpha_{1,SL} = 35\%$ of $\alpha_{1,UL} \cdot N_{UL}$ PRBs PRBs for V2X service in uplink direction). The traffic generation associated to each eMBB UE at a random position assumes that services generate sessions following a Poisson process with rate λm, required bit rate $R_b = 1$ Mb/s and average session duration of 120 s. The gNB supports a cell with a channel organized in 200 RBs composed by 12 subcarriers with subcarrier separation $\Delta f = 30$ kHz, which corresponds to one of the 5G NR numerologies defined in [18]. The actions specify the fraction of resources for

Table 1. Simulation parameters

Parameter	Value	Parameter	Value
Cell radius	500 m	Path loss model	The path loss and the LOS probability for cellular mode are modeled as in [20]. In sidelink mode, all V2V links are modeled based on freeway case (WINNER+B1) with hexagonal layout [ITU-R] [21]
Number of RBs per cell	$N_{UL} = N_{DL} = 200$ RBs		
Base station antenna gain	5 dB		
GBR (Rreq)	1 Mb/s	Number of actions	20
λ_a	1 UE/s		$\alpha_{1,x}$: varies from 0.05 to 1 in steps of 0.05 $\alpha_{2,x}$: varies from 1 to 0.05 in steps of 0.05
Shadowing standard deviation	3 dB in LOS and 4 dB in NLOS	Length of the street	Freeway length = 1 km
Frequency	2.6 GHz	Lane width	4 m
Average session duration	120 s	Size of cluster	250 m
		Number of lanes	3 in one direction
Vehicle speed	80 km/h	Number of clusters	4
Learning rate α	0.1	Vehicular UE height	1.5 m
Temperature parameter τ	0.1	Safety message size (Sm)	300 bytes
		Time window T	30 s
Spectral efficiency model to map SINR	Model in section A.1 of [19]. The maximum spectral efficiency is 8.8 b/s/Hz	Average generation rate	Slice 1: $\lambda v = 1$ [packets/s] Slice 2: λm = varied from 0.2 to 1.2 sessions/s

V2X and eMBB and they are defined such that action $a_{k,x}$ corresponds to $\alpha_{1,x}(k) = 0.05 \cdot k$ and $\alpha_{2,x}(k) = (1 - 0.05 \cdot k)$ for $k = 1,...,20$, $x \in \{UL, DL\}$. All relevant system and simulation parameters are summarized in Table 1. The presented evaluation results intend to assess and illustrate the performance of the proposed solutions in terms of network capacity, throughput, and network congestion. As a reference for comparison, we assume a simpler RAN slicing strategy denoted as "Proportional Scheme", in which the ratio of RBs for each slice is proportional to its total traffic rate (in Mb/s). Figure 2 presents the RB utilization for V2X and eMBB slices in the UL as a function of the session generation rate (λ_m) for eMBB users.

From the presented results, we notice that our proposed model with off-line Q-learning maintains high resource utilization compared to the proportional strategy in different load scenarios. This is due to the RL-based slicing strategy that inherently tackles slice dynamics by selecting the most appropriate action considering the resource utilization in the reward. It is clearly observed that, as the arrival rate of requests increases, the RB utilization of the system increases gradually. For the proposed off-line Q-learning, when the arrival rate for the traffic of slice_ID = 2 is 1.2, the system utilizes around 79% of radio resources. For the proportional approach, the utilization is only about 73% of radio resources. Figure 3 depicts the throughput delivered in Mbits/sec for both eMBB and V2X slices in the sidelink and uplink. The figures illustrate with two lines the behavior of the proposed solution and the proportional scheme. Here, we can observe that the off-line Q-learning outperforms the proportional scheme in terms of throughput. The proposed scheme with off-line Q-learning achieved maximum throughput of 120 Mb/s in uplink when the eMBB arrival rate is 1.2 sessions/s, whereas in case of the proportional strategy model, the maximum throughput is only reached 114 Mb/s in uplink. The reasons are two-fold. First, when the arrival rate λ_m of eMBB UEs is increased, more users will use the network and this will increase the number of eMBB sessions and request more RBs to be used in transmissions. Second, as the number of eMBB sessions increases, requiring more radio resources, the proposed off-line Q- learning approach ensures more RBs which can be used to transmit data, while the proportional approach provides a lower number of available RBs for use in data transmissions.

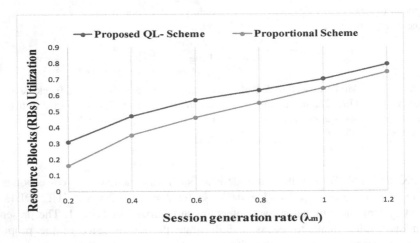

Fig. 2. Uplink RB utilization as a function of the eMBB session generation rate λ_m (sessions/s).

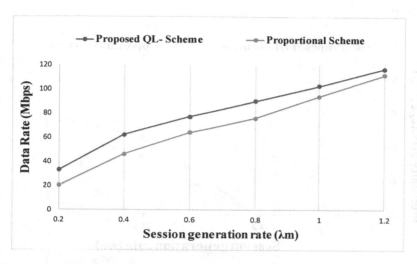

Fig. 3. Aggregated throughput experienced by both slices in uplink as a function of the eMBB session generation rate λ_m (sessions/s).

Figure 4 presents the RB utilisation for the sidelink. It shows that the proposed scheme with off-line Q-learning is able to improve the resource utilization compared to the reference model in different load scenarios. The proposed scheme with off-line Q-learning achieved maximum RB utilization of 93% in sidelink when the V2X arrival rate is 5 packets/s while in case of the proportional strategy model, the maximum utilization of RBs is only reached 72%.

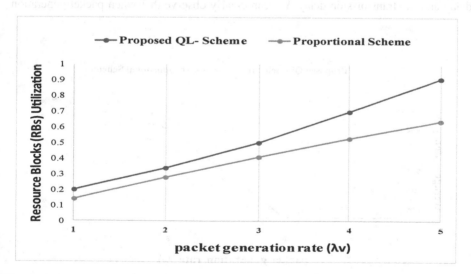

Fig. 4. Resource blocks utilization for sidelink transmissions as a function of the V2X UEs packet generation rate λ_v (packets/s).

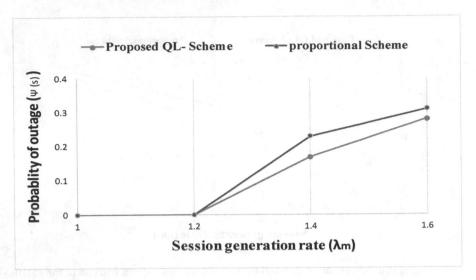

Fig. 5. Outage probability as a function of the eMBB session generation rate λ_m (UEs/s).

In Fig. 5, we investigate the probability of having congestion due to the lack of radio resources at a certain point of time. The outage probability of the proposed and proportional strategy is plotted against the eMBB session generation rate λ_m. As shown in the figure, increasing the traffic load leads to an increase in the outage probability of service. It can be also noted that our proposed scheme with off-line Q-learning can substantially reduce the congestion probability.

Figure 6 depicts the average latency for V2X service caused by channel access delay and the transmission delay. We can clearly observe that when packet generation

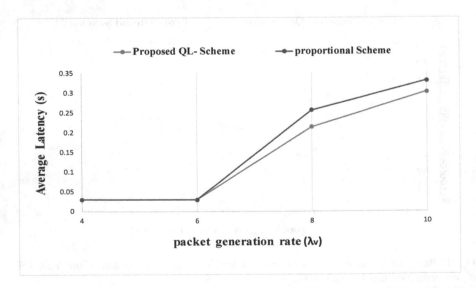

Fig. 6. Average latency as a function of the V2X UEs packet generation rate λ_v (packets/s).

rate λ_v is increased, more vehicles will use the network and request RBs to be used for the transmissions. This cause an increase in the waiting time and therefore increase the latency. We notice that our proposed model with off-line Q-learning approach reduces the latency compared to the proportional model and this is due to the fact that the proposed solution guarantees higher availability of resources avoiding outage situations.

5 Conclusions and Future Work

In this paper, we have investigated the splitting of radio resources into multiple RAN slices allocated to support V2X and eMBB services in uplink, downlink and sidelink (direct V2V) communications. We proposed a new RAN slicing strategy based on off-line Q-learning to determine the split of resources assigned to eMBB and V2X slices. This strategy has been compared against a reference scheme that makes an allocation of resources in proportion to the traffic rate of each slice. Extensive simulations were conducted to validate and analyze the performance of our proposed solution. Simulation results show the capability of the proposed algorithm to allocate the resources efficiently and improve the network performance. From the presented results, we notice that our proposed scheme outperforms the proportional scheme in terms of resource utilization, data rate, latency and congestion probability.

Acknowledgements. This work has been supported by the Spanish Research Council and FEDER funds under SONAR 5G grant (ref. TEC2017-82651-R).

References

1. Report ITU-R M.2410-0: Minimum Requirements Related to Technical Performance for IMT - 2020 Radio Interface(s), November 2017
2. 3GPP TR 38.912 V15.0.0: Study on New Radio (NR) access technology (Release 15), June 2018
3. 5G Americas: Network Slicing for 5G Networks and Services. Technical report, November 2016. http://www.5gamericas.org/files/3214/7975/0104/5GAmericasNetworkSlicing11.21Final.pdf
4. 3GPP TS 22.261 v16.0.0: Service requirements for the 5G system; Stage 1 (Release 15), June 2018
5. Soliman, H.M., Leon-Garcia, A.: QoS-aware frequency-space network slicing and admission control for virtual wireless networks. In: IEEE GLOBECOM (2016)
6. Sama, M.R., An, X., Wei, Q., Beker, S.: Reshaping the mobile core network via function decomposition and network slicing for the 5G era. In: Proceedings of IEEE Wireless Communications Networking Conference Workshops (WCNCW), pp. 90–96, April 2016
7. Hoang, D.T., Niyato, D., Wang, P: Optimal cross slice orchestration for 5G mobile services. arXiv:1712.05912v1, 16 December 2017
8. Aijaz, A.: Hap-SliceR: a radio resource slicing framework for 5G networks with haptic communications. IEEE Syst. J. **12**, 2285–2296 (2017)
9. Jiang, M., Condoluci, M., Mahmoodi, T.: Network slicing in 5G: an auction-based model. In: IEEE ICC (2017)

10. Caballero, P., Banchs, A., de Veciana, G., Costa-Pérez, X.: Network slicing games: enabling customization in multi-tenant networks. In: IEEE INFOCOM (2017)
11. Tang, L., Tan, Q., Shi, Y., Wang, C., Chen, Q.: Adaptive virtual resource allocation in 5G network slicing using constrained Markov decision process. IEEE Access **6**, 61184–61195 (2018)
12. Sciancalepore, V., Zanzi, L., Costa-Perez, X., Capone, A.: (PDF) ONETS: online network slice broker from theory to practice, January 2018
13. 3GPP TS 36.300 v15.0.0: Evolved Universal Terrestrial Radio Access (E-UTRA) and Evolved Universal Terrestrial Radio Access Network (E-UTRAN); Overall description; Stage 2 (Release 15), December 2017
14. 3GPP TR 38.885 v1.0.0: NR; Study on Vehicle-to-Everything (Release 16), November 2018
15. 3GPP TS 38.401 v15.2.0: NG-RAN; Architecture description (Release 15), June 2018
16. Albonda, H.D.R., Pérez-Romero, J.: An efficient mode selection for improving resource utilization in sidelink V2X cellular networks. In: IEEE (CAMAD) Workshops, Barcelona, Spain, September 2018
17. Sutton, R.S., Barto, A.G.: Reinforcement Learning: An Introduction. MIT Press, Cambridge (1998)
18. 3GPP TS 38.211 v15.2.0: NR; Physical channels and modulation (Release 15), June 2018
19. 3GPP TR 36.942 v15.0.0: Radio Frequency (RF) system scenarios, September 2018
20. Report ITU-R M.2135: Guidelines for evaluation of radio interface technologies for IMT-Advanced (2009)
21. WINNER II Channel Models, D1.1.2 V1.2. http://www.cept.org/files/1050/documents/winner2%20%20final%20report.pdf

Assessment of Spectrum Management Approaches to Private Industrial Networks

Pekka Ojanen[1(✉)] and Seppo Yrjölä[2]

[1] Co-Worker Technology Finland Oy, Turku, Finland
pekka.ojanen@co-workertech.com
[2] Nokia, Oulu, Finland

Abstract. Ubiquitous connectivity is one of the foundational technologies enabling data sharing amongst participating components of an industrial internet of things (IIoT) system. The growing pressure to open the mobile market for location specific networks has resulted in new regional licensing and sharing-based models for spectrum access, to allow the emergence of local networks to serve different verticals. While the development of technical solutions for network performance is progressing, less attention has been paid to the spectrum management approaches for the new industrial networks and location specific service offerings. This paper examines solutions to problems currently faced by industry in acquiring spectrum to support the IIoT along with introducing a framework that should be accounted for when assessing the feasibility of the spectrum management approaches. The aspects of how spectrum is currently allocated and how it addresses the IIoT needs were assessed in six selected countries: Canada, Finland, France, Germany, Netherlands, UK and US.

Keywords: Industrial internet of things · LTE · Private networks · Spectrum management · Regulation · 5G

1 Introduction

The fourth industrial revolution, "Industry 4.0", is the next era in industrial production, aiming at improving significantly the flexibility, versatility, usability and efficiency of industrial manufacturing through integrating the internet of things (IoT) and related services in [1]. Intelligent networking can be used by companies for flexible production, optimized logistics, advanced use of data on various production processes. Ubiquitous connectivity is one of the foundational technologies enabling data sharing amongst participating components of an industrial IoT (IIoT) system [2].

Wireless communication, and in particular, the 5th generation mobile networks (5G), is an important means of achieving the required flexibility and efficiency of production [3]. The 5G will not only expand the broadband capabilities of mobile networks but has been designed to provide advanced wireless connectivity suitable for all vertical industries, such as the manufacturing, logistics, transportation, automotive and agricultural sectors. To achieve this, 5G supports three usage scenarios: enhanced mobile broadband (eMBB), massive machine-type communication (mMTC), and ultra-reliable low-latency communications (URLLC) [4]. In some industrial use cases the

© ICST Institute for Computer Sciences, Social Informatics and Telecommunications Engineering 2019
Published by Springer Nature Switzerland AG 2019. All Rights Reserved
A. Kliks et al. (Eds.): CrownCom 2019, LNICST 291, pp. 277–290, 2019.
https://doi.org/10.1007/978-3-030-25748-4_21

required network performance and characteristics may be relatively similar to those of public mobile networks, while for some use cases the requirements may be significantly different, particularly related to required network availability, security and privacy [5]. Implementation and deployment of the distinct requirements of industrial applications can be delivered by private networks deployed directly by industrial organizations themselves, or by external service providers [6]. There are recent indications that industry players may prefer to set up their own networks, e.g., the German car manufacturers have shown a preference to operate their own private 5G networks, as they are reluctant to entrust their digitized operations and data to third parties, such as public mobile network operators (MNOs) [7]. If an industrial or a private network is deployed by an organization other than current mobile license holder, e.g., by a microoperator [8], the organization needs to get access to suitable spectrum. [9] addressed the use of private long term evolution (LTE) networks for mission and business critical mobile broadband communications, analyzed the three plausible service delivery models, and highlighted the role of the spectrum management in assigning dedicated spectrum. Spectrum management aims at allocating spectrum to the right use, and high overall efficiency by assigning it to those who value it most. Ultimately, the spectrum management decisions are about maximizing the value of spectrum, its efficient utilization, and its benefits to society [10].

Traditionally, private mobile radio (PMR) systems have been deployed in specific, rather narrow bands, as the main service has been voice and narrowband data. Due to historical· reasons the bands have been very much country specific. However, the spectrum demand has changed towards wider bandwidths to facilitate higher network capacities and higher bitrates. Wireless local area networks, (RLANs) have been used widely in industry environment as a solution to provide wideband local connectivity, but utilization of unlicensed bands and limitation to very short ranges do not often provide the required performance and quality of service (QoS). Instead, employment of standardized wide area technologies, such as LTE, and later 5G could better meet the performance requirements of industrial networks [11]. To date, most third generation partnership project (3GPP) defined LTE bands have been and are being made available for nationwide public mobile networks through competitive awarding, i.e. auctions, and the same approach has dominated the 5G spectrum release so far [12, 13]. Only a few countries have introduced regulatory frameworks for making spectrum available for individually authorized, locally deployed private LTE or 5G networks [14]. As one solution to overcome the spectrum scarcity, variants of 3GPP technologies have been developed suitable for deployment in the license exempt bands, especially to the 5 GHz RLAN bands. However, use of shared bands does not provide sufficient spectrum certainty and quality for all applications and use cases [15].

While there are already some studies on suitable regulatory frameworks for local 5G deployments [16], more detailed studies on locally and temporary shared spectrum [17] and micro licensing and the associated regulatory framework [15], there is no preceding work and method to assess spectrum management approaches in the context of locally deployed IIoT networks. Furthermore, to the best knowledge of the authors, summary of the recent spectrum management approaches in this context has not been presented in the literature. This paper examines, what would be suitable spectrum management approaches facilitating deployment and operation of private and industrial networks to meet the requirements towards the 5G era. The focus is on use cases where private networks are deployed by industrial organizations for their own use and thus they need to get access to suitable spectrum on affordable terms. Information on studied spectrum management approaches have been collected both from public sources and directly from the regulators. This research addresses the planning and authorization processes which are key parts of the overall spectrum management process [18].

The rest of the paper is organized as follows. In Sect. 2, applications and their spectrum and regulatory requirements are discussed, and assessment framework introduced. Next, selected recent spectrum management approaches relevant for private industrial networks are described, assessed and regulatory recommendation given. Finally, conclusions are drawn in Sect. 4.

2 Spectrum and Regulatory Requirements of Private Industrial Networks

The 3GPP has analyzed use cases of vertical industries and defined a set of functional requirements and system parameters related to communication services for each use case in each domain [5, 19]. Several of the developed service performance requirements have an impact on preferred spectrum management approach. High communication service availability can be reached through exclusive access to dedicated spectrum assignments and through protection from harmful interference. Access to wide bandwidths is needed. The required service areas are typically geographically limited, covering one or several, local or regional areas, ranging from indoor coverage, up to few km^2. This means that frequency ranges below 4 GHz with sufficient transmit powers are preferred if outdoor coverage is required. Depending on the application, traffic may range from symmetric up to very asymmetric, in either direction requiring uplink/downlink ratio (UL/DL) flexibility from the technology, the deployment and the band regulation. Use of time division duplex (TDD) technology can provide the required duplex flexibility, though adjacent networks may need to be synchronized, which would limit the applicability. 5G Alliance for Connected Industries and Automation (5G-ACIA) [3] addresses major challenges of 5G, highlighting spectrum and operator models. In order to meet extremely demanding latency and reliability requirements, licensed spectrum and protection from harmful interference are highly preferred.

Investment cycles of vertical industries differ from cycles of the telecom industry: cycles for media and entertainment are typically shorter, ranging between 2–3 years, for automotive industry 7–8 years, energy, manufacturing and mechanical industries 25 years, and for oil & gas from 10 to 25 years [11]. Partly due to this difference, vertical industries may prefer to deploy their own networks. Furthermore, the timing for investing in wireless communications depends solely on their own business plans. For example, in the Netherlands, UtilityConnect [20] has deployed its own dedicated network for smart energy grid applications rather than relying on 3G or 4G networks which were considered too short-term solutions.

Vertical industries require the assurance that for their networks there will be a continuity of service, without unjustified price increases, spectrum re-farming or technology upgrades over their planned life span. On the other hand, deploying and operating a wireless networks for IIoT are not their core business, but an enabler for optimizing operations and productivity, enhancing security and safety, and improving planning and decision making.

This all means that the cost of spectrum should be affordable, suitable authorization process would be application based, and that the applications should be allowed to be submitted any time, based on the business need. It also means, that the license duration should be comparable to the investment cycle, and that overall regulatory certainty is needed for years to come. The key elements of the spectrum management approaches and the preferences are identified and summarized in Table 1.

Table 1. Spectrum management assessment framework.

Spectrum management framework element	Private industrial network preference
Band type	Dedicated bands for applications requiring high spectrum quality, harmonized 3GPP bands for economies of scale
Bandwidth	Applicant defined: sufficient for wideband deployments
Spectrum availability	Full time, guaranteed
Harmful interference protection	Yes
Interference coordination	By regulator, through authorization conditions or automatic through technical solution
Sharing conditions	Stable, pre-defined
Technical or operational restrictions	Pre-defined, no substantial restrictions on network deployment or services
Location and overage area	Applicant defined: outdoor local or regional, indoor. One or multiple areas
Maximum transmit power	Sufficient for required local/regional outdoor and/or indoor coverage
Technology	Technology neutral approach, flexible duplex technology such as TDD allowed
Authorization method	Application based, time-to-air critical
Application timing	Submission any time, based on business need
Authorization duration	In line with investment cycle, typically long, e.g. 10 years
Cost/pricing	Fee, administrative one-off or annual fee, per license/area
Regulatory certainty	High, beyond authorization duration

3 Spectrum Management Approaches for Private Industrial Networks

In a few countries the spectrum management frameworks are already in place or are being defined for making dedicated spectrum available for wideband private networks. Furthermore, shared, license exempt bands are also widely available. The selected exemplary cases present spectrum management approaches in several countries on 2.3 GHz, 2.5 GHz and 3.5 GHz licensed frequency bands considered for LTE or 5G. The license exempt usage of the 5 GHz RLAN bands is also addressed for comparison. Information on studied spectrum management frameworks have been collected both from public national regulatory authority (NRA) sources and interviews.

3.1 Country Specific Spectrum Management Approaches

The C-Band in the Netherlands (NL)

The bands 3410–3500 MHz and 3700–3800 MHz are used by a military satellite earth station in the northern part of the country. In order to protect the incumbent users, those frequencies cannot be used for mobile services north from the Amsterdam – Zwolle line, while the rest of the C-band is available for commercial, nationwide mobile use. The bands are available for local broadband networks, including private LTE networks south from the line, with certain operational limitations [21]. For example, the emitted power towards north must be limited [22]. Deployment and operation of networks in those bands requires a license from the regulator. For the moment, the licenses are temporary, with the end date in 2022 for the higher band and in 2026 for the lower band. For every base station a separate license is needed with the maximum bandwidth of 40 MHz. This opportunity has been popular, more than 150 licenses have been issued and in some areas, like the Rotterdam harbor area, the bands are getting very occupied. As the authorization process is based on the first-come-first-served principle, in most popular areas it may not be possible to get a license. There are plans to remove the earth station from the band, and at the same time reorganize the use of the whole C-band. It seems that one option would be to re-farm the local networks to the 3700–3800 MHz band as planned in Germany and Sweden.

The C-Band in Germany (GER)

In Germany, the band 3400–3700 MHz will be auctioned in 2019 for public mobile networks whereas the band 3700–3800 MHz will be made available for regional and local assignments, based on application. The process allows applying for regional and local assignments not only at the time when the new regulatory frameworks comes to force, but also at a later date, flexibly and in line with the demand. There is currently a large amount of regional and local assignments in the 3400–3700 MHz band, with licenses expiring latest at the end of 2022. Those networks will be relocated to the 3700–3800 MHz band. The German regulator BNetzA has published the planned application procedure and the rules for the band 3700–3800 MHz [23]. Access to the band is based on application and requires an individual authorization from BNetzA. Operators having current access to licensed spectrum in 700–3700 MHz already are only eligible for a temporary access, in case there are unused parts in the 3700–3800 MHz band. License duration is 10 years, and the licenses are transferable.

According to the proposed rules the whole band 3700–3800 MHz will be available for indoor deployments. There are technical restrictions for ensuring that no harmful interference is created outside the facility. The band 3700–3780 MHz is reserved for outdoor regional deployments, and the 3780–3800 MHz for outdoor local deployments.

There are rules also for outdoor deployments to ensure that no harmful interference is created outside the defined coverage area. The locations and the area of the region can be defined based by the applicants. All assignments are based on 10 MHz blocks. The approach is service and technology neutral, though TDD is the only allowed duplex technology, and networks must be synchronized. National roaming is not mandated, but allowed. Efficient use of the assignment is required, also throughout a region, with a principle use-it-or-lose-it. There is a fee for the spectrum use.

The 3.6 GHz WBS Band in Canada (CAN)

The band 3650–3700 MHz is designated in Canada for wireless broadband service (WBS) and fixed and mobile systems fulfilling regulator ISED's technical requirements [24, 25] are permitted. Deployed technologies include WiMax and LTE, which have been adjusted to comply with the specific requirements of the band. Licenses are issued on all-come all-served basis for Tier 4 service areas and there is a service area specific annual fee for the license. The licenses will expire on March 31 of each year, and will generally be renewable upon payment of required fees. The eligibility is not restricted.

The band is shared between all the WBS licensees within the service area, and the licensees are expected to cooperate to identify and resolve possible interference problems by themselves. To assist in facilitating cooperation and coordination, ISED has developed a publicly accessible spectrum management system database (SMS) showing both current license and site-specific data. Licensees will be required to upload their information to the database at least six weeks before putting a site to service and keep it up to date. In addition to coordination between the WBS licensees, they must ensure that no harmful interference is created towards incumbents deployed in the adjacent band and meet coordination distance requirements from the Canada-US border. A public consultation was conducted in 2018 [26], which addressed among other things also the possible need to develop the WBS band regulation further. To date, conclusions based on the responses have not been published.

The 2.6 GHz in France (FRA)

The 3GPP band 38 (2570–2620 MHz) had been planned for public mobile networks, but was not licensed because of the lack of market demand. Therefore, its suitability for wideband PMR networks was investigated through several trials and two public consultations were held. The trials showed that the band 38 is suitable for private LTE networks, and the first consultations concluded that a 40 MHz sub-band, 2575–2615 MHz, could cover the spectrum needs of superfast PMR. It seems that regulator ARCEP has to look in more detail to the case where demand exceeding the supply, particularly what would be the regulatory approach if there is a need to deploy several networks in the same geographical area, aiming to utilize the same band. According to the working assumptions [27] the aim is to grant access to blocs of 10, 15 or 20 MHz, in limited geographical areas and for maximum ten years licenses. Each applicant would have to specify the requested coverage area and justify the spectrum needs within that area. Compliance with the technical conditions of the EU would be required [28], e.g. the use of TDD would be required and the maximum field level at the edge of coverage area is restricted.

The 2.3 GHz band in Finland (FIN)
The 2.3 GHz band is identified globally for IMT [29] and the European regulation by the CEPT and the EU is in place [30]. However, in many European countries the band is used by various incumbents, and removing them would be difficult or impractical. As there are still spectrum resources unused by the incumbents the CEPT has defined a possible regulatory approach for administrations that would prefer to deploy public mobile networks in the band in shared manner employing the licensed shared access (LSA) concept [31].

In Finland the 2.3 GHz band is not designated for public mobile use. The main incumbent primary usage in the band is operation of wireless cameras. The use of the cameras is individually authorized, and the license allows operation of wireless cameras anywhere in the country. The number of cameras is very limited, so there are unused spectrum resources all over the country, in places where the cameras are not in use. Mobile networks could access the band on a secondary basis, but the public mobile network operators in Finland have not shown interest towards accessing the band, especially since the areas to be protected could change over the time and that the remaining areas available for mobile networks would also change consequently. Moreover, a rather complex LSA solution should be employed and still there could be restrictions on the spectrum availability for the mobile network. Private LTE networks could possibly be deployed and operated in rural and remote locations without complex coordination or severe operational restrictions.

Shared Access Mobile Bands in the UK (UK)
The UK regulator Ofcom has published a consultation on shared access in the 3.8–4.2 GHz band, in a portion of the 2.3 GHz band (2390–2400 MHz) and in the 1800 MHz shared spectrum (1781.7–1785 MHz paired with 1876.7–1880 MHz) [32]. The proposed approach is to provide spectrum for local networks in locations unused by other licensed users. The Ofcom proposes a single authorization approach for all three bands. As the bandwidth in the 1800 MHz band is rather limited, this paper addresses only the first two bands. The consultation includes also an interesting regulatory proposal to allow access to the unused parts of the bands awarded to public mobile networks, but that case is not analyzed in this paper further.

The 3.8–4.2 GHz band is currently being used by several incumbent services, but in addition the band could be used for private networks as there are unused spectrum resources. The band is next to the 3.4–3.8 GHz band, which has been identified as a pioneer 5G band in Europe and several countries are looking to expanding the 5G deployments also to the 3.8–4.2 GHz band. The highest 10 MHz of the 2.3 GHz band are used currently for military applications, and also there is room for local network deployments. The 2.3 GHz band 40 is already widely used for 4G deployments, especially in Asia, and therefore, LTE equipment is widely available.

The access to the bands is planned to be individually authorized, allowing operation in a certain location. Two types of licenses are defined: a low power license allowing operation of a low power base station within a 50 m radius circle and a medium power license for operating a medium power base station in a rural area. The applicants would need to specify the bands they would like to access, as well as the planned locations. The Ofcom would then assess for each application the interference to and from other licensees in the band, based on coordination methodology and parameters proposed by them and make the assignments on a first come first served basis. This approach would provide certainty for the spectrum access and a possibility to provide QoS. The license fees would be cost based administrative fees, charged annually on a per area based or on a per base station basis, amount depending on the used bandwidth: the original proposal being £80 per 10 MHz. In the current proposals the Ofcom would deal with coordination between the licensees, but they intend to explore the potential for introducing dynamic spectrum access (DSA) in the proposed three shared access bands.

The 3.5 GHz CBRS Band for PAL Use in the Unites States (US)
In the US the band 3550–3700 MHz is being made available for citizens broadband radio service (CBRS) [33]. The band is currently used by several incumbent services, but there are unused spectrum resources in the band. The CBRS users comprise of two tiers of users: priority access license (PAL, tier 2) users and general authorized access (GAA, tier 3) users. Spectrum for PAL users is auctioned in 10 MHz pieces and one licensee can hold up to 4 PAL channels, i.e. up to 40 MHz. The authorizations are regional. The PAL authorization allows PAL licensees to access the spectrum resources available from the incumbents, while the incumbents must be protected from harmful interference. The rules allow the PAL licensees to lease their spectrum within their PAL area, which is beyond their deployment coverage. For example, in an industrial area a PAL holder may lease one or more of their channels to industrial enterprises. Spectrum not used by incumbents or by a PAL licensee is available for general authorized access (GAA) users on an unlicensed, shared basis. The amount of GAA spectrum may vary based upon variations incumbent and PAL usage; furthermore, unlicensed GAA users may experience harmful interference from other GAA users.

For both GAA and PAL, the base stations (CBSD) must register with a spectrum access system (SAS) and request a spectrum grant. The SAS will identify suitable spectrum for the CBSDs while ensuring that higher tier users are protected from harmful interference from lower tier users. The CBRS band could be suitable for IIoT, because the PAL licenses may allow a path for acquiring exclusive spectrum for regional use and GAA allows for a no cost option for non-mission critical services. The leasing rules for CBRS PALs provide also a potentially lower cost option for enterprises to lease spectrum at their facility.

The 5 GHz RLAN Bands (5G RLAN)
The International Telecommunication Union, Radiocommunication Sector (ITU-R) has identified the bands 5150–5350 MHz and 5470–5725 MHz globally for wireless access systems (WAS), including radio local area networks (RLAN). There are also several incumbent applications deployed in those bands, and the deployment of WASs must not cause harmful interference to the incumbent applications. Therefore, a number of technical and operational restrictions have been defined, such as maximum transmit power, restriction to indoor use in part of the bands and restrictions on antenna pattern [34] as well as required interference mitigation methods, such as dynamic frequency selection (DFS), transmit power control (TPC) and listen-before-talk protocol (LBT) to facilitate coexistence both with incumbent applications and among WASs sharing the band [35]. The global regulation is reflected, e.g. to Europe, and similar restrictions apply. Systems compliant with the ETSI standards [36] are allowed to be deployed in

the 5 GHz RLAN bands within the EU and the CEPT. Also, the FCC has defined their technical requirements. LTE-U [37] and MulteFire [38] are LTE based technologies that are designed to be compliant with the regulatory requirements in the 5 GHz bands. RLAN networks can be deployed in the 5 GHz bands under a general authorization, and this applies in most countries also to compliant LTE based networks for private and industrial applications.

3.2 Spectrum Management Approach Summary and Assessment

This section compares and assesses selected spectrum management approaches based on the analysis framework introduced in Sect. 2. A summary is depicted in Table 2. The key areas of the proposed spectrum management assessment framework are authorization, spectrum assignment, co-ordination, cost/pricing and regulatory certainty.

Table 2. Assessment of country specific spectrum management approaches.

Spectrum management framework element	Netherlands C-band	Canada WBS band	Germany 3.7 GHz	France 2.6 GHz	UK Shared 3.8 and 2.3 GHz	Finland 2.3 GHz	US CBRS PAL/GAA	RLAN 5 GHz
Band type	+	+	+	+	+	+	+/+	−
Bandwidth	+	+	+	+	+	+	+/+	+
Spectrum availability	+	−	+	+	+	−	−/−	−
Protection from harmful interference	+	−	+	+	+	−	+/−	−
Interference coordination	+	−	+	+	+	−	+/+	+
Sharing conditions	+	+	+	+	+	−	+/+	−
Technical or operational restrictions	+	+	+	+	+	−	+/−	−
Location and coverage area	+	+	+	+	−	+	+/+	+
Maximum transmit power	+	+	+	+	−	+	+/+	−
Technology	+	+	+	+	+	+	+/+	+
Authorization method	+	+	+	+	+	+	−/+	+
Application timing	+	+	+	+	+	+	−/+	+
Authorization duration	−	+	+	+	−	TBD	+/+	+
Cost/pricing	+	+	+	TBD	+	+	−/+	+
Regulatory certainty	−	−	+	+	+	−	+/+	+

In all presented country specific cases there are specific bands designated for private networks and the access is **individually authorized.** In some cases, there are incumbents having primary rights to the band, which means that private networks as secondary users are not allowed to cause harmful interference to the primary users (UK, FIN, US). In the unlicensed 5 GHz RLAN case, there are also incumbents requiring protection, which is reflected in technical and operational restrictions on the private networks. Furthermore, there are several other technologies allowed to share the 5 GHz bands horizontally with the private networks, which may restrict the spectrum availability and certainty. In all example cases, there is room for sufficient **bandwidths.** Dedicated access to a band together with protection from harmful interference provides the highest **spectrum availability** (NL, GER, UK).

Several **coordination approaches** are employed: only the first entrant may get the access (NL, GER, UK), the regulator may do the coordination when a new entrant submits an authorization application (UK), the entrant may need to prove that there is no harmful interference towards incumbents (FIN), the licensees have to coordinate among them in case of interference (CAN) or coordination is done by technical means (US). In Germany the designated band is wide, and it is partitioned for regional and local outdoor deployments, both shared with indoor deployments. This approach allows geographically parallel networks. For 5 GHz RLAN unlicensed bands the coordination is covered by technical and operational restrictions for vertical sharing and by LBT protocol for horizontal sharing.

The **maximum transmit power** is sufficient for various coverage requirements and deployments in the 3GPP defined LTE bands, except in the shared use case (UK). In parts of the 5 GHz RLAN bands the transmit power is limited to 200 mW, and operation is limited to indoor environment (not applicable in the US).

UL/DL flexibility is possible in all cases, as the bands are unpaired and the use of TDD is either possible or required (GER, FRA). In most cases adjacent networks may need to be synchronized, which would limit the UL/DL flexibility. In Canada and in Finland the networks may be located a significant distance apart from each other, so that synchronization may not be needed.

Application based individual **authorization** without timing restrictions for submission is widely employed (NL, CAN, GER, FIN, UK). If the principle is first-come-first-served, late applicants may be left without access to spectrum, especially if the available bandwidth does not support deployment of several geographically overlapping networks (NL, FRA, UK). If the principle is all-come-all-served, there may be coexistence problems due to overlapping networks, unless there is a specific coordination approach or process (CAN). The coordination process could be performed by the regulator (UK), or be on the responsibility of the licensees (CAN).

The **license duration** should be rather long, and in line with verticals' investment cycles. The longest offered license periods are 10 years (GER, FRA). On the other hand, in one case the license period is only one year, but the **license renewal** is done simply by paying the next year's fees (CAN). In the US CBRS band licenses are for ten years with possibility of renewal. In other cases the renewal approach is not defined, which undermines the regulatory certainty.

Spectrum cost should be affordable, and therefore nationwide authorizations and competitive award methods are not suitable for authorization of locally deployed private networks. In most of the presented cases, there is a fee for the spectrum access, which is a preferred choice. The fee can depend on coverage area population (CAN), or on used bandwidth (UK). In one case auctions are used (US CBRS PAL), but the unused spectrum resources can be leased. Moreover, the cost of additional environmental sensing capability (ESC) monitoring and SAS control equipment provided by the SAS operator, and possibly by a third-party ESC operator will be passed on to end users through higher service fees.

The **regulatory certainty** is highest if there is a newly confirmed spectrum management framework and licenses are granted for a long period, preferably longer than the investment cycle of the licensee (GER, US). In some cases there is a spectrum management framework in place, but at the same time ongoing or planned process to review the framework (NL, CAN). Also, the licenses may be granted for shorter period, e.g. three years or one year, which may reduce the certainty. In two cases early proposals for a spectrum management framework have been published, but major changes are possible (FRA, UK).

3.3 Discussion

Based on the assessment results, several recommendations can be drawn. The heterogeneity of industrial use cases, applications and requirements leads to a flexibility requirement in spectrum award and use. For mission critical applications, and applications requiring high QoS, individual authorizations and protection from harmful interference are required. The authorizations should be based on applications, and submissions should be allowed any time, based on the business need. The applicant should be allowed to define the required bandwidth and coverage area. The preferred license duration should be aligned with the licensee's investment cycle, and for industrial networks 10 years would be desirable with renewal option. Only reasonable administrative fees should be applied, on a yearly basis or per transmitter where practical. Spectrum should be made available from harmonized 3GPP bands in a suitable frequency range, for most cases below 5 GHz, on exclusive basis, and the bandwidth of the designation should allow deployment of more than one network in a location. Band segmentation, e.g. between local and regional deployments, and sharing between indoor and outdoor deployments should be considered. In all cases the maximum field strengths at the coverage edge could be defined as a coordination method. The complexity of the spectrum monitoring and control system should be minimized as well as the related services fees. UL/DL asymmetry should be facilitated through employment of TDD duplexing, and network synchronization should be required only in locations where it is required for coexistence reasons. Where synchronization is needed, the UL/DL ratio should be defined in a manner that takes into account the requirements of all affected networks.

4 Conclusions

The emergence of digital automation and enabling local high-quality private networks have the potential of improving the productivity and efficiency of vertical industries, promoting innovation and competition in the market and advancing society in totally new ways. There is an increasing interest for the deployment of local high-quality networks by different stakeholders to complement MNOs' networks especially for industrial vertical specific service delivery. This paper has highlighted the importance of understanding the different approaches for spectrum management in the context of the upcoming IIoT networks, whose deployment will be location specific to complement the previous generations that addressed wide-area coverage.

We have defined an assessment framework to consider the specifics arising from local IIoT networks operating in regionally licensed and shared spectrum bands, and assessed lately introduced spectrum management approaches in eight selected real-life cases. To do so, we have first identified the key elements of spectrum management approached based on a literature review focusing on the spectrum planning and authorization processes from the views of the vertical industries and a local private operator. It is critical to consider the authorization, spectrum assignment, coordination, cost/pricing approaches as well as the regulatory certainty. After that we have provided a case study of the recent spectrum authorization decisions in different countries and analyzed the decisions from the viewpoints of the identified key elements of spectrum management framework. Our analysis has shown that different countries have adopted different spectrum management approaches, which ultimately define who can enter the IIoT market and benefit from the related business models. Some of the first decisions have been taken to facilitate networks for verticals in regional license areas and to promote vertical specific service delivery through the possibility to establish local private networks by different stakeholders. Results showed that the, most promising approaches being proposed by Germany and France.

From technology perspective future work is needed on the reducing the complexity of coexistence management while improving the accuracy and certainty e.g. through propagation modelling and 3D clutter databases. On the policy and business future research could consider the valuation and pricing of spectrum for local and regional assignments.

References

1. Federal Ministry for Economic Affairs and Energy: Platform Industrie 4.0. https://www.plattform-i40.de/I40/Navigation/EN/Industrie40/WhatIsIndustrie40/what-is-industrie40.html. Accessed 24 Jan 2019
2. Industrial Internet Consortium homepage. https://www.iiconsortium.org/members.htm. Accessed 24 Jan 2019
3. 5G Alliance for Connected Industries and Automation (5G-ACIA): White Paper, 5G for Connected Industries and Automation (2018)
4. ITU-R Recommendation M.2083: IMT Vision – Framework and overall objectives of the future development of IMT for 2020 and beyond (2015)

5. 3rd Generation Partnership Project; Technical Specification Group Services and System Aspects; Study on Communication for Automation in Vertical Domains (Release 16), 3GPP TR 22.804, V16.1.0 (2018)
6. Zander, J.: Beyond the ultra-dense barrier: paradigm shifts on the road beyond 1000x wireless capacity. IEEE Wirel. Commun. 24(3), 96–102 (2017)
7. Handelsblatt. https://www.handelsblatt.com/today/companies/closer-connections-carmakers-want-their-own-5g-networks/23583750.html?ticket=ST-889835-xNQgjhppS99d6fTefpal-ap2. Accessed 24 Jan 2019
8. Matinmikko, M., Latva-aho, M., Ahokangas, P., Yrjölä, S., Koivumäki, T.: Micro operators to boost local service delivery in 5G. Wirel. Pers. Commun. 95(1), 69–82 (2017)
9. Ferrus, R., Sallent, O.: Extending the LTE/LTE-A business case: mission- and business-critical mobile broadband communications. IEEE Veh. Technol. Mag. 9(3), 47–55 (2014)
10. Beltran, F.: Accelerating the introduction of spectrum sharing using market-based mechanisms. IEEE Commun. Stand. Mag. 1(3), 66–72 (2017)
11. 5GPPP: 5G empowering vertical industries: Roadmap paper. The 5G Infrastructure Public Private Partnership, Brussels (2016)
12. Cramton, P.: Spectrum auction design. Rev. Ind. Organ. 42(2), 161–190 (2013)
13. Hazlett, T.W., Muñoz, R.E.: A welfare analysis of spectrum allocation policies. RAND J. Econ. 40(3), 424–454 (2009)
14. Matinmikko-Blue, M., Yrjola, S., Seppänen, V., Ahokangas, P., Hämmäinen, H., Latva-aho, M.: Analysis of spectrum valuation approaches: the viewpoint of local 5G networks in shared spectrum bands. In: IEEE International Symposium on Dynamic Spectrum Access Networks (DYSPAN), Seoul (2018)
15. Matinmikko, M., Yrjölä, S., Latva-aho, M.: Micro operators for ultra-dense network deployment with network slicing and local spectrum micro licensing. In: IEEE VTC Spring 2018. The 2018 IEEE 87th Vehicular Technology Conference, Porto (2018)
16. European Commission: RSPG18-005 Strategic spectrum roadmap towards 5G for Europe: RSPG second opinion on 5G networks. Radio Spectrum Policy Group (RSPG) (2018)
17. Guirao, M.D.P., Wilzeck, A., Schmidt, A., Septinus, K., Thein, C.: Locally and temporary shared spectrum as opportunity for vertical sectors in 5G. IEEE Netw. 31(6), 24–31 (2017)
18. International Telecommunication Union (ITU): the ICT regulation toolkit project by ITU and the World Bank's Information for Development Program (infoDev). http://www.ictregulationtoolkit.org/toolkit/5. Accessed 24 Jan 2019
19. GPP: TS 22.261, v16.6.0 - Technical Specification Group Services and System Aspects; Service requirements for the 5G system; Stage 1 (Release 16) (2018)
20. Utility Connect Homepage. https://www.utilityconnect.nl/. Accessed 24 Jan 2019
21. Agentschap Telecom: Internetverbinding verbeteren en 3,5 GHz-band (2019)
22. Staatscourant: nr 8903 (2011)
23. BNetzA: Anhörung zur lokalen und regionalen Bereitstellung des Frequenzbereichs 3.700 MHz bis 3.800 MHz für den drahtlosen Netzzugang (2018)
24. ISED: Radio Standards Specification RSS-197—Wireless Broadband Access Equipment Operating in the Band 3650–3700 MHz (2018)
25. ISED: SRSP-303.65—Technical Requirements for Wireless Broadband Services (WBS) in the Band 3650–3700 MHz
26. ISED: Consultation on Revisions to the 3500 MHz Band to Accommodate Flexible Use and Preliminary Consultation on Changes to the 3800 MHz Band (2018)
27. ECC: Draft ECC Report 292 (2018)
28. EC: Decision 2008/477/EC (2008)
29. ITU-R: Radio Regulations, Volume 1: Articles (2016)

30. ECC: Dec (14)02. Harmonised technical and regulatory conditions for the use of the band 2300-2400 MHz for Mobile/Fixed Communications Networks (MFCN) (2014)
31. ECC: Report 205 (2014)
32. Ofcom: Enabling opportunities for innovation. Consultation (2018)
33. FCC: Part 96 - Citizens Broadband Radio Service (2015)
34. ITU-R: Resolution 229. Use of the bands 5150–5250 MHz, 5250–5350 MHz and 5470–5725 MHz by the mobile service for the implementation of wireless access systems including radio local area networks (2012)
35. ITU-R: Recommendation M.1450-5. Characteristics of broadband radio local area networks (2014)
36. ETSI: EN 301 893. 5 GHz RLAN; Harmonised Standard covering the essential requirements of article 3.2 of Directive 2014/53/EU (2017)
37. LTE-U Forum homepage
38. MulteFire alliance homepage. http://www.multefire.org/. Accessed 24 Jan 2019

Unlocking the Potential of QoS-Aware Pricing Under the Licensed Shared Access Regime

Vaggelis G. Douros$^{(\boxtimes)}$, Andra M. Voicu, and Petri Mähönen

Institute for Networked Systems, RWTH Aachen University,
Kackertstrasse 9, 52072 Aachen, Germany
{vaggelis.douros,avo,pma}@inets.rwth-aachen.de

Abstract. We present a techno-economic analysis of a cellular market that operates under the licensed shared access (LSA) regime, consisting of a mobile network operator (MNO) that leases spectrum to a number of Programme Making and Special Events (PMSE) users. The MNO offers two quality-of-service (QoS) classes (high and low), differentiating the price based on the QoS class. The key question that we address is whether and to which extent the MNO has incentive to adopt this form of QoS-aware pricing. The first step is to model the parameters that are controlled by each PMSE user: (i) the way to choose between the two QoS classes and (ii) the available budget per QoS class. The second step is to compute the maximum revenue of the MNO. Our analysis reveals that the MNO can always tune the prices so as to maximise its revenue for the scenario where all users belong to the high QoS class. This is a consistent result throughout our study, that holds for any considered set of user-controlled parameters and of technical parameters. We conclude that the adoption of QoS-aware pricing in the LSA market generates a tussle between the MNO and the regulator. The MNO has incentive to support fewer users but with high QoS and charge them more, which is not aligned with the regulator's goal for social welfare maximisation.

Keywords: Techno-economics · Mobile network operators ·
Programme Making and Special Events

1 Introduction and Related Work

Licensed shared access (LSA) [6] has been adopted in Europe as a promising paradigm to dynamically share licensed spectrum between different networks and technologies. LSA proposes a two-tier approach where the initial target use case considered mobile network operators (MNOs) leasing spectrum in the 2.3–2.4 GHz band from incumbent technologies like Programme Making and Special Events (PMSE) [7]. However, recent initiatives from industry and spectrum regulators have proposed a symmetric use case, where PMSE users could lease

© ICST Institute for Computer Sciences, Social Informatics and Telecommunications Engineering 2019
Published by Springer Nature Switzerland AG 2019. All Rights Reserved
A. Kliks et al. (Eds.): CrownCom 2019, LNICST 291, pp. 291–305, 2019.
https://doi.org/10.1007/978-3-030-25748-4_22

spectrum from MNOs, targeting reliable short-term use of spectrum for concerts, conferences, etc. [13].

Though the adoption of LSA brings significant benefits from a technical perspective, a number of business challenges arise for the key stakeholders of the market (*i.e.*, regulator, incumbent spectrum user, and LSA licensee). These include the MNO's costs of additional infrastructure and the required modifications of the existing systems to support and manage the sharing procedure, as well as the license fees [15]. Thus, the stakeholders must perform a techno-economic analysis in order to assess whether LSA is worth the investment. However, business research on LSA is scarce [4,5,10] and focuses on the qualitative domain, without offering quantitative results on whether LSA schemes are techno-economically attractive.

The work closest to ours is [16], where an MNO that operates under the LSA framework leases spectrum to a number of PMSE users that belong to two distinct quality-of-service (QoS) classes, admitting either low or high QoS requirements. As in [16], we study scenarios where all users have either high or low QoS requirements, as well as mixed QoS requirements (*i.e.*, some users have low and some users have high QoS requirements). We extend the approach of [16], aiming at unlocking the potential of QoS-aware pricing in this LSA market, where we adopt price differentiation based on the QoS class. Our key contributions are the following. From the perspective of the PMSE users, we model the behaviour of the users regarding how they choose between the two QoS classes, as well as their available budgets for the two QoS classes. Through this process, we are able to predict the distribution of the users between the two QoS classes for each possible combination of considered prices.

From the perspective of the MNO, we identify the prices that correspond to the maximum revenue that can be achieved for each QoS scenario. A consistent result arises independently of (i) the distribution of the budgets, (ii) the way that the users choose between the QoS classes, and (iii) the values of the technical parameters. The MNO can always tune the prices so that the maximum revenue for the high QoS scenario is the highest, followed by the mixed QoS scenario and finally by the low QoS scenario. This result highlights the potential of QoS-aware pricing for the MNO, since the MNO has motivation to sacrifice some of the users with low QoS in order to support more users with high QoS and charge them more. This is also interesting from a regulatory point of view, since we identify a constant tussle in the LSA market, where the goal of the MNO (*i.e.*, revenue maximisation) is not aligned with the goal of the market regulator (*i.e.*, social welfare maximisation). Finally, we quantify the impact of the budget parameters on the revenue of the QoS scenarios, providing insights for which markets have the potential to be more profitable for the MNO.

2   The Techno-Economic Problem

We first summarise the techno-economic input from [16] that we are going to use for our analysis. Then, we introduce our extensions. We assume a monopolistic market with one MNO and N PMSE users that are interested in leasing spectrum from the unique MNO. Consistent with one of the business models in [13], the PMSE users also utilise the network infrastructure of the MNOs. Furthermore, the PMSE users are classified into two distinct QoS classes: there are at most N_L PMSE users with low QoS requirements (e.g., audio speech applications) and at most N_H PMSE users with high QoS requirements (e.g., high definition audio productions). We are interested in analysing from a techno-economic point of view the following three QoS scenarios:

- *Low QoS Scenario*: The MNO can support at most N_L users, where all of them have the same low QoS requirements Q_L.
- *High QoS Scenario*: The MNO can support at most N_H users, where all of them have the same high QoS requirements Q_H.
- *Mixed QoS Scenario*: The MNO supports users with mixed QoS requirements, *i.e.*, at most $N_{L,M}$ users with Q_L and at most $N_{H,M}$ users with Q_H.

Given the maximum number of supported PMSE users for the three QoS scenarios, the goal of the MNO is to define a pricing policy and choose the scenario that will maximise its revenue. Among the four pricing policies that have been considered in [16], we apply QoS-aware pricing, where the differentiation in the price is based on the QoS class that each user belongs to [8]. Depending on the assumptions and the model, QoS-aware pricing may maximise e.g. the revenue of the MNO or the social welfare [14,17].

We adopt a type of QoS-aware pricing which corresponds to an application of the *second degree of price discrimination* [9]. In this form of discrimination, there are at least two distinct prices, which correspond to at least two different types of services. Any customer who wants the same type of service will pay the same price. In our case, we propose that the discrimination is based on the QoS class that each PMSE user belongs to; each user that targets Q_L pays $P_L \in [P_{L,\min}, P_{L,\max}]$, whereas each user that targets Q_H pays P_H. We also define parameter $K = \frac{P_H}{P_L}$ which is always above 1. Then, the revenue of the MNO for each of the three QoS scenarios is:

$$\text{Low QoS Scenario: } N_L P_L, \tag{1}$$

$$\text{High QoS Scenario: } N_H P_H = N_H K P_L, \tag{2}$$

$$\text{Mixed QoS Scenario: } N_{L,M} P_L + N_{H,M} P_H = N_{L,M} P_L + N_{H,M} K P_L. \tag{3}$$

Clearly, the scenario that maximises the MNO's revenue can be computed by the following formula: $\max\{N_L, N_H K, N_{L,M} + N_{H,M} K\}$.

In [16], there has been an extensive study of the revenue for the three QoS scenarios. For different values of the technical parameters including carrier frequency f, propagation environment, base station (BS) transmit power level, and bandwidth, the maximum number of supported PMSE users for the three QoS scenarios has been computed. Then, the revenue after the application of QoS-aware pricing has been estimated for a fixed value of P_L and a range of values of P_H. A key assumption during the whole analysis was that the MNO always serves the maximum number of users that can be technically supported.

We generalise this study towards the following two directions. First, we introduce an additional degree of freedom studying markets with different values of P_L. Second, we relax the assumption that the market always performs at its maximum capacity by proposing a methodology to compute the exact number of PMSE users that will be admitted in each QoS scenario. In order to do so, we need to model the behaviour of the users. Initially, we need to model how a user chooses between the two QoS classes. Therefore, we introduce a metric w that quantifies the preference of each user i for each QoS class by weighing the importance that the user gives to the price and the QoS. For the high QoS class, w is defined as follows:

$$w_{H,i} = a_i \frac{P_L}{P_L + P_H} + (1 - a_i) \frac{Q_H}{Q_L + Q_H},$$

where the user-specific parameter a_i follows a uniform distribution in (0,1). When a_i is above 0.5, user i considers as the most important factor the price that it has to pay, otherwise the most decisive factor is the QoS that it gets. We note that we use fractions for a relative comparison of the two factors that influence the decision of the user, which is why w also ranges between 0 and 1.

Similarly, for the low QoS class, w is defined as:

$$w_{L,i} = a_i \frac{P_H}{P_L + P_H} + (1 - a_i) \frac{Q_L}{Q_L + Q_H}.$$

Note that $w_{H,i} + w_{L,i} = 1$, meaning that each user i needs to compute just one of them. If $w_{H,i}$ is higher than 0.5, then user i prefers the high QoS class. Otherwise, it prefers the low QoS class.

Another aspect that was not modelled in [16] is the user's available budget for each QoS class. Though we are not aware of specific studies for the distribution of the budgets of the PMSE users, we expect that it follows a (variation of the) normal distribution. This is in accordance with adjacent telecommunication markets [9]. More specifically, we model the distribution of the budget for the low QoS B_L as a truncated normal distribution with minimum value $P_{L,min} = \$10$ [11]. We need a minimum value, otherwise a user can never get access to this QoS class, so it is not of interest for this market. We study 6 cases for B_L, where the mean $\mu_L = \{0.5, 0.7, 0.9\} P_{L,max}$ and the standard deviation $\sigma_L = \{0.2, 0.4\} P_{L,max}$, with $P_{L,max} = \$120$ [11].

Then, we model the distribution of the budget for the high QoS B_H as a truncated normal distribution with minimum value B_L. The motivation for this minimum threshold is that the user's budget for the high QoS class should be at least equal to its budget for the low QoS class. For B_H, we also consider 6 cases, where the mean $\mu_H = \{0.2, 0.4, 0.6\} \frac{Q_H}{Q_L} B_L$ and the standard deviation $\sigma_H = \{0.2, 0.4\} \frac{Q_H}{Q_L} B_L$. The quantity $\frac{Q_H}{Q_L} B_L$ is used as a benchmark, since, as we know from adjacent markets [9], a typical user is expected to be willing to spend at most $\frac{Q_H}{Q_L}$ times more to get the class Q_H instead of the class Q_L. Moreover, since the budget of the users for more expensive services is expected to be tighter, the coefficients of μ_H are typically lower than the ones of μ_L.

Maximum Number of PMSE Users

Table 1 summarises the values of the technical parameters from [16] used to estimate the maximum number of PMSE users that can be technically supported. Each PMSE user has either high or low QoS requirements. We define the QoS requirements in terms of the target Application-layer throughput R, where high QoS and low QoS correspond to 4.61 Mbps and 150 kbps, respectively. These values are consistent with the highest and lowest PMSE audio throughput requirements in [3,13], where low throughput values correspond to audio speech applications, while high throughput values are required for high definition audio productions [12]. Based on these values of the technical parameters, Table 2 summarises from [16] the maximum number of users that can be supported for the three QoS scenarios. Since the number of users for the carrier frequencies of 2600 MHz and 3800 MHz are quite similar, we analyse only three cases: (i) 800 MHz for the indoor propagation environment, (ii) 800 MHz for the outdoor propagation environment, and (iii) 3800 MHz for the indoor propagation environment.

Table 1. PMSE user QoS requirements and technical parameters.

Parameter	Value		
	Low QoS scenario	High QoS scenario	Mixed QoS scenario
PMSE user QoS requirements as application-layer throughput R	150 kbps [3,13]	4.61 Mbps [3,13]	4.61 Mbps for 50% of the users in the high QoS scenario and 150 kbps for other users
Bandwidth C	20 MHz [2]		
Carrier frequency f	800, 2600, 3800 MHz [2]		
BS transmit power T	30 dBm [1,2] (same for all BSs)		
Propagation environment	Indoor, outdoor		

3 Revenue Analysis: A Case Study

In this section, we illustrate the evolution of the revenue for the three QoS scenarios for the example of the carrier frequency $f = 3800$ MHz and the indoor propagation environment. We assume that the market consists of 41 PMSE users so that, provided that all of them have the necessary budget to pay for the prices P_L and P_H, the maximum number of supported users can be admitted (*i.e.*, either $N_L = 37$, or $N_H = 4$). For a given set of prices P_L and P_H, we assume that the users follow a so-called *non-strict* version for the choice of the QoS class. In this non-strict version, a user initially applies for getting access to the QoS

Table 2. Max. number of users that can be supported for the three QoS scenarios for the different values of the technical parameters.

Frequency, Environment	Scenario			
	Low QoS	High QoS	Mixed QoS	
	Users N_L	Users N_H	Users $N_{L,M}$	Users $N_{H,M}$
$f = 800\,\mathrm{MHz}$, indoor	65	6	21	3
$f = 800\,\mathrm{MHz}$, outdoor	7	2	4	1
$f = 2600\,\mathrm{MHz}$, indoor	36	4	13	2
$f = 2600\,\mathrm{MHz}$, outdoor	31	4	12	2
$f = 3800\,\mathrm{MHz}$, indoor	37	4	13	2
$f = 3800\,\mathrm{MHz}$, outdoor	33	4	12	2

class that it prefers more based on the value of the weighted metric w. It gets access to this QoS class provided that the following two conditions hold: (i) it can afford to pay the price that the MNO has announced and (ii) the MNO has not reached the maximum number of PMSE users that it can support for this QoS class. If the user does not get access to the QoS class of its first choice, then it applies for the other QoS class and it gets admitted provided that the same conditions hold. In the following section, we also consider a *strict* version for the choice of the QoS class, where each user applies for only one QoS class, *i.e.*, the one that corresponds to the highest value of the weighted metric w.

After deciding whether a user will be admitted and, if so, in which QoS class, the MNO computes the revenue for the three QoS scenarios. We consider four values of P_L, corresponding to 30, 60, 90, and 120 \$ for 48-hour access [11]. For a given P_L, we apply QoS-aware pricing where $P_H = K P_L$, with parameter $K \in \{2, 3, \ldots, \left\lfloor \frac{Q_H}{Q_L} \right\rfloor = 30\}$.

Figure 1 shows the evolution of the revenue for the three QoS scenarios for the four values of P_L. Each subfigure corresponds to the revenue as a function of parameter K, for a given P_L. The results are averaged based on the simulation of 1000 markets, each consisting of 41 users. As we notice from Fig. 1(a), when parameter K is below 7, the low QoS scenario generates the highest revenue. This is justified since the price differentiation between Q_H and Q_L is small enough to not overcome the difference between the actual number of users that are supported for Q_H and Q_L. For higher values of K, the high QoS scenario generates the highest revenue, followed by the mixed QoS scenario. Also, the revenue for both the high QoS and the mixed QoS scenario increases linearly with K. This is expected from the corresponding Eqs. (2) and (3) provided that the number of users N_H and $N_{H,M}$ does not change with K. Finally, for the low QoS scenario, the revenue does not change with K, so any fluctuation is due to changes in the number of users.

Figure 1(b) shows the revenue for $P_L = \$60$, where we notice some differences in the trends. First, though P_L was doubled compared to Fig. 1(a), the revenue

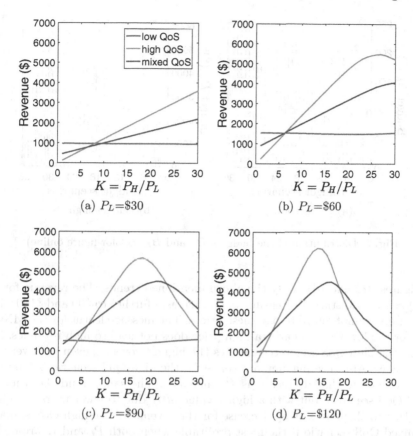

Fig. 1. Evolution of the revenue for the three QoS scenarios. Technical parameters: carrier frequency f=3800 MHz and indoor propagation environment. Parameters of the distribution of the budgets B_L and B_H: $\mu_L = 0.7P_{L,\max}$, $\sigma_L = 0.4P_{L,\max}$, $\mu_H = 0.4\frac{Q_H}{Q_L}B_L$, $\sigma_H = 0.2\frac{Q_H}{Q_L}B_L$. The choice of the QoS class is non-strict.

for the low QoS scenario was not doubled. This means that the budget B_L of some of the users is below \$60 and, therefore, they cannot afford to pay for this QoS class. Due to this, the high QoS scenario generates the highest revenue starting with a smaller value of K (it is for $K > 6$, whereas for $P_L = \$30$ it was for $K > 7$). Moreover, for high values of K, the revenue for the high QoS scenario starts increasing sub-linearly and then it decreases. This is again due to budget constraints, this time for the budget B_H. The trend of a sub-linear increase is also noticed for the mixed QoS scenario, though it starts for higher values of K compared to the high QoS scenario. This is expected since, for the mixed QoS scenario, the maximum number of users with high QoS that can be admitted is 2 instead of 4 for the high QoS scenario (see Table 2). Therefore, for higher values of K, it is easier to find 2 instead of 4 users with Q_H.

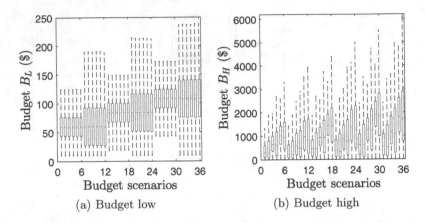

(a) Budget low   (b) Budget high

Fig. 2. Distribution of the budgets B_L and B_H. (Color figure online)

Figures 1(c) and (d) verify the above mentioned trends. The revenue for the low QoS scenario starts decreasing as P_L increases further to \$90 and \$120, since many users cannot afford to pay these prices. The message learnt for the MNO is that, for the low QoS scenario, a high price does not lead to high revenues. Due to this, the high QoS scenario generates the highest revenue, even with very low values of K. Also, the maximum revenue for the high QoS scenario is admitted for a value of K that decreases as P_L increases. The same trends hold for the mixed QoS scenario, but with a higher value of K due to fewer users with high QoS. Due to this and a steep decrease for the revenue of the high QoS scenario, the mixed QoS scenario is the most profitable when both P_L and K are high.

4   Revenue Analysis: General Results

Through the detailed analysis of the previous section, we are able to compute the expected revenue of the three QoS scenarios for every possible combination of the techno-economic parameters. Though this methodology provides a fine-grained view for each case, we need to extract general conclusions. Indeed, for a given set of techno-economic parameters, the ultimate challenge for the MNO is to choose the prices P_L and P_H so that its revenue will be maximised. Therefore, we can consider this fine-grained analysis as an internal process for the MNO to compute: (i) the value of P_L that maximises its revenue for the low QoS scenario, (ii) the value of P_H, *i.e.*, parameter K and P_L, that maximises its revenue for the high QoS scenario, and (iii) the values of P_L and P_H that maximise its revenue for the mixed QoS scenario. Then, the MNO can choose which QoS scenario maximises globally its revenue.

Though the MNO controls the technical parameters and the price, the distribution of the users' budgets as well as the users' preferences for the two QoS classes are private information. The complementary problem of how to estimate this piece of information is not addressed in this paper. However, we present a broad number of scenarios for the parameters that each user controls, so as to estimate the revenue for the three QoS scenarios under different users' behaviours.

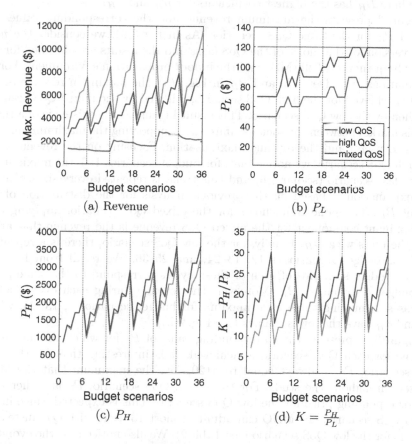

Fig. 3. Max. revenue and the corresponding values for P_L and P_H for the three QoS scenarios. Technical parameters: $f = 3800$ MHz, indoor. The choice of the QoS class is non-strict.

Initially, we generalise the results of the previous section where we consider 36 budget scenarios for the distribution of the users' budgets B_L and B_H. The number of budget scenarios arises since the 4-tuple $\{\mu_L, \sigma_L, \mu_H, \sigma_H\}$ can get $3 \cdot 2 \cdot 3 \cdot 2 = 36$ possible values. Figure 2 represents the evolution of the budget distribution. We progressively update the elements of the 4-tuple in four loops, with the following order from the outermost loop to the innermost loop: (i) μ_L, (ii) σ_L, (iii) μ_H, and (iv) σ_H. Due to this, as we can see from Fig. 2(a), μ_L, depicted as a red line, increases every 12 budget scenarios, remaining the same for scenarios 1–12, 13–24, and 25–36. Let us consider scenarios 1–12: due to a higher value of σ_L, scenarios 7–12 have higher upper quartiles and whiskers than scenarios 1–6. For the case of B_H (Fig. 2(b)), we notice that every 6 scenarios where μ_L and σ_L are fixed (*i.e.*, scenarios 1–6, 7–12, etc.), the upper quartile increases. Moreover, the maximum upper whiskers correspond to scenarios 6, 12, etc., where B_H has the highest coefficients for μ_H and σ_H.

Figure 3 presents the maximum revenue and the corresponding values for P_L and P_H for the three QoS scenarios. As in Fig. 3(a), we consider the *non-strict* version for the choice of the QoS class and the results are obtained for the carrier frequency $f = 3800$ MHz and the indoor propagation environment. For all combinations of budgets B_L and B_H in Fig. 3(a), the maximum revenue of the MNO is achieved for the high QoS scenario, followed by the mixed QoS scenario and then by the low QoS scenario. This result highlights the existence of a tussle for this market between the social welfare (*i.e.*, supporting the maximum number of PMSE users) and the revenue maximisation. Focusing on the revenue from the high QoS scenario, we notice that, for budget scenarios 1–6, the maximum is for the last scenario (scenario 6) and this trend is repeated every six scenarios. The explanation is based on the previous analysis for the distribution of the budget B_H. The same trend holds for the mixed QoS scenario, implying that the dominant component for the mixed QoS revenue is the revenue that arises from the users with Q_H. Finally, for the low QoS scenario, there is a repeating trend for budget scenarios 1–12, 13–24, and 25–36. We recall from Fig. 2(a) that all budget scenarios of each of these cycles correspond to the same μ_L of the budget distribution B_L. Moreover, the revenue during each cycle slightly decreases, admitting three local maxima for budget scenarios 1, 13, 25, where μ_H and σ_H have the lowest values (see Fig. 2(b)).

Figure 3(b) presents the corresponding value of P_L for which the maximum revenue for each QoS scenario is achieved. It is interesting that for the high QoS scenario, P_L is always equal to $120, *i.e.*, the maximum that the MNO can set throughout the study. For the mixed QoS scenario, P_L is higher than the corresponding price for the low QoS scenario. This is expected, since in the mixed QoS scenario, the MNO can admit at most 13 users with Q_L, instead of 37 users for the low QoS scenario (see Table 2). We also notice that the evolution of P_L is similar for both low and mixed QoS scenarios, with the highest values being for budget scenarios 31–36, where μ_L and σ_L get the highest values (see Fig. 2(a)).

Then, we show in Fig. 3(c) the corresponding value of P_H. As expected, it is higher for the mixed QoS scenario where at most 2 users with Q_H can be supported than for the high QoS scenario where $N_H = 4$. Moreover, the curves follow the same trend with the revenue. Finally, Fig. 3(d) depicts the evolution of parameter $K = \frac{P_H}{P_L}$, where the trends are similar with the trends for P_H. Clearly, there is room for the MNO to apply higher price differentiation for the case of the mixed QoS scenario compared to the high QoS scenario. Our analysis suggests that in budget scenarios where μ_H and σ_H get the highest values, the MNO has motivation to charge the mixed QoS users with Q_H at the maximum level of price differentiation, *i.e.*, 30 times more than the users with Q_L.

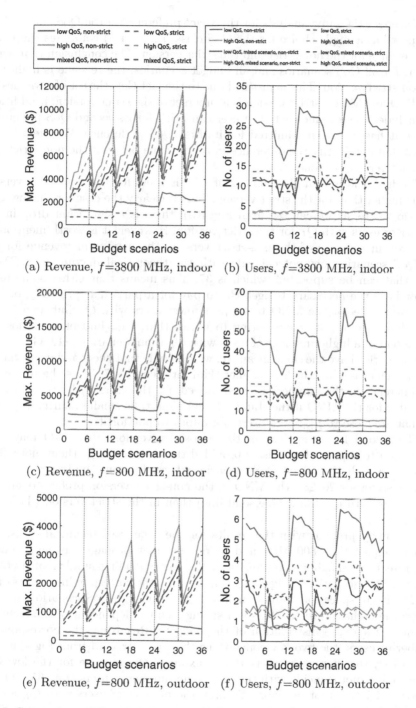

Fig. 4. Comparison of the max. revenue and the corresponding number of users for the non-strict and the strict choice of the QoS class. (Color figure online)

We repeat the same analysis for the *strict* preference of the QoS class, where each user has a single choice for the QoS class. Figure 4(a) compares the maximum revenue for the non-strict and the strict version. The conclusion that arises is that, for all QoS scenarios and all budget scenarios, the revenue is higher for the non-strict version. This is justified due to the fact that the set of revenues for the MNO for the non-strict version is a superset of the strict version: it additionally includes the revenue that each user can bring for its second QoS preference in case it has not been admitted for its first QoS preference. We identify the factors that can justify the difference in the revenue between the non-strict and the strict version, as follows.

The first one is that the number of PMSE users for the non-strict version can be higher than for the strict version. This is clearly the case for the low QoS scenario where, as we can see from Fig. 4(b), there is a significant drop in the number of users with Q_L for the strict version. However, it is worth mentioning that even in the case of the non-strict version, the maximum revenue for the low QoS scenario does not coincide with the theoretical maximum of PMSE users that can be supported, which is 37. This means that either some users do not have the necessary budget B_L to pay for a particular price P_L, or it is more profitable for the MNO to support fewer users with Q_L but at a higher price. Furthermore, it is interesting to notice that, e.g., budget scenarios 1–6 correspond to a higher number of users with Q_L than scenarios 7–12. Given that these scenarios have the same mean μ_L, we conclude that the standard deviation σ_L for scenarios 1–6, which is smaller than for scenarios 7–12, is the reason for the difference in the number of users. Indeed, for the users with Q_L, it is more profitable for the MNO if the standard deviation σ_L is smaller, since, for prices P_L that are close to μ_L, more users can afford to pay for it.

The second factor is that, in the non-strict version, the MNO may have motivation to support fewer users provided that it can charge them more. This is the case with the mixed QoS scenario, where, for some budget parameters (budget scenarios 26–28), the MNO in the non-strict version prefers to support fewer users with Q_L (dark blue solid line) than in the strict version (dark blue dashed line).

We finally proceed with the results for the other two technical cases, *i.e.*, carrier frequency f=800 MHz and indoor/outdoor propagation environment. We present the maximum revenue and the corresponding number of users for the three QoS scenarios in Figs. 4(c)–(f), omitting the corresponding values of P_L and P_H due to space constraints. As in Fig. 4(a), the high QoS scenario generates always the highest revenue. This is a strong result independent of the technical parameters and the distribution of the budgets. Regarding the corresponding number of users, the two key conclusions that we extracted from Fig. 4(b) still hold. First, the number of users that maximises the revenue for the low QoS scenario does not coincide with the maximum number of users (*i.e.*, 65 users for indoor and 7 users for outdoor). Second, the number of users with Q_L for the mixed QoS scenario is in general lower for the non-strict version compared to the strict version, since the MNO has motivation to support fewer users with

Q_L in order to admit more users with Q_H and charge them with high values of K. This trend becomes clearer in Fig. 4(f), where the non-strict version of the mixed QoS scenario (dark blue solid line) is almost always below the strict version of the mixed QoS scenario (dark blue dashed line).

5   Conclusions and Outlook

The goal of this work was to unlock the potential of QoS-aware pricing for an MNO that operates under the LSA regime. The business model for the MNO was to lease spectrum to PMSE users, differentiating their prices based on whether they belong to the high or the low QoS class. We analysed three QoS scenarios: (i) all users have the same low QoS requirements, (ii) all users have the same high QoS requirements, and (iii) a mixed QoS scenario.

From the perspective of the PMSE users, we made two contributions. First, we modelled the behaviour of the users regarding how they choose between the two QoS classes, quantifying the importance that each user gives to the QoS class versus the price that it has to pay. Second, we modelled the distribution of the budget of the users for the two QoS classes. The added value of these models is that we were able to perform a fine-grained analysis, predicting the distribution of the users between the two QoS classes for each possible combination of considered prices.

From the perspective of the MNO, the challenge was to choose the prices P_L and P_H so as to compute the maximum revenue that can be achieved for each QoS scenario. Our analysis revealed a consistent result that holds independent of (i) the distribution of the budgets, (ii) the way that the users choose between the QoS classes, and (iii) the values of the technical parameters. The MNO can always tune the prices so that the maximum revenue for the high QoS scenario is the highest, followed by the mixed QoS scenario and finally by the low QoS scenario. This result highlights the potential of QoS-aware pricing for the MNO. For the high and mixed QoS scenarios where QoS price differentiation can be applied, the MNO can consistently generate higher revenue than for the low QoS scenario. This is also interesting from a regulatory point of view, since the MNO has motivation to support few users charging them at a higher price instead of supporting more users at a lower price. Therefore, we identified a constant tussle in the LSA market, where the goal of the MNO (*i.e.*, revenue maximisation) is not aligned with the goal of the market regulator (*i.e.*, social welfare maximisation).

Through the analysis of the revenues for the different budget scenarios, we identified the impact of the budget parameters on the revenue of the QoS scenarios. The revenue for the high and mixed QoS scenarios admits local maxima when both the mean and the standard deviation of the budget distribution B_H are high (budget scenarios 6, 12, etc.). On the other hand, the revenue for the low QoS scenario admits local maxima when the mean of the budget distribution B_L and both parameters of the budget distribution B_H are small (budget scenarios 1, 13, 25). These trends hold for any values of the technical parameters. We argue that they are useful in particular for an MNO who evaluates the business opportunities in different markets before entering into them since they provide insights for which markets have the potential to be more profitable.

Finally, we conclude with two key messages extracted from our study for the mixed QoS scenario. First, there is higher room for price differentiation for the mixed QoS scenario, since fewer users with Q_H can be admitted compared to the high QoS scenario. Second, for the non-strict version of the choice of the QoS class, the MNO usually prefers to sacrifice some of the users with Q_L in order to support more users with Q_H and charge them more. Both conclusions reinforce the message learnt, $i.e.$, that the application of QoS-aware pricing unlocks significant revenue opportunities.

As future work, it is interesting to extend this study by introducing an additional (intermediate) QoS class and evaluate the robustness of the results. This also requires a modification for the way that the users choose among the three QoS classes. Another interesting direction is to consider an oligopoly market with two or three MNOs, analysing the churn of the users and the evolution of the revenue as the MNOs update their pricing policies.

Acknowledgment. The authors would like to thank Shaham Shabani who conducted the simulations in ns-3 for estimating the maximum number of PMSE users for all QoS scenarios.

References

1. 3GPP: Evolved Universal Terrestrial Radio Access (E-UTRA); Further advancements for E-UTRA physical layer aspects. TR 36.814, V9.2.0, March 2017
2. 3GPP: LTE; Evolved Universal Terrestrial Radio Access (E-UTRA); Base Station (BS) radio transmission and reception. TS 36.104, V14.5.0, October 2017
3. 3GPP: Study on communication for automation in vertical domains. TR 22.804, V1.2.0, May 2018
4. Ahokangas, P., et al.: Business scenarios for incumbent spectrum users in licensed shared access (LSA). In: Cognitive Radio Oriented Wireless Networks and Communications (CROWNCOM). Oulu, Finland, June 2014
5. Ahokangas, P., et al.: Business models for mobile network operators in Licensed Shared Access (LSA). In: IEEE Dynamic Spectrum Access Networks (DYSPAN). McLean, VA, USA, April 2014
6. ECC: Licensed Shared Access (LSA). Report 205, February 2014
7. ECC: Guidance for the implementation of a sharing framework between MFCN and PMSE within 2300–2400 MHz. Recommendation (15)04, July 2015
8. Huang, J., Gao, L.: Wireless network pricing. Synth. Lect. Commun. Netw. **6**(2), 1–176 (2013)
9. Maillé, P., Tuffin, B.: Telecommunication Network Economics: From Theory to Applications. Cambridge University Press, Cambridge (2014)
10. Matinmikko-Blue, M., Yrjölä, S., Seppänen, V., Ahokangas, P., Hämmäinen, H., Latva-aho, M.: Analysis of spectrum valuation approaches: the viewpoint of local 5G networks in shared spectrum bands. In: IEEE Dynamic Spectrum Access Networks (DYSPAN). Seoul, South Korea, October 2018
11. Ofcom: PMSE fees, August 2018. https://www.ofcom.org.uk/manage-your-licence/radiocommunication-licences/pmse/fees

12. Pilz, J., Holfeld, B., Schmidt, A., Septinus, K.: Professional live audio production: a highly synchronized use case for 5G URLLC systems. IEEE Netw. **32**(2), 85–91 (2018)
13. PMSE-xG: White paper: PMSE and 5G, March 2017
14. Shetty, N., Schwartz, G., Walrand, J.: Internet QoS and regulations. IEEE/ACM Transact. Netw. (TON) **18**(6), 1725–1737 (2010)
15. Tehrani, R.H., Vahid, S., Triantafyllopoulou, D., Lee, H., Moessner, K.: Licensed spectrum sharing schemes for mobile operators: a survey and outlook. IEEE Commun. Surv. Tutor. **18**(4), 2591–2623 (2016)
16. Voicu, A.M., Shabani, S., Douros, V.G., Simić, L., Mähönen, P.: Techno-economics of licensed shared access with mobile network operators leasing spectrum to PMSE users. In: Research Conference on Communications, Information and Internet Policy (TPRC). Washington, DC, USA, September 2018
17. Wang, Y.C., Tsai, T.Y.: A pricing-aware resource scheduling framework for LTE networks. IEEE/ACM Transact. Netw. (TON) **25**(3), 1445–1458 (2017)

Location Dependent Spectrum Valuation of Private LTE and 5G Networks in Europe

Topias Kokkinen[1], Heikki Kokkinen[1(✉)], and Seppo Yrjölä[2]

[1] Fairspectrum, Turku, Finland
{info,heikki.kokkinen}@fairspectrum.com
[2] Nokia, Oulu, Finland
seppo.yrjola@nokia.com

Abstract. Emerging private LTE and 5G services and applications have created need for local radio spectrum licensing. The existing pricing models for licenses do not work well in this context. This paper introduces three new location dependent valuation methods that aim to produce more accurate pricing for local licenses. We use FICORA Frequency Fee as our base-case general spectrum valuation model, and we replace the population density based location coefficient with proxies such as employee density, value added per employee, and rent prices. By comparing the differences in the prices yielded by the models, we show that the new models can in some cases identify high demand areas like hospitals and industrial districts better than the original population density based model. Additionally, we conclude that the original population density based model and the new employee density based model could be used together to capture both the consumer and the industrial spectrum demand simultaneously.

Keywords: Private LTE · 5G · Spectrum pricing · Valuation

1 Introduction

1.1 Motivation

Private LTE and 5G networks serve enterprise business, government or education using mobile network technology. The studied mobile networks operate in the radio spectrum bands, which are defined by 3GPP in specification TS 36.101 (2018b). The value of the mobile spectrum for private LTE and 5G networks is dependent on multiple factors, including the bandwidth, duration, area, and location. In this study, the specific interest is in the location of the network area. For services targeted to consumers, the valuation follows the population density, but as no one lives in factories, ports, or in shopping malls, other location proxies should be found for spectrum valuation of private LTE and 5G networks.

Mobile Network Operators (MNOs) have spent billions on gaining exclusive access to spectrum assets and deploying network infrastructure needed to meet current and future consumer demand. LTE evolution and future 5G networks target location specific solutions to meet stringent wireless connectivity needs of distinct vertical use

© ICST Institute for Computer Sciences, Social Informatics and Telecommunications Engineering 2019
Published by Springer Nature Switzerland AG 2019. All Rights Reserved
A. Kliks et al. (Eds.): CrownCom 2019, LNICST 291, pp. 306–319, 2019.
https://doi.org/10.1007/978-3-030-25748-4_23

cases in the capacity bands. The 5G technologies bring drastic changes to how mobile spectrum is used (5 GPPP, 2016). A significant driver for this is the change in the demand for the licenses of mobile spectrum. During LTE era, the demand has mainly consisted of Mobile Network Operators, which provide Mobile Broadband (MBB) services. However, the potential user-base of 5G is much more diverse. For example, 5G serves the specific communication needs of industries like manufacturing, logistics, and education (Cave and Nicholls, 2017).

The industry and site specific networks require local, private network deployments in contrast to the nation-wide public networks of MNOs. Because of this, there is also a need for a change in the supply side. The recently proposed micro licensing model (Matinmikko, et al. 2017a and 2017b, private LTE and 5G (Ferrus and Sallent, 2014), and network slicing (Alliance, 2016) are concepts for creating customized mobile communications services. Of these methods, micro licensing and private LTE and 5G allow the transaction of spectrum rights in a dynamic way that caters the needs of different user types. In practice, the allocation could be done through a marketplace that works as a centralised, efficient secondary market of private LTE and 5G licenses (Kokkinen et al., 2017) or network slicing (Lemstra, 2018). Cramton and Doyle (2015) state that an open access market for spectrum would increase competition and make the process more efficient, transparent, fair, and simple.

The possibility to deploy private LTE and 5G networks is highly dependent on the spectrum availability. The spectrum could become available through local licenses in the mobile spectrum bands or by allowing unlicensed access to 5G bands (European Commission, 2017). Furthermore, all commercial 5G licenses are advised to be subject to trading or leasing (European Commission, 2018). The novel regulatory approaches include locally licensed mobile spectrum (The Federal Network Agency Germany, 2018; The Swedish Post and Telecom Authority, 2018; Radiocommunications Agency Netherlands, 2018; Ofcom, 2018) wholesale spectrum provisioning, and the secondary market of spectrum. Especially, the 2.3 GHz, 3.5 GHz, and 24 GHz frequency bands are likely to require different approaches to authorization, as they are expected to be the enablers for private LTE and 5G services and applications. The regulators foresee the need for more flexibility in 5G spectrum authorization approaches including the commons approach (general authorization, unlicensed), licensed shared use between different users, geographical sharing, or dynamic spectrum sharing in time, frequency, and location (European Commission, 2018). Sharing-based spectrum management approaches facilitate more efficient spectrum use by allowing two or more radio systems to operate in the same frequency band (Beltran, 2017). Prominent sharing concepts under standardization and trials are the European Licensed Shared Access (LSA) (ECC, 2014), the US based Citizens Broadband Radio Service (CBRS) (FCC, 2016), and the unlicensed LTE technologies: LAA (3 GPP, 2015), LTE-U (LTE-U Forum, 2016), and MulteFire (MulteFire, 2016). 5G convergence with IEEE family of technologies using unlicensed spectrum particularly indoors and dense urban area introduce new opportunities for the co-existence of the 3GPP and the IEEE Wi-Fi ecosystems (Abinader et al., 2014). The 3GPP study item "Study on New Radio (NR) based Access to Unlicensed Spectrum" determines a global solution for NR-based access to unlicensed spectrum (3 GPP, 2018a).

Spectrum management aims at effectiveness by allocating spectrum to the right use, and efficiency by assigning to those what value it the most (Beltran, 2017). Regulators aim at making the best value of spectrum in their decisions, but assessing the value of spectrum is a complex process with multiple perspectives (Bazelon and McHenry, 2013; Mölleryd et al. 2012; ITU-R, 2012). Different wireless services, such as mobile broadband (MBB) communications, Private Mobile Radio (PMR), broadcast, and military use, have different basis for their value due to their distinct business models, technologies, and role in society. Ultimately, the spectrum management decisions are about maximizing the value of spectrum, its efficient utilization, and its benefits to society (Beltran, 2017).

The mobile communication market has traditionally been centered around a small number of MNOs that have been granted long-term exclusive spectrum licenses, most recently through auctions with high up-front payments (Cramton, 2013). While auctions have resulted significant income for the governments in many countries, their impact on society goes beyond auction revenues and has turned out to be a complicated topic to analyze (Cramton, 2013; Hazlett and Muños, 2009; Cave and Nicholls, 2017; Klemperer, 2002). For example, competition, which will ultimately lead to greater innovation and better and cheaper services, will likely generate greater governmental revenues in the long term compared to the sole auction revenues (Cramton, 2013). LTE evolution and future 5G networks are expected to change the mobile communication market structure and be increasingly locally deployed by new entrant stakeholders. Facility owners' role as a local operator serving MNOs' customers is highlighted by Zander (2017) and Ahmed, Markendahl and Ghanbari (2013) as a feasible solution for the deployment of building specific ultra-dense networks. Furthermore, local high-quality 5G wireless networks are gaining increasing attention as the solution to deliver guaranteed quality of service, particularly concerning the low latency requirements, in various use cases of vertical sectors and enterprises (Guirao et al., 2017). Private mobile communication networks as stand-alone solutions or collaboratively serving MNOs' customers are particularly envisaged to operate in shared spectrum bands (ETSI, 2018). Spectrum options for local indoor network deployments by local operators were assessed by Ahmed, Markendahl and Ghanbari (2013) for different spectrum allocation options where the local operators were either collaborating closely with the MNOs or deploying their own independent networks.

1.2   Contribution

In Finland since 2009, the spectrum price of the mobile spectrum bands consists of auction price and yearly frequency fee, which is called FICORA frequency fee in this paper (Note that Finnish Communications Regulatory Authority merged with the Finnish Transport Agency, and as of 1.1.2019, the organisation is called TRAFICOM). In this paper, we use FICORA frequency fee with the modification that the population coefficient is replaced with employee density and with employee density factored with the industry specific value of an employee in the employee-dense areas. The final value of the spectrum is the higher one of FICORA frequency fee and our employment based FICORA frequency fee.

The allocation of local licenses raises an interesting question about how they should be valuated. A marketplace requires a cost-effective and accurate method for valuating the licenses. Traditionally, the mobile spectrum licenses have been sold in large nation-wide bundles to (MNOs) through auctions. There has been less research on how the value of the licenses is distributed on a local level. The demand and value of licenses can drastically change between different locations. Moreover, the type of use also affects the price. While the current mobile licenses are primarily used for Mobile Broadband, 5G technology can be used for diverse use cases, including private net-works. These should be taken into consideration in the valuation of the licenses.

There are several challenges related to spectrum market, one of which is the val-uation of the licenses. In this paper, we research different methods of valuation and compare the pricing results that the methods yield. The aim of this paper is to develop a location dependent valuation method of private LTE and 5G spectrum for industrial users. This study contributes to the literature the need to extend the private LTE and 5G spectrum valuation methods with an employment based, geographic distribution of the spectrum price. The reminder of this paper will continue as follows: Spectrum valu-ation for private LTE and 5G is introduced in Sect. 2, Sect. 3 discusses the proposed valuation model and data to validate the model, the results are described in Sect. 4, and the conclusions can be found in Sect. 5.

2   Spectrum Valuation for Private LTE and 5G

2.1   Conventional Valuation Methods

The commercial value is the price of a private LTE and 5G radio spectrum license should it be for sale. We focus on the commercial value of local licenses for private LTE and 5G networks. The value is determined by the expected future cash flows that the license generates to its holder. These cash flows are generated through additional increases in revenues or reductions in costs. The license can for example be used to offer a new product to increase revenues, or to implement a new manufacturing method to reduce costs. Licenses can also be used defensively to limit competition. By blocking competition, the license holders can use their improved market power to increase cash flows. Additionally, Marks et al. (2009) note that radio spectrum licenses hold a significant option value.

The valuation of licenses should be based on the above mentioned underlying economic factors. Conventionally, the price of a mobile license is determined through auctions. Given a sufficient number of buyers, auctions result in relatively accurate pricing as the buyers can use sophisticated, context specific financial models to esti-mate the economic factors. However, due to the small value and illiquidity of licenses for private LTE and 5G networks, auctions are often not a viable method of pricing. One local license can for example only cover the area of a single factory or port, which means that the license has only one potential buyer. Small number of buyers makes auctions inefficient (Tonmukayakul and Weiss, 2008). Additionally, the context specific financial models used by the buyers in auctions, as well as in the bidding process, can be very resource intensive. These costs can be relatively large compared to the commercial value generated by the trade of a local license for private LTE and 5G networks.

Benchmarking is another market-based approach for pricing radio spectrum licenses. Benchmarking would solve the problem of high valuation costs as it can gather large amounts of market information cost effectively. However, the markets for local 5G licenses do not currently exist in large extent, so due to the immaturity of the market, benchmarking is not viable in the early stages of 5G. Furthermore, even if the market was more mature, it could prove to be very difficult to find sufficiently similar comparables for the licenses as the value of the license is determined by many context specific factors. Thus, if benchmarking was to be used, it should be adjusted to account for these factors.

2.2   General Spectrum Valuation Model

As the market-based valuation methods accommodate the market for local 5G licenses poorly, we seek to find a general valuation model that uses the characteristics of the spectrum to determine a price for the license. In this paper we study whether some set of intrinsic characteristics can be used as proxies to estimate the commercial value with a sufficient accuracy.

Typically, general spectrum valuation models used by regulators consist of the following variables: the opportunity cost for a given band and location, the amount of spectrum used, the type of service, the frequency band, and the location (Marks et al., 2009). Kokkinen et al. (2018) used the frequency fee developed by the Finnish Communications Regulatory Authority to price spectrum licenses. The Table 1 shows that the formula used by FICORA (Finnish Ministry of Transport and Communications, 2017) fits well to the general model. Both models measure similar intrinsic commercial value drivers.

Table 1. Fitting FICORA frequency fee to the general model.

General model	FICORA frequency fee
Opportunity cost for a given band and location	P basic fee
Amount of spectrum used	B0 relative bandwidth
Type of service provided	S basic fee coefficient (Type of radio equipment used) C6b system coefficient (Scaled number of transmitters used)
Frequency band	C1 frequency band coefficient
Location	Cinh population coefficient

FICORA fee uses population density as a measure for location value. However, as noted by Kokkinen et al. (2018), population density might not accurately estimate the commercial value of local 5G licenses as these are often used in industrial districts and sites such as factories and ports. These locations typically have a low population density even though the willingness to pay and demand for licenses might be high. Thus, using population density might underestimate the value of local 5G licenses. In this paper, we present alternative proxy measures to estimate the location value for local 5G licenses.

We sought to find alternative measures that are based on globally available open data. We research potential measures that would be based on proxies such as land prices, density of business activity, and value added locally. However, we selected employee density and commercial property rental prices as proxies for location value because the availability of data and our hypothesis that including either one or both of these measures in the valuation formula would improve its accuracy in valuation of local 5G licenses.

3 Valuation Model and Data

3.1 Approach

We use the FICORA Frequency Fee formula as our base case model. The formula uses population density as a measure for location value. In this section, we substitute the population density with other measures, namely employee density, adjusted employee density, and commercial property rental prices. The total value of the spectrum is obtained using the FICORA Frequency Fee, and it is then redistributed using other valuation methods. Thus, this paper mainly studies the relative prices yielded by different valuation methods. The absolute prices of licenses would change if other methods such as benchmarking would be used to calculate the total base value of the spectrum. The total value of licenses is the same in all models. However, the way in which this total value is distributed to different locations changes.

If a combination of two methods is used, such as the max function of two valuations, the total value for the spectrum will be higher than the total base value of the spectrum. The base value reflects the spectrum value when it is used only for consumer services.

3.2 Data

We use population density, employee density, and area data from Official Statistics of Finland (2018a). The database includes statistics for all 3030 postcode areas and it uses 2018 postal area classification. The database includes the most recent population and employee data, from 2016 and 2015 respectively. The value added per employee by industry data used for Adjusted Employee Based valuation is from Official Statistics of Finland (2018b). The commercial property rental prices are obtained from City of Helsinki (2018). The rental price data used a different area classification so the data was matched to postcode areas as closely as possible.

3.3 Base Case: FICORA Frequency Fee Using Population Density

FICORA Frequency Fee formula is currently used in Finland to determine the annual frequency fee for all spectrum licenses in Finland. It is based on factors such as availability, usability, and number of frequencies in the license (Finnish Ministry of

Transport and Communications, 2017). The formula fits well to the general model as seen from Table 1.

$$FICORA\ Frequency\ Fee = C_1 * C_{inh} * C_{6b} * B_0 * S * P \tag{1}$$

Table 2. FICORA frequency fee coefficients.

Coefficient name	Coefficient	Value
Frequency band coefficient	C_1	0.4
Population coefficient	C_{inh}	Variable
System coefficient	C_{6b}	1
Relative bandwidth	B_0	2000
Basic fee coefficient	S	0.018
Basic fee	P	1295.5 €

The constant values on Table 2 are set by FICORA (Finnish Ministry of Transport and Communications, 2017) for a 1-year public mobile network license with a bandwidth of 10 MHz and area of 1 km in the 3.5 GHz frequency band. The values are also the same as used in Kokkinen et al. (2018). The population coefficient is calculated for each postal code using the Eq. (2).

$$C_{inh} = \frac{POP_{PC}}{POP_{FIN}} * \frac{1\,km}{A_{PC}} \tag{2}$$

Where POP_{PC} is the population of the postal code area, POP_{FIN} the total population of Finland, $1\,km$ the constant area of the license, and A_{PC} the area of the postcode.

3.4   Location Coefficient Variation: Employee Density

The Employee Density formula is the same as FICORA formula, with the exception that employee density data is used instead of population density data. The Employee Coefficient Cemp is obtained by dividing the number of employees working in the license area by the number of employees in Finland.

$$Employee\ Density\ Valuation = C_1 * C_{emp} * C_{6b} * B_0 * S * P \tag{3}$$

$$C_{emp} = \frac{EMP_{PC}}{EMP_{FIN}} * \frac{1\,km}{A_{PC}} \tag{4}$$

Where EMP_{PC} is the number of employees working in the postal code area, EMP_{FIN} the total number of employees in Finland, $1\,km$ the constant area of the license, and A_{PC} the area of the postcode.

3.5   Location Coefficient Variation: Adjusted Employee Density

The Adjusted Employee Density formula is the same as the Employee Density formula, with the exception that employees from different industries different weights. The weights are based on the industry value added per employee. The rationale behind this is that the number of employees might not be comparable between different industries. For example, some industries are more automated than others. By using weights, locations where employees work in high value adding industries are more expensive

Each industry's employee weights (EW) are calculated by dividing the average employee value added in that industry by average employee value added in Finland (Eq. 7). If there was no data on a particular industry's value added, the average for whole Finland (weight of 1) was used for that industry.

$$Adjusted\ Employee\ Density\ Valuation = C_1 * C_{emp,adj} * C_{6b} * B_0 * S * P \qquad (5)$$

$$C_{emp,adj} = \frac{\sum_i EW_i * EMP_{PC,i}}{EMP_{FIN}} * \frac{1\ km}{A_{PC}} \qquad (6)$$

$$Employee\ Weight(EW_i) = \frac{Average\ Employee\ Value\ Added\ in\ Industry\ i}{Average\ Employee\ Value\ Added\ in\ Finland} \qquad (7)$$

3.6   Location Coefficient Variation: Rent Based

The Rent Based valuation uses commercial office rental prices per square meter to calculate the location value. For this research, data was available for selected areas in the Helsinki region. We calculated an average price for a 1-year public mobile network license with bandwidth of 10 MHz and area of 1 km in the 3.5 GHz Frequency Band in the selected areas using the Employee-Based Valuation. The average price was calculated using employee-weighted average, i.e. the relative number of employees in the area was used as the weight. We then used relative rent prices as a coefficient to evaluate licenses in different areas.

$$Rent\ Based\ Valuation = License\ Value * \frac{Rent_{PC}}{Rent_{ALL}} \qquad (8)$$

Where License Value is the employee-weighted average of a license in the selected areas according to the Employee-Based Valuation. $Rent_{PC}$ is the average rent in the postcode area, and $Rent_{ALL}$ is the employee-weighted average of rent in the selected areas.

The data used is from selected areas from the Helsinki region and it uses office rental prices. Data from wider area using industrial rental prices exists but was not available for this research (KTI Property Information Ltd. 2018). This more extensive data could be used to increase the accuracy of this method.

4 Results

All prices in this paper have been calculated for a 1-year license with bandwidth of 10 MHz in the 3.5 GHz frequency band. In the Table 3 and Fig. 1, we show descriptive statistics of prices obtained by different valuation methods for all postcode areas and selected areas in the Helsinki region. The rent based valuation is calculated only for the selected areas in the Helsinki region. The pricing results including the rent based method are shown in Table 4 and Fig. 3.

4.1 Comparison of FICORA, Employee Based, and Adjusted Employee Based Prices

In this section, we have selected 100 postcode areas that have the highest valuation based on the Employee Based method as these types of areas are most relevant for local 5G licenses.

Table 3. Comparison of prices yielded by different valuation methods. The 100 highest priced areas using the employee density method are included.

	Min (€)	Max (€)	Mean (€)	Employee-weighted mean (€)	Median (€)
FICORA	0.00	73.39	11.82	14.10	9.17
Employee based	10.30	364.09	48.07	63.58	24.21
Adjusted employee Based	10.20	366.34	48.36	64.86	24.73

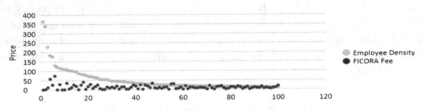

Fig. 1. Comparison of prices yielded by the employee density and the FICORA Fee method. The 100 highest priced areas using the employee density method are included.

The Employee Based valuation methods generate significantly higher values than the FICORA Frequency Fee for certain areas. This is explained by the fact that employment is concentrated more than residency. Areas such as commercial and industrial districts have a very high employee density compared to the population density of even the most populated residential areas. Conversely, as residency is more spread out, the population based prices of for example rural and residential areas are typically higher than employee based. Interestingly, there is a group of postcode areas that have no residents but a high employee density. Examples of these locations are the hospital area of Joensuu, the office park of Ilmala, Turku University of Applied Sciences, and industrial district of Martinlaakso (Fig. 2).

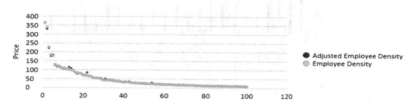

Fig. 2. Comparison of prices yielded by the employee density and the adjusted employee density method. The 100 highest priced areas using the employee density method are included.

Employee Density and Adjusted Employee Density based valuation methods yield very similar prices. However, differences occur in areas where the employees work dominantly in industries that have either relatively high or low value added per employee. Examples of locations where the Adjusted Employee Density yields higher results are postcode areas with large powerplants (Olkiluoto, Tahkoluoto), some industrial districts (Martinlaakso industrial area), and university campuses (Otaniemi).

4.2 Comparison of Prices, Including the Rent Based Method

In this section, we show the prices for selected areas in the Helsinki Region. The reason for selecting these areas was the availability of detailed commercial rent prices.

Table 4. Comparison of prices yielded by different valuation methods, including the rent based method. Selected areas in the Helsinki Region are included.

	Min (€)	Max (€)	Mean (€)	Employee-weighted mean (€)	Median (€)
FICORA	2.71	73.39	16.77	18.00	11.18
Emp. based	3.87	228.05	54.50	90.57	25.75
Adj. emp. based	3.86	221.47	56.33	94.66	24.54
Rent based	56.84	126.34	81.14	90.57	76.43

Using the Rent Based valuation, the license prices are significantly more evenly distributed than using the other methods. Because population density and employee density can vary significantly between areas, the prices based on these methods also vary significantly and they can even be close to zero. However, office rents do not have this same characteristic and thus license prices based on rents are distributed more evenly.

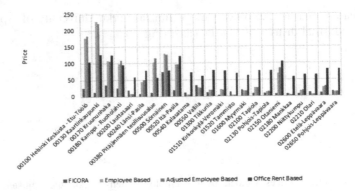

Fig. 3. Comparison of prices yielded by different valuation methods, including the rent based method. Selected areas in the Helsinki Region are included.

5 Conclusion

As the 5G utilising technology and use cases for local licenses develop, there is a growing need to evaluate local 5G licenses. In the introduction, we presented arguments why conventional valuation methods, namely auctions and benchmarking might not work in this context. Additionally, we argued that existing population density based general valuation models might not accurately proxy the underlying drivers of location value in commercial local licenses. In this paper, we proposed two alternative proxies for location value drivers: employee density and commercial rental prices.

As seen from the results, the four valuation methods used distribute the total value of spectrum differently. The FICORA Frequency Fee is based on population density, and it is a good measure of location value for mobile broadband, where the customers are mainly private consumers. However, we can see from the results that this valuation method is not always sufficiently accurate in the context of commercial local licenses. There exist many postcode areas that have a population of zero but that have potentially high demand for local licenses. These areas include, for example, industrial districts and hospital areas. Using the Frequency Fee based valuation, the prices for licenses in these areas are very low, which does not accurately reflect reality. This problem would be even more noticeable if we were to use areas smaller than postcode areas.

Kokkinen et al. (2018) summarises this problem with a sentence: "No one lives in factories or ports." We might add: "But many work there". The two proposed employee density based methods are able to identify areas with low population density but high potential demand for spectrum. The methods also show that employment is more concentrated than residency. Companies tend to group up in small areas, which locally increases the demand for spectrum. This is reflected in the prices of licenses: the highest prices using employee density are significantly higher than the ones the population density based method yields. Conversely, prices for low demand areas are lower with employee density valuation.

In the basic Employee Density valuation, all employees drive the spectrum value equally. However, this might not reflect the reality as different types of employees, companies, and industries have different demand and willingness to pay for spectrum. Because of this we introduced the Adjusted Employee Density valuation, where employees from different industries were weighted based on their average value added. This method distinguished for example that the employees of energy companies such as nuclear plants have a very high value added per employee and thus areas with energy plants were given a higher license price per employee. The industry categories we used were very broad. For example, all manufacturing companies were consolidated in the same category. With more detailed categories, the usability of this method would increase significantly. It could, for example, be very useful to distinguish smart factories as their own category, as these factories typically have a very low ratio of employees to value added. Additionally, value added per employee is not the only, nor necessarily the best way to weight employees. Some industries might have a high value added but no demand for local licenses. Further research on how new 5G technologies will benefit different industries, especially in monetary terms, would improve the pricing of licenses between industries.

Value of Mobile Broadband licenses is very much location dependent. If an area has no users for the service, the price of that local license is close to zero. The value of the license increases in the number of users as the potential revenues also increase. The Employee Density based valuation makes the same assumption on commercial local licenses and it might not always reflect reality. Often a factory makes similar profits no matter if it is surrounded by other businesses or not. A new factory under construction might not have many employees working at the site yet but still its willingness to pay for license can be high. In the light of these arguments, it is possible that the Employee Density based method generates too stark differences between locations.

Because of this, we introduced another proxy for location value, commercial rental prices. Average rents change based on the location, but not as abruptly as employee density. Rent prices do not drop to zero even in very rural areas. As seen from the results, the rent based method yields more -evenly distributed prices than the other methods. Still, it ranks different locations very similarly to the employee based method as the prices generated by these two methods have a strong correlation. Because of the limited availability of data for this research, we used only office rent prices for selected areas in the Helsinki region. There exist more extensive databases, which could be used in further research for more accurate pricing.

The valuation method should enable an allocation where the party with the highest willingness to pay gets the license. To achieve this, a combination of different valuation methods could be used. For example, the license price could be the maximum of the population density based valuation and the employee density based valuation. If the location has a high population density compared to employee density, the most efficient allocation is most likely to use the license for mobile broadband. By using max function, the license will always be sold at the higher price, which in this case incentives mobile broadband use.

Spectrum pricing should follow demand and be based on transparent methods and easily available data. The demand for consumer services such as mobile broadband follows population density but this is not necessarily true for industrial demand. We recommend the use of employee density as a measure for the industrial demand. This can be adjusted to reflect the differences between industries. Optionally, rent based valuation can also be used to price licenses. A combination of these methods, such as the max function of population density and employee density valuation, allows the consideration of both consumer and industrial demand simultaneously. This method will always price the licenses to match the highest willingness to pay.

References

3GPP TR 36.889: Feasibility Study on Licensed-Assisted Access to Unlicensed Spectrum. 3rd Generation Partnership Project (2015)

3GPP TR 38.889 V0.0.1: Study on NR-based Access to Unlicensed Spectrum (2018a)

3GPP TS 36.101. User Equipment (UE) radio transmission and reception (2018b)

5GPPP 5G empowering vertical industries: Roadmap paper. The 5G Infrastructure Public Private Partnership (2016)

Abinader, F.M., et al.: Enabling the coexistence of LTE and Wi-Fi in unlicensed bands. IEEE Commun. Mag. 52(11), 54–61 (2014)

Ahmed, A.A.W., Markendahl, J., Ghanbari, A.: Evaluation of spectrum access options for indoor mobile network deployment. In: 2013 IEEE 24th International Symposium on Personal, Indoor and Mobile Radio Communications (PIMRC Workshops), London, pp. 138–142 (2013)

Alliance, N.G.M.N.: Description of network slicing concept. NGMN 5G P1 (2016)

Bazelon, C., McHenry, G.: Spectrum value. Telecommun. Policy 37, 737–747 (2013)

Beltran, F.: Accelerating the introduction of spectrum sharing using market-based mechanisms. IEEE Commun. Stan. Mag. 1(3), 66–72 (2017)

Cave, M., Nicholls, R.: The use of spectrum auctions to attain multiple objectives: Policy implications. Telecommun. Policy 41(5–6), 367–378 (2017)

City of Helsinki: Toimitilamarkkinat Helsingissä ja Pääkaupunkiseudulla 2017, (2018). https://www.hel.fi/hel2/tietokeskus/julkaisut/pdf/18_01_23_Tilastoja_01_Henriksson.pdf

Cramton, P.: Spectrum auction design. Rev. Ind. Organ. 42(2), 161–190 (2013)

Cramton, P., Doyle, L.: An Open Access Wireless Market. White paper (2015)

ECC: Report 205 Licensed Shared Access (LSA) (2014)

ETSI: Feasibility study on temporary spectrum access for local high-quality wireless networks. ETSI TR 103 588, p. 30 (2018)

European Commission: Study on spectrum assignment in the European Union. Doc. SMART 2016/0019 (2017)

European Commission: Strategic spectrum roadmap towards 5G for Europe: RSPG second opinion on 5G networks. Radio Spectrum Policy Group (RSPG), RSPG18-005 (2018)

FCC: The Second Report and Order and Order on Reconsideration finalizes rules for innovative Citizens Broadband Radio Service in the 3.5 GHz Band. FCC 16–55 (2016)

Ferrus, R., Sallent, O.: Extending the LTE/LTE-a business case: mission- and business-critical mobile broadband communications. IEEE Veh. Technol. Mag. 9(3), 47–55 (2014)

Finnish Ministry of Transport and Communications: Act 1028/2017. Liikenne- ja viestintäministeriön asetus hallinnollisista taajuusmaksuista ja Viestintäviraston taajuushallinnollisista suoritteista perittävistä muista maksuista (2017). https://www.finlex.fi/fi/laki/alkup/2017/20171028

Guirao, M.D.P., Wilzeck, A., Schmidt, A., Septinus, K., Thein, C.: Locally and temporary shared spectrum as opportunity for vertical sectors in 5G. IEEE Netw. 31(6), 24–31 (2017)

Hazlett, T.W., Muñoz, R.E.: A welfare analysis of spectrum allocation policies. RAND J. Econ. 40(3), 424–454 (2009)

ITU-R: Exploring the value and economic valuation of spectrum. ITU Telecommunication Development Sector, Report, p. 35 (2012)

Klemperer, P.: What really matters in auction design. J. Econ. Perspect. **16**(1), 169–189 (2002)

Kokkinen, H., Yrjölä, S., Engelberg, J., Kokkinen, T.: Pricing private LTE and 5G Radio licenses on 3.5 GHz. In: Moerman, I., Marquez-Barja, J., Shahid, A., Liu, W., Giannoulis, S., Jiao, X. (eds.) CROWNCOM 2018. LNICST, vol. 261, pp. 133–142. Springer, Cham (2019). https://doi.org/10.1007/978-3-030-05490-8_13

Kokkinen, T., Kokkinen, H., Yrjölä, S.: Spectrum broker service for micro-operator and CBRS priority access licenses. In: Marques, P., Radwan, A., Mumtaz, S., Noguet, D., Rodriguez, J., Gundlach, M. (eds.) CrownCom 2017. LNICST, vol. 228, pp. 237–246. Springer, Cham (2018). https://doi.org/10.1007/978-3-319-76207-4_20

KTI Property Information Ltd: Market Information (2018). https://kti.fi/en/services/market-information/. Accessed on 18 Jan 2018

Lemstra, W.: Leadership with 5G in Europe: two contrasting images of the future, with policy and regulatory implications. Telecommun. Policy **42**(8), 587–611 (2018)

LTE-U Forum (2016). http://www.lteuforum.org/index.html

Marks, P., Pearson, K., Williamson, B., Hansell, P., Burns, J.: Estimating the commercial trading value of spectrum. Report for Ofcom, Plum Consulting (2009)

Matinmikko, M., Latva-aho, M., Ahokangas, P., Seppänen, V.: On regulations for 5G: micro licensing for locally operated networks. Telecommun. Policy **42**(8), 622–635 (2017a)

Matinmikko, M., Latva-aho, M., Ahokangas, P., Yrjölä, S., Koivumäki, T.: Micro operators to boost local service delivery in 5G. Wirel. Pers. Commun. **95**(1), 69–82 (2017b)

MulteFire: MulteFire alliance (2016). http://www.multefire.org/

Mölleryd, B.G., Markendahl, J., Mäkitalo, Ö., Werding, J.: Mobile broadband expansion calls for more spectrum or base stations-analysis of the value of spectrum and the role of spectrum aggregation. In: 21st European Regional ITS Conference, Copenhagen, Denmark, p. 22 (2010)

Ofcom: Enabling opportunities for innovation Shared access to spectrum supporting mobile technology. Ofcom, UK (2018). https://www.ofcom.org.uk/__data/assets/pdf_file/0022/130747/Enabling-opportunities-for-innovation.pdf

Official Statistics of Finland: Open Data by Postal Code Area. [e-publication] (2018a). http://pxnet2.stat.fi/PXWeb/pxweb/en/Postinumeroalueittainen_avoin_tieto/

Official Statistics of Finland: Structural business and financial statement statistics. [e-publication] (2018b). http://www.stat.fi/til/yrti/index_en.html

Radiocommunications Agency Netherlands: Internetverbinding Verbeteren (2018). https://www.agentschaptelecom.nl/onderwerpen/internetverbinding-verbeteren. Accessed 18 Jan 2018

The Federal Network Agency Germany: Regionale und locale Netze (2018). https://www.bundesnetzagentur.de/DE/Sachgebiete/Telekommunikation/Unternehmen_Institutionen/Frequenzen/OeffentlicheNetze/RegionaleNetze/regionalenetze-node.html. Accessed 18 Jan 2018

The Swedish Post and Telecom Authority: The Swedish Post and Telecom Authority's intents for the assignment of frequencies for 5G, following referral (2018). https://www.pts.se/globalassets/startpage/dokument/icke-legala-dokument/rapporter/2018/radio/preliminary-study-consultation-pts-responses.pdf. Accessed 18 Jan 2018

Tonmukayakul, A., Weiss, M.B.: A study of secondary spectrum use using agent-based computational economics. Netnomics **9**(2), 125–151 (2008)

Zander, J.: Beyond the ultra-dense barrier: Paradigm shifts on the road beyond 1000x wireless capacity. IEEE Wirel. Commun. **24**(3), 96–102 (2017)

Spectrum Allocation Options for Point-to-Multipoint Services in 5G

Heikki Kokkinen[1](✉) ⓘD, Juha Kalliovaara[2] ⓘD, and Tero Jokela[2] ⓘD

[1] Fairspectrum, Turku, Finland
heikki.kokkinen@fairspectrum.com
[2] Turku University of Applied Sciences, Turku, Finland

Abstract. If the same content is to be delivered to a large number of users at the same time, point-to-multipoint transmissions (broadcast/multicast) could present a more efficient delivery mechanism when compared to point-to-point transmission schemes (unicast). Therefore point-to-multipoint is considered to be an essential feature for many 5G applications. Different 5G applications differ greatly in terms of coverage, bit rate and quality of service they require. This paper considers spectrum allocation options for different use cases in the following 5G vertical sectors: Media & Entertainment, Public Warning, Automotive and Internet of Things. This study analyses the spectrum allocation options for each use case in different spectrum bands and with different spectrum allocation methods, ranging from exclusive licensing to spectrum sharing and unlicensed spectrum. The different analyses map use cases to spectrum bands, spectrum bands to allocation options, use cases to allocation options, and all these are brought together in use case - spectrum band - allocation option - operator mapping.

Keywords: 5G · Point-to-multipoint · Spectrum · Allocation · Media and Entertainment · Public warning · Automotive · IoT

1 Introduction

In scenarios where a very large number of users consume the same data, such as popular media content, emergency messages and software updates, broadcast/multicast technologies can be used to deliver the same content to potentially unlimited number of users within the network coverage area with a defined and stable quality of service [1]. The broadcast/multicast services are a point-to-multipoint (PTM) scheme, while the traditional unicast services are a point-to-point (PTP) scheme. The use of PTM instead of PTP can be very cost-effective and high-quality delivery mechanism when data needs to be delivered to a large number of users, while the networks using PTP data delivery can potentially get congested in such scenarios as the data needs to be sent to each user individually.

5G-Xcast is a 5G-PPP Phase 2 project which develops an architecture for PTM capabilities in 5G. The project has identified and defined six different use cases which could benefit from PTM capabilities. This paper describes spectrum allocation options for these six different 5G PTM service use cases. Three of the use cases are in the field

© ICST Institute for Computer Sciences, Social Informatics and Telecommunications Engineering 2019
Published by Springer Nature Switzerland AG 2019. All Rights Reserved
A. Kliks et al. (Eds.): CrownCom 2019, LNICST 291, pp. 320–330, 2019.
https://doi.org/10.1007/978-3-030-25748-4_24

of media and entertainment, one in public warning, one in automotive and in Internet of Things.

The use cases differ greatly in terms of required coverage, bit rate and quality of service. This paper studies and analyses the spectrum allocation options for each use case in different spectrum bands and with different spectrum allocation methods, ranging from exclusive licensing to spectrum sharing and unlicensed spectrum. We also study the type of operator who would have most benefit in the selected combination of use case and spectrum assignment.

This paper analyses spectrum allocation and usage options under the following categories and allocation options: use cases (M&E1, M&E2, M&E3, PW1, Auto1, IoT1), spectrum bands (470–694 MHz, 700 MHz, 2.3 GHz, 3.5 GHz, 3.8–4.2 GHz, 6 GHz, 26 GHz and above), allocation/usage options (Nation-wide long-term licenses, Local and temporary licenses, CBRS, Licensed Shared Access, Concurrent Shared Access, Unlicensed spectrum.), and operator (MNO, broadcaster, other).

The uses cases are presented in Sect. 2, the spectrum bands, the allocation and usage options and the operators in Sect. 3. Section 4 presents the analysis and results of mapping different use cases, spectrum bands, and allocation options while Sect. 5 concludes the paper.

2   Use Cases

PTM transmissions present a more efficient delivery mechanism compared to PTP when the same content is to be delivered to multiple users at the same time. Therefore PTM is considered to be an essential feature for many 5G applications. Use cases evaluated in 5G-Xcast project cover the following vertical sectors: Media & Entertainment (M&E), Public Warning (PW), Automotive (Auto) and Internet of Things (IoT) [2]. The use cases are briefly described in the following sections. More detailed descriptions are available in [2].

2.1   Use Case M&E 1 – Hybrid Broadcast Service

Users have access to any combination of linear and non-linear audio-visual content in addition to social media. Content and services can be delivered over a combination of several networks and types of networks simultaneously. Continuity of the users' experience is preserved when switching between different access networks, possibly operated by different operators. The population of concurrent users may be very large (i.e. millions of viewers of a popular live event) and may substantially change over short periods of time.

2.2   Use Case M&E 2 – Virtual/Augmented Reality Broadcast

Virtual Reality (VR) is a technology that creates a perception of a user's physical presence in a rendered environment, real or imagined, leading to an immersive experience and may allow for user interaction. Virtual realities artificially create sensory experience, which in principle can include sight, touch, hearing, and smell. Augmented

Reality (AR) is a technology that composites multimedia or other types of content in the user's view of the real world. In this use case, large amount of users should be able to receive high-quality VR/AR content over the air simultaneously.

2.3   Use Case M&E 3 – Remote Live Production

In order to support the workflows in a typical production environment, it is often required that different people have access to the same video feed at the same time (e.g. directors, editors, commentators, those that create metadata). Given the potentially very high bit-rates in use, it is not practicable to carry multiple unicast copies of content. Instead, a feed from the production equipment such as cameras and microphones is carried over a unicast uplink connection (e.g. multilink) to the infrastructure access point and distributed via multicast to enable concurrent viewing by multiple users. The most important capability is to receive the video feed continuously, without breaks, freezes, artefacts or other issues. M&E use cases are illustrated in Fig. 1.

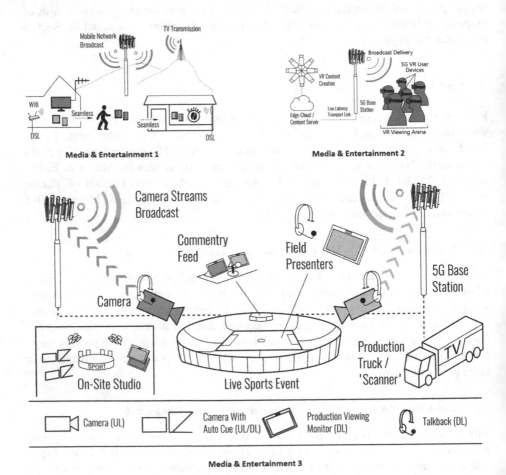

Fig. 1. Media & Entertainment use cases 1–3 [2].

2.4 Use Case PW 1 – Multimedia Public Warning Alert

In the Multimedia public warning alert, users are notified with alerts carrying multimedia and manifold information, which improves the effectiveness and reactivity of the users' responses. The alert is send to a targeted location. Within that targeted location, all users need to be notified promptly.

2.5 Use Case Auto 1 – V2X Broadcast Service

Various V2X applications like road safety, signage, mapping and autonomous driving require information delivered from the Intelligent Transport System (ITS) infrastructure (such as ITS roadside units and sensors) to the vehicle. The delivery of information that would benefit multiple recipients concurrently could utilize a point-to-multipoint service.

2.6 Use Case IoT 1 – Massive Software and Firmware Updates

IoT devices such as smart water-metering are installed deep indoors and wake up once or twice a day to send the consumption reports to the water metering network that is regularly extended. Based on the growing amount of data, the system configuration is adjusted, requiring the delivery of small configuration updates to all metering devices. Moreover, the water metering manufacturer regularly provides non-critical software updates. Figure 2 illustrates the public warning, automotive and IoT use cases.

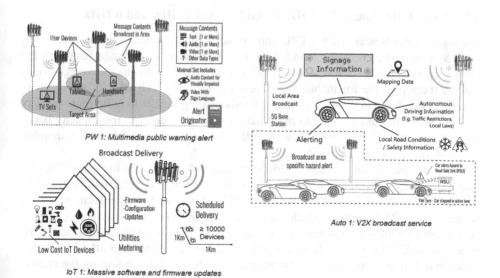

Fig. 2. Public warning, automotive and IoT use cases [2]

3 Spectrum Bands, Allocation Options and Operators

This section discusses the different spectrum bands, spectrum allocation methods and types of operators for the considered use cases. The spectrum bands are divided into three groups: coverage bands below 1 GHz, mid-capacity bands between 1 to 6 GHz and high capacity bands above 6 GHz. Sections 3.1, 3.2 and 3.3 discuss these bands, Sect. 3.4 the options for spectrum allocation and Sect. 3.5 the operator types.

3.1 Sub-1 GHz Bands: 470–694 MHz & 700 MHz

The frequency bands of mobile networks are traditionally divided to coverage bands and capacity bands. The coverage bands are below 1 GHz. They propagate well over long distances and it is economical for a mobile operator to build a nation-wide coverage using the coverage bands. The bandwidth in the coverage bands is narrow, and due that it is difficult to provide broadband connectivity or have many data-hungry applications in the same cell simultaneously. All capacity bands have been utilized tens of years and in order to get coverage bands for 5G, the bands have to be cleared from the existing use. The coverage bands are difficult to share with other types of spectrum users and the primary spectrum assignment method for coverage bands is exclusive licensing. In practise, the coverage bands have been used by the terrestrial television. The pioneer 5G coverage band globally is 700 MHz band, and it may be extended to cover lower digital TV UHF bands 470–694 MHz in the future.

3.2 1 to 6 GHz Range: 2.3 GHz, 3.5 GHz, 3.8–4.2 GHz, and 6 GHz

The capacity bands begin from 1 GHz and extend to higher frequencies. In some cases, the frequency bands between 1 and 2 GHz may be used as coverage bands by the mobile operators. The coverage bands offer wider bandwidths than coverage bands making them possible for mobile broadband services. The cell sizes of the capacity bands are smaller than those of coverage bands making it easy to build high capacity network areas, but uneconomic to build nation-wide coverage. As the capacity bands are not expected to be deployed with full coverage, spectrum sharing with other spectrum users becomes feasible.

The mid-band of the capacity bands is limited to 6 GHz in the high end. The first pioneer capacity mid-band is 3.5 GHz. It will be extended to cover 3.4–4.2 GHz. Also the LSA band 2.3 GHz will be used for 5G and 6 GHz is being harmonized for unlicensed use. The countries which are able to clear the band before assigning them to 5G can assign nation-wide licenses or in some cases a part of the spectrum is dedicated to private LTE/5G networks. Most countries will not be able to clear all mid-capacity bands and different spectrum sharing methods will be used depending on the characteristics of the incumbent spectrum user. For static incumbents, static sharing using license terms is the prevailing method and for the dynamic incumbents, dynamic spectrum sharing is required.

3.3   Above 6 GHz: 26 GHz and Above

The high-frequency capacity bands are above 6 GHz. Although, the band naming begins on 6 GHz, the pioneer band is 26 GHz, and it will be followed by even higher frequencies. They are often called millimeter waves. The bandwidths are very wide compared to any other communication system allowing gigabit/second-level wireless bitrates. The connectivity between the base station and user equipment requires a line of sight, the cell sizes are very small and the beams can be very directive. The millimeter wave bands are very suitable for spectrum sharing. Italy is the first European country, which included club use-type of spectrum sharing as a part of the 26 GHz auction rules.

3.4   Spectrum Allocation Options

The considered allocation options are exclusively licensed spectrum, nation-wide long-term licenses, local and temporary licenses, and shared spectrum. Following the 5G Spectrum Position Paper [3] of GSMA, the primary spectrum management approach for 5G remains exclusively licensed spectrum. Practically, all mobile network bands are currently on exclusively licensed bands. Spectrum sharing and unlicensed bands complement that. An option for assigning spectrum for industrial users are local licenses for private LTE/5G networks.

Furthermore, GSMA Public Policy Position [4] about Spectrum Sharing from November 2018 states that the commonly discussed spectrum sharing frameworks include Citizens Broadband Radio Service (CBRS), Licensed Shared Access (LSA), and Concurrent Shared Access (club licensing). CBRS has been developed for the band 3550–3700 MHz in the US. It is based on the Federal Communications Commission (FCC) regulation Part 96 [5]. LSA has been specified by ETSI [6] and CEPT [7], and it has been decided to be used in the 2300–2400 MHz band by the European Commission [8]. Concurrent access is defined in the GSMA Public Position Paper as club-use, meaning that the band is divided to the license holders. The license holders have the national priority for their own part of the band, but if they do not deploy the network in a part of the country, the other license holders have a right to use it on secondary basis.

Unlicensed spectrum is the most difficult one for sharing with incumbents. A part of 5 GHz unlicensed band is shared with radars and all devices using the shared part of the spectrum have to support Dynamic Frequency Selection (DFS). The spectrum users have felt the DFS use cumbersome and the channels requiring DFS are very little used. The next significant unlicensed band allocation will be 6 GHz and it will require fixed link and satellite ground station incumbent protection in a few countries as well. FCC has proposed a central interference protection system called Automated Frequency Coordination (AFC) for that band.

3.5   Operator Types

Mobile network operators (MNOs) are the entities operating mobile networks that are used to transmit audio and mobile broadband services. They traditionally use exclusive spectrum licenses. The typical mobile network transmissions are low tower low power (LTLP), as the transmitter heights are typically below 50 m.

Broadcasters are the entities which operate broadcast networks that are used to deliver audiovisual TV content over digital terrestrial television (DTT) networks, traditionally using exclusive spectrum licenses. The existing infrastructure used to broadcast DTT transmissions is characterized by very tall towers and high transmission powers. Transmissions of this type are known as high tower high power (HTHP). The height of HTHPs is typically around 300 m and the transmitter coverage around 100 km.

Other types of operators could include for example virtual network operators and local private network operators. The number of private LTE/5G networks is growing significantly at the moment as the regulators begin to assign radio licenses also to local networks on LTE and 5G bands.

4   Analysis and Results

This section analysis the mapping of different use cases with the selected spectrum assignment options. The different mappings are: use case to spectrum band, spectrum band to allocation option, use case to allocation option, and all these are brought together in mapping use case – spectrum band – allocation option – operator.

4.1   Mapping Use Cases to Spectrum Bands

The linear TV services have been offered on UHF terrestrial TV band on 470–862 MHz for tens of years and it not surprise that the same frequency band is recommended also in this study. M&E1 shares the same basic characteristics as linear TV and due to that the mapping of linear TV and M&E1 are generally the same. Virtual and augmented reality require very high bitrate and due to that they fit best to the highest capacity bands. Remote live production benefits from high uplink capacity. On the other hand, live production in a remote location needs coverage. The coverage is best achieved on the coverage bands and utilization of the current primary PMSE camera link band, 2.3 GHz for shorter communication distances, could be a practical combination. Public warning should reach as many people as possible, so coverage bands are preferred. The media services require more capacity than the coverage bands can offer, so the mid capacity bands could be used for providing them. The use case spectrum band mapping can be found in Table 1.

Table 1. Use case to spectrum band mapping

MHz	Linear TV	M&E1	M&E2	M&E3	PW1	Auto1
470–694	X	X				
700	X	X		X	X	
2300				X		
3400–3800			X			X
3800–4200			X			
6000			X			X
26000			X			

4.2 Mapping Spectrum Bands to Allocation Options

The combinations of allocation options and spectrum bands are not directly related to the requirements of the use cases, but in practice spectrum bands and the allocation options are tied together. Nation-wide licenses require clearing the band and should be possible on coverage bands and at least in a few countries in the 3.5 and 3.7 GHz bands. Local licenses for private LTE/5G networks have been allocated in the mid-capacity bands and they will most like be allocated on millimeter waves, as well. At the moment we do not expect CBRS to be deployed in Europe in near or medium term. The dynamic spectrum access option for Europe is LSA and dynamic incumbents can be found in the mid-capacity bands. The concurrent shared access as club licensing has been proposed for 26 GHz. The next unlicensed band for broadband communication is expected to be 6 GHz, see Table 2.

Table 2. Allocation option to spectrum band mapping

MHz	Nation- wide	Local, temporary	CBRS	LSA	Concurrent	Unlicensed
470–694	X					
700	X					
2300		X		X		
3400–3800	X	X		X		
3800–4200		X		X		
6000						X
26000		X			X	

4.3 Mapping Use Cases to Allocation Options

All use cases can be used with nation-wide exclusive licenses, which are the primary spectrum allocation option for 5G and other mobile networks. The main question here is which services could also be operated using the other allocation options. Linear TV, hybrid broadcasting and public warning need nation-wide coverage and do not tolerate any service breaks; the nation-wide exclusive licenses are the only recommended option for them. Virtual reality services will be offered largely also in Wi-Fi networks, which only provide opportunistic access. The M&E2 services could be provided with any allocation option which can offer high bitrates. The remote video production requires more guarantees for service coverage and capacity than unlicensed bands can offer. Local, temporary licenses, and LSA could complement well the exclusive licensing. Video services for automotives could also use any capacity available. Concurrent use was left out, because it is expected to be used mainly on 26 GHz, which has too small cells to provide media services to moving vehicles. The results of use case to allocation option analysis are collected in Table 3.

Table 3. Use case to allocation option mapping

MHz	Linear TV	M&E1	M&E2	M&E3	PW1	Auto1
Nation-wide	X	X	X	X	X	X
Local, temp			X	X		X
CBRS						
LSA			X	X		X
Concurrent			X			
Unlicensed			X			X

4.4 Mapping Use Cases, Spectrum Bands, Allocation Options and Operators

Table 4 combines the results of the mappings in the previous tables. Linear TV and hybrid broadcasting fit best to the similar spectrum use as the TV services have been using for decades. The coverage bands below 1 GHz, nation-wide exclusive licenses having either broadcaster or MNO as the operator would work best considering also that societies have been using them for TV broadcasting. The virtual and augmented reality services require very high bitrates, which can only be provided on the highest capacity bands beginning from around 3 GHz. All spectrum allocation options are feasible. The operator for the services is most likely MNO, but other local operators can provide them in private LTE/5G networks, as well. Remote video production has two sides: one is remoteness and the other is bandwidth requirements of video. Remote can easily be translated to coverage band, i.e. 700 MHz and video production to 2.3 GHz which is used for that purpose by broadcasters and production companies. Any allocation method providing even a little bit higher availability than unlicensed should be considered. The spectrum license holder can be MNO, broadcaster or a private LTE/5G license holder. Public warning system requires highest coverage and availability limiting the choices to nation-wide exclusive licenses on 700 MHz and provided by MNO or broadcaster. Media services to vehicles could be provided in the 3.5 GHz or 6 GHz bands using any other allocation method but concurrent, which is expected here to be available only on 26 GHz. The media services to cars could be provided by broadcaster, MNO and other companies dedicated to roadside communications.

Table 4. Use case – spectrum band – allocation option – operator

	Linear TV	M&E1	M&E2	M&E3	PW1	Auto1
Band	<1 GHz	<1 GHz	>3 GHz	700, 2300 MHz	700 MHz	3.5, 6 GHz
Allocation	Nationwide	Nationwide	All	Nationwide, local, LSA	Nationwide	All, but concurrent
Operator	Broadcaster, MNO	MNO	MNO, other	MNO, Broadcaster, other	Broadcaster, MNO	Broadcaster, MNO, Other
Notes						

5 Conclusion

PTM transmissions (broadcast/multicast) could present a more efficient delivery mechanism in many scenarios when compared to PTP transmission schemes (unicast). 5G-Xcast project develops an architecture for PTM in 5G and has identified different use cases, or use case families, which cover the scenarios where the highest benefits of 5G PTM could potentially be achieved. The use cases belong to the following 5G vertical sectors: Media & Entertainment, Public Warning, Automotive and Internet of Things. Different 5G use cases and applications differ greatly in terms of coverage, bit rate and quality of service they require. Thus, the combination of spectrum bands and spectrum quality they need is different in each use case.

This paper has analysed spectrum allocation options in different frequency bands for the six different PTM use cases. The use cases have been analysed against the spectrum bands they could use, then the spectrum bands have been analysed against the different allocation options (ranging from exclusive licensing to spectrum sharing and unlicensed spectrum), and the use cases were analysed against the allocation options. Finally, all of these were brought together in use case - spectrum band - allocation option - operator mapping.

The analyses of this paper provide valuable information on what needs to be considered when choosing a combination of frequency band and allocation option for a service or use case. Though this paper considers only PTM use cases, the same methodologies can also be applied to PTP use cases.

Acknowledgement. We would like to acknowledge the support of the European Commission under the 5GPPP project 5G- Xcast (H2020-ICT-2016-2 call, grant number 761498). The views expressed in this contribution are those of the authors and do not necessarily represent the project.

References

1. Gomez-Barquero, D., Navratil, D., Appleby, S., Stagg, M.: Point-to-multipoint communication enablers for the fifth generation of wireless systems. IEEE Commun. Stand. Mag. 2(1), 53–59 (2018). https://doi.org/10.1109/mcomstd.2018.1700069
2. 5G-Xcast Deliverable 2.1 v1.1 Definition of Use Cases, Requirements and KPIs. http://5g-xcast.eu/wp-content/uploads/2018/07/5G-Xcast_D2.1_v1_1_web.pdf
3. GSMA 5G Spectrum Public Policy Position, November 2018. https://www.gsma.com/spectrum/wp-content/uploads/2018/12/5G-Spectrum-Positions-1.pdf
4. GSMA: Spectrum Sharing Public Policy Position, November 2018. https://www.gsma.com/spectrum/wp-content/uploads/2018/11/Spectrum-Sharing-Positions.pdf
5. CFR Title 47 Part 96 Citizens Broadband Radio Service
6. ETSI TS 103 235: System architecture and high level procedures for operation of Licensed Shared Access (LSA) in the 2300–2400 MHz band. https://www.etsi.org/deliver/etsi_ts/103200_103299/103235/01.01.01_60/ts_103235v010101p.pdf

7. CEPT Report 55, CEPT Report 56 and CEPT Report 58 in response to the Mandate from the European Commission on 'Harmonised technical conditions for the 2300–2400 MHz ('2.3 GHz') frequency band in the EU for the provision of wireless broadband electronic communications services'
8. ECC Decision (14)02: Harmonised conditions for MFCN in the 2300–2400 MHz band

Wireless Network Virtualization with Long-Term Device-to-Device Communication

Zhengyu Su[1,2](✉) and Biling Zhang[1,2]

[1] School of Network Education, Beijing University of Posts and Telecommunications,
Beijing, People's Republic of China
{zhengyusu,bilingzhang}@bupt.edu.cn
[2] The State Key Laboratory of Integrated Services Networks, Xidian University,
Xi'an, People's Republic of China

Abstract. To reduce the transmission energy and latency when servicing the users who are interested in a common popular content, the base station (BS) chooses to deliver the content to the users nearby with less power. After these users receive and cache the required content, they can act as relay users (RUEs) to serve those who are far away from the BS by means of Device-to-Device (D2D) and thus are called D2D users (DUEs). In such a scenario, how to classify the users into RUEs and DUEs and associate the DUEs to the RUEs is an important but not trivial problem. In this paper, we formulate the joint RUEs selection and DUEs association problem from a long-term perspective. To find a low complexity computational solution to the problem, we first propose an algorithm to select the RUEs based on the set criteria, and then a coalition formation game based algorithm is proposed for the DUEs association. We further prove that the proposed algorithm is convergent. Numerical results demonstrate that the algorithms we proposed yield notable gains compare with short-term optimal scheme and non-cooperative scheme.

Keywords: Device-to-Device communication ·
Long-term users association · Coalition formation game

1 Introduction

Industry foresee a gigabit experience with zero latency for the next generation mobile communication systems beyond 2020. The fifth generation wireless system brings together novel advanced services to offer a solid user experience, such as tactile internet, high resolution (4K) video streaming, advanced sensing and monitoring, autonomous driving. Besides throughput enhancements and reduction of cost and power requirements of devices, latency and reliability requirements are among the key performance indicators to meet the targeted in the 5G.

© ICST Institute for Computer Sciences, Social Informatics and Telecommunications Engineering 2019
Published by Springer Nature Switzerland AG 2019. All Rights Reserved
A. Kliks et al. (Eds.): CrownCom 2019, LNICST 291, pp. 331–343, 2019.
https://doi.org/10.1007/978-3-030-25748-4_25

In order to meet the low-latency communication requirement in the 5G, many approaches have been proposed. In [1], the authors investigate the latency performance of content delivery networks with the aid of edge-caching, in which a data center is serving the users via a shared wireless medium. The latency-minimization resource allocation for a multi-user mobile edge computation offloading system is studied in [2]. The authors in [3] study a minimum delay routing problem in the context of distributed networks and propose novel predetermined path routing algorithms. In [4], the authors investigate the latency-minimization problem in a multi-user time-division multiple access mobile-edge computation offloading system. In [5], the authors study delivering delay-sensitive data to a group of receivers with minimum latency which consists of the time that the data spends in overlay links as well as the delay incurred at each overlay node.

In the 5G network, all users are cognitive which can perceive the changing network conditions and optimize end-to-end performance. Device-to-Device (D2D) communication as an underlay to heterogeneous cellular network can facilitates the direct communication between the cellular users [6]. Since users within communication range can directly communicate with each other without relying on the base station, D2D communication can provide several benefits for the heterogeneous networks in reusing spectrum resources and achieving high bit rate. A number of works studied content distribution problems in mobile wireless networks with D2D communication [7–9]. In particularly, the typical resource allocation problem has been addressed under versatile content distribution scenarios including relay networks [7], social networks [8], as well as mmWave cellular networks [9]. Many comprehensive studies on the co-existence of large-scale deployed cellular and D2D networks are pursued in [10,11]. In [10], underlay and overlay D2D communications are considered along with users access selections. The transmission capacity region in D2D integrated cellular networks when two prevalent interference management techniques, power control and Successive Interference Cancellation are utilized is studied in [11].

However, aforementioned works focus only on the strategies that are suitable for short-term period, e.g., one time slot. When the network condition changes in the next time slot, the optimal strategy has to be changed, which may lead to instability problem. What's more, users' association schemes change between time slots will cause switching cost. In such a case, the existing studies cannot make full use of existing resources. In this paper, to minimize the average downloading duration from a long-term perspective, we use the existing resources more efficiently by integrating long-term D2D communication into cognitive wireless virtual network (WVN) where users from different operators can perform D2D communication and be served by a same base station (BS). It must be mentioned that in the long-term communication scenario we studied, the users can only associate one access node until a complete content is received. The considered scenario is shown in Fig. 1 where users from different operators require a popular content simultaneously. To reduce the average transmission latency of the network, the BS first chooses to deliver the content to the users nearby.

After these users receive the content, they can act as relay users (RUEs) to serve those who are far away from the BS by means of D2D and thus are called D2D users (DUEs). Different from other scenarios, the users will change their locations between time slots but remain stationary in one time slot. If the problem is solved by traversal algorithm, the optimal solution can be obtained but the complexity of the algorithm will increase exponentially as the number of users increases. To find a low complexity computational solution to the problem, we propose an algorithm to select the RUEs and a coalition formation game based algorithm for the DUEs association. We further prove that the coalition formation game based algorithm is convergent. Finally, our proposed scheme and algorithms are proved to achieve a great performance by simulation when compared to the short-term optimal scheme and non-cooperative scheme.

The remainder of this paper is organized as follows. Section 2 introduces the system model of the long-term D2D communication in cognitive WVN and the selection criteria for all users. The RUEs selection algorithm and the coalition formation game based algorithm that is used to determine the DUEs served by each RUE are described in Sect. 3. Experiment setup and numerical results are presented in Sect. 4. Finally, we conclude this study with future work in Sect. 5.

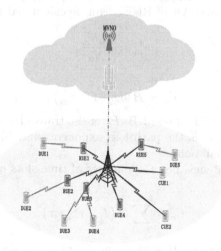

Fig. 1. A downlink cognitive wireless virtual network.

2   System Model

As shown in Fig. 1, we present a downlink 5G system with a single BS managed by a single mobile virtual network operator (MVNO), where users requiring the same popular content with size L, form the set $\mathcal{U}$ with size $|\mathcal{U}|$. To characterize the users' quality of experience (QoE), we assume each user has a same latency threshold, i.e., the maximum number of time slots it can tolerate according to its QoE constrain t_{max}. How t_{max} is obtained will be introduced in Sect. 2.2.

To reduce the transmission energy and service latency, the BS chooses to deliver the content to the users nearby with less power. After these users receive and cache the required content, they can act as RUEs to serve the users far away from the BS by means of D2D that are called DUEs. The other users that served by the BS but are not suitable to serve the DUEs are called cellular users (CUEs).

2.1 Achievable Rates

Before we introduce the algorithms for selecting RUEs and associating them with the DUEs, the achievable rate (AR) and the latency of each type of user is defined first.

Suppose K resource blocks (RBs) are set aside in the BS, each of which is assigned to one RUE and then reused by DUEs that the RUE served in the D2D communication stage. In such a case, $M(M \leq K)$ users will be selected from $\mathcal{U}$ as RUEs and form a set $\mathcal{U}^{\mathcal{R}}$.

Let $h_m^t = d\left(w_0 - w_m^t\right)^{-\alpha}$ be the path loss of wireless channel between the BS and RUE_m at time slot t. Here, w_0 and w_m^t are the locations of the BS and RUE_m at time slot t, respectively, and $d\left(w_0 - w_m^t\right)$ is the distance between them. Then the SNR and AR of RUE_m can be calculated as

$$\gamma_m^t = \frac{P_0 h_m^t}{N_0}, \tag{1}$$

and

$$r_m^t = B \log_2 \left(1 + \gamma_m^t\right), \tag{2}$$

where B is the bandwidth of one RB, P_0 is the transmission power the BS uses when it serves RUEs, α is the path loss exponent, and N_0 is the power of the additive white Gaussian noise.

As a result, the latency, i.e., the number of time slots used by RUE_m to get the content, is

$$t_m = \sum_{t=0}^{t_{\max}} O\left(L - r_m^t * \tau\right), \tag{3}$$

where τ is the duration of a time slot, and $O\left(x\right)$ is an indicative function whose value equals to 1 when x is greater than 0, and 0 otherwise.

For those users in $\mathcal{U} \setminus \mathcal{U}^{\mathcal{R}}$, if a user can get the content from one RUE with the latency not exceeding t_{max}, then it can be one DUE. Otherwise, it has to appeal to the BS and becomes one CUE. Let $\mathcal{U}^{\mathcal{D}}$ and $\mathcal{U}^{\mathcal{C}}$ be the set of DUEs and CUEs, respectively. Suppose the number of DUEs is $N(N \leq |\mathcal{U}| - M)$, then the number of CUEs will be $|\mathcal{U}| - M - N$. Obviously, $\mathcal{U}^{\mathcal{R}} \cup \mathcal{U}^{\mathcal{D}} \cup \mathcal{U}^{\mathcal{C}} = \mathcal{U}$.

We assume that one DUE can only connect with one RUE until the DUE receive the content while one RUE can serve multiple DUEs simultaneously. One RUE and the DUEs it serves forms a coalition, and the collection of the coalitions is represented as $\mathcal{S}^{\mathcal{R}} = \{s_1^r, \ldots, s_M^r\}$ where s_m^r represents the coalition constructed by RUE_m and the DUEs it serves.

If DUE_n is served by RUE_m, the path loss of the channel between them is denoted as $h_{m,n}^t = d\left(w_m^t - w_n^t\right)^{-\alpha}$. Assuming all RUEs have equal transmission power P_r, then the SNR DUE_n receives at time slot t will be

$$\gamma_{m,n}^t = \frac{P_r h_{m,n}^t}{N_0},\tag{4}$$

and the corresponding AR will be

$$r_{m,n}^t = \frac{B}{|s_m^r| - 1} \log_2\left(1 + \gamma_{m,n}^t\right),\tag{5}$$

where $|s_m^r| - 1$ indicates the number of DUEs that served by RUE_m.

Hence, the latency of DUE_n served by RUE_m will be

$$t_{m,n} = \sum_{t=t_m+1}^{t_{max}+t_m} O\left(L - r_{m,n}^t * \tau\right),\tag{6}$$

For the CUEs, since their QoE constraint cannot be satisfied if they are served by the RUEs, they will be served also by the BS directly with a low guaranteed rate λr_{th} introduced in Sect. 2.2.

Therefore, the number of time slots of one CUE getting the content is

$$t_c = \frac{L}{\lambda r_{th}}\tag{7}$$

2.2 Quality-of-Experience Model

In order to reasonably optimize the average delay of all users receive the content in our proposed scenario, we present one QoE model for the users considering their data rates and device types in this section. In the literature, the mean opinion score (MOS) model [12] is usually used to measure the QoE of wireless users as shown in Fig. 2, where QoE is categorized into five levels according to the users' sensitivity to the AR. The mapping between a user's AR and his/her MOS value is given by

$$\bar{r} = \frac{\lambda r_{th} - r_{min}}{r_{max} - r_{min}},\tag{8}$$

where $\bar{r}$ is the MOS value, r_{th} is the lowest average rate that guarantees users' QoE, λ is the diameter length of the user's device which also influence the user's

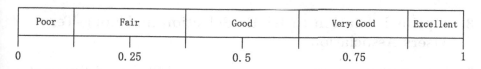

Fig. 2. Mean opinion score model

MOS value, $r_{\max}$ and $r_{\min}$ are the maximum and minimum average rates to get the content in the system, respectively.

Before the users are classified into RUEs, DUEs and CUEs, the minimum and maximum average rates of the system can be calculated as follows.

$$r_{\min} = \min\left\{\min_{m} \frac{1}{t_m}\sum_{t=1}^{t_m} r_m^t \; , \; \min_{m,n} \frac{1}{t_{m,n}}\sum_{t=t_m+1}^{t_{\max}+t_m} r_{m,n}^t\right\}, \tag{9}$$

$$r_{\max} = \max\left\{\max_{m} \frac{1}{t_m}\sum_{t=1}^{t_m} r_m^t \; , \; \max_{m,n} \frac{1}{t_{m,n}}\sum_{t=t_m+1}^{t_{\max}+t_m} r_{m,n}^t\right\}. \tag{10}$$

In Eqs. (9) and (10), the users represented by m m and n will traverse the set $\mathcal{U}$.

Then given $\bar{r}$ and λ, r_{th} can be expressed as

$$r_{th} = \frac{\bar{r}r_{\max} + (1-\bar{r})r_{\min}}{\lambda}. \tag{11}$$

The maximum number of time slots to get the content guaranteeing users' QoE can be expressed as

$$t_{\max} = \frac{L}{\lambda r_{th}} = \frac{L}{\bar{r}r_{\max} + (1-\bar{r})r_{\min}}. \tag{12}$$

2.3   Problem Formulation

Given this system model, our goal is to minimize the total latency of all the users acquiring the content while guarantee the QoE of each user. Suppose that after the users submit the content requirement to the MVNO, their current locations are also reported. By using machine learning, the MVNO can estimate their locations in a period in the future. Then the problem includes RUEs selection and the DUEs association problems can be formulated as follows.

$$\min_{\mathcal{U}^{\mathcal{R}},\mathcal{S}^{\mathcal{R}}} \frac{1}{|\mathcal{U}|}\left(\sum_{m\in\mathcal{U}^{\mathcal{R}}} t_m + \sum_{s_m^r\in\mathcal{S}^{\mathcal{R}}}\sum_{n\in s_m^r} t_{m,n} + \sum_{c\in\mathcal{U}^{\mathcal{C}}} t_c\right)$$

$$\text{s.t.} \quad r_m^t \geq \lambda r_{th}, \forall t \in [1, t_m], \tag{13}$$

$$r_{m,n}^t \geq \lambda r_{th}, \forall t \in [t_m + 1, t_n^{\max} + t_m].$$

In (13), the constraints guarantee that in each time slot the ARs of RUEs and DUEs must meet their QoE.

3   Optimal Solution to RUEs Selection and Long-Term Users Association

The optimal solution of the joint RUE selection and DUE association problem in Eq. (13) can only be obtained by traversal. However, as the number of users increases, the complexity of the traversal algorithm will explode. In order to find a sub-optimal solution with low complexity, we divide the problem solving process into two steps. First, we select a subset of users to act as RUEs based on the set criteria. Then, the DUEs served by each RUE and the CUEs served by the BS are determined by the coalition formation game based algorithm.

3.1 Selection of RUEs

To guarantee the QoE of users, we set the first criterion for selecting RUEs: the users that cannot meet their QoE when BS transmits the content to them with P_0 and a proprietary RB are classified as CUEs. Moreover, the RUEs should have high potential to help minimize the total latency of the system. In other words, a potential RUE should be the one that close to BS during the reception of the content and close to the users it will serve after caching the content. Therefore, in algorithm for RUEs selection, we use the sum number of time slots used by users to receive the content from BS and distribute it to other users as the other criterion for selecting RUEs.

The proposed RUEs selection algorithm is mainly composed of three steps: (i) apply the first selection criterion to all users and divide some users into CUEs; (ii) BS directly serves remaining users with P_0 and a proprietary RB if the number of remaining users doesn't exceed K, otherwise the next step is continued; (iii) apply the second selection criterion to the remaining users and select K users with the best expected performance to act as RUEs. The algorithm for RUEs selection is shown in Algorithm 1.

Algorithm 1. RUEs Selection

---

for u $\in \mathcal{U}$ **do**

    Calculating the value of t_u which is the number of time slots BS used to transmit the content to every user with P_0 and a proprietary RB.

    if $l_u > \frac{L}{\lambda_r^{th}}$ **then**

        The user will be divided into $\mathcal{U}^C$ and served by BS directly.

    end if

end for

if $|\mathcal{U}| - |\mathcal{U}_C| \leq \mathcal{K}$ **then**

    The remaining users will be divided into $\mathcal{U}^R$ and served by BS with P_0 and an exclusive RB.

else

    for u $\in \mathcal{U} \backslash \mathcal{U}_C$ **do**

        Calculating the value of $t_u + \frac{1}{|\mathcal{U} \backslash \mathcal{U}_C| - 1} \sum\limits_{n \in \mathcal{U} \backslash \mathcal{U}_C \backslash u} \sum\limits_{t=t_u+1}^{t_{max}+t_u} O(L - B \log_2(1 + \gamma_{u,n}^t) * \tau)$.

    end for

    The K users that with minimal value are selected as RUEs.

end if

3.2 DUEs Association

After RUEs are selected and received the content, they will select users to serve according to the distance between them during the process of transmission. Here, coalition formation game, which is a powerful analytical tool to study the behavior of rational players when they may cooperate with each other, is used to solve the DUEs association problem. A coalition $s_m^r \subseteq S^R$, which is consisted of one RUE and some DUEs, is a non-empty subset of S^R. RUEs are divided into disjoint coalitions and one DUE can only access one RUE until him/her gets the content. Thus, $\forall i, i' \in \{1, 2, \ldots, M\}$ and $i \neq i'$, it is clear that $\cup s_i^r = S^R$ and $s_i^r \cap s_{i'}^r = \emptyset$.

The coalition formation game can be defined as a triplet $(\mathcal{T}, \mathcal{P}, \mathcal{V})$, where $\mathcal{T}$ represents the users set defined as $\mathcal{U}^R \cup \mathcal{U}^D$, $\mathcal{P}$ represents a collection of coalitions which can be seen as a partition of $\mathcal{T}$, and $\mathcal{V}$ represents the value of coalitions. The utility of DUE and coalition are denoted as $\mathcal{V}_n$ and $\mathcal{V}(s_m^r)$, respectively. According to the MOS model that we mentioned in Sect. 2, user's QoE is closely related to delay, and the reduction of delay will increase user's QoE. Thus we define $\mathcal{V}(n)$ as the reciprocal of delay to receive the content, which can be calculated as

$$\mathcal{V}_n = \frac{1}{t_{m,n}}, \tag{14}$$

where $t_{m,n}$ has been calculated in Eq. (6).

Similarly, we define $\mathcal{V}(s_m^r)$ as the reciprocal of the average delay of users in coalition s_m^r, which can be calculated as

$$\mathcal{V}(s_m^r) = \begin{cases} \dfrac{|s_m^r| - 1}{\displaystyle\sum_{n \in s_m^r} t_{m,n}}, & \text{if } |s_m^r| \geq 1, \\ 0, & \text{otherwise.} \end{cases} \tag{15}$$

The necessary condition for users to join a coalition is that the users can get the content from RUE in the coalition in up to $t_{\max}$ time slots. Intuitively, we can find the optimal coalition structure by traversing all possible formation. However, the complexity of such a primitive calculation will show an exponential growth trend as the increases of DUEs. Therefore, we proposed a suboptimal solution to solve the coalition formation game using the concept of defection order. The defection order of the DUE_n served by RUE_m can be represented as

$$s_{m'}^r \rhd s_m^r = \begin{cases} \mathcal{V}_n\left(s_{m'}^{r,new}\right) > \mathcal{V}_n\left(s_m^r\right), \\[2mm] t_{m,n'} \leq t_{\max}, \forall DUE_n \in s_{m'}^r, \\[2mm] \mathcal{V}\left(s_{m'}^{r,new}\right) + \mathcal{V}\left(s_m^{r,new}\right) \\ > \mathcal{V}\left(s_{m'}^r\right) + \mathcal{V}\left(s_m^r\right), \end{cases} \tag{16}$$

where $s_{m'}^r$ is the coalition in S^R except for s_m^r; $\mathcal{V}_n(s_m^r)$ indicates the utility of DUE_n in coalition s_m^r; $s_{m'}^{r,new} = s_{m'}^r \cup DUE_n$ while $s_m^{r,old} = s_m^r \backslash DUE_n$; $\rhd$ indicates $s_{m'}^r$ is better than s_m^r for DUE_n.

The defection order lists three necessary conditions based on users' QoE constraint: (i) DUE_n in $s_{m'}^r$ is more profitable than in s_m^r; (ii) the participation of DUE_n will not violate the QoE constraint of DUEs originally in $s_{m'}^r$; (iii) the defects of DUE_n will increase the utility of $\mathcal{S}^{\mathcal{R}}$. The corresponding algorithm, which MVNO uses to find the optimal association between RUEs and DUEs, is proposed in Algorithm 2.

Algorithm 2. The Long-term Coalition Formation Game

Initial Stage

The network is partitioned by RUEs, CUEs and candidate users which may become DUEs or CUEs in a non-cooperative way.

Coalition Formation Stage

repeat

    for u $\in \mathcal{U} \backslash \mathcal{U}^{\mathcal{C}} \backslash \mathcal{S}^{\mathcal{R}}$ **do**

        Calculating the value of $t_{m,n}$ defined in Eq. (6) for each $m \in \mathcal{U}^{\mathcal{R}}$.

        if $min\ t_{m,n} > t_{\max}$ **then**

            The user will be divided into $\mathcal{U}^{\mathcal{C}}$.

        else

            The user submit application to the RUE that can transmits the content to it with minimal value of $t_{m,n}$.

        end if

    end for

    The user which won't cause the users connected it cannot satisfy their QoE and maximize the value of $\mathcal{V}(s_m^r)$ defined in Eq. (15) will join in s_m^r.

until All users in $\mathcal{U} \backslash \mathcal{U}^{\mathcal{R}}$ served by RUEs or BS.

repeat

    for All DUEs in $\mathcal{S}^{\mathcal{R}}$ **do**

        if The defection order defined in Eq. (16) is satisfied **then**

            The DUE transfers to the coalition that ensures its maximum utility.

        end if

    end for

until The defection of one DUE can only cause the change of $\mathcal{S}^{\mathcal{R}}$ doesn't exceed ε which is equal to 0.01.

The goal of Algorithm 2 is to reach a stable coalition partition, where no DUE will defect. The proposed long-term coalition formation game is composed of two stages: initial stage and coalition formation stage. Initially, the network is partitioned in a non-cooperative way. The coalition formation procedure is performed in the second stage: (i) all users who have not determined their access schemes send access requests to the access point that can provide maximum average receiving rate, and then each RUE will associate with users that maximize utility of the coalition from the users sent access requests to him/her; (ii) DUEs traverse all coalitions and join the coalition that can maximize its utility according to defection order. After the long-term coalition formation game, users will decide to become DUEs or CUEs according to the distance to RUEs in future time slots.

In the following Theorem 1, we prove the convergence of the coalition formation game.

Theorem 1. *The proposed long-term coalition formation game based algorithm is convergent.*

Proof. The convergence of the long-term coalition formation game based algorithm is guaranteed due to the fact that: (i) the total number of possible partitions with overlapping coalitions is finite when the total number of users is finite in this system; (ii) the transition from a coalition to another coalition leads to the increase of individual utility; (iii) the game contains mechanism to prevent the users to re-visit a previously formed coalitional structure with the set of ε.

4   Simulation Results and Analysis

In this section, the proposed algorithms in Sect. 3 are evaluated based on a data-driven numerical simulation, and the results are compared with two other baseline schemes. The first baseline scheme is called short-term optimal scheme, which MVNO will select RUEs and corresponding DUEs according to their locations at the beginning of each time slot. The second baseline scheme is called non-cooperative scheme, which all users are served by BS directly.

4.1   Experiment Setup

We assume some users that require same popular content are uniformly and randomly distributed in a $500\,\text{m} \times 500\,\text{m}$ square area. There is one BS at the center of the square area. At the beginning of each time slot, users move to new locations and unmoved during the time slot. We set all users' transmission power to 6 dBm, the BS's transmission power to 43 dBm, the power of background additive white Gaussian noise to an average of -120 dBm (at all access points), the λ to 1, the locations prediction interval to 1 s [13], the duration of one time slot to $\tau = 1$ ms, the bandwidth of one RB to 180 kHz, the path loss exponent to $\alpha = 3$.

4.2   Numerical Results

Based on the setup, we compare the average number of time slots used by users with the variety of L when $|\mathcal{U}| = 30$ and the number of RBs is 20, the results of which are depicted in Fig. 3. Obviously, the advantage of Long-term Optimal Algorithm (LOA) we proposed isn't obvious when the value of L is small. As the value of L increases, the average number of time slots used by D2D users in our algorithm increase slowly and the advantage of the algorithm is outstanding when the value of L is large. It's due to the fact that, users can receive the content in very few time slots when the value of L is small and LOA can't show its advantage. When the value of L is relatively large, the users in Short-term

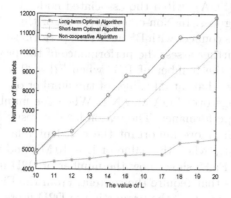

Fig. 3. The average number of time slots varies with the value of L

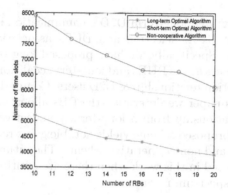

Fig. 4. The average number of time slots varies with the number of RBs

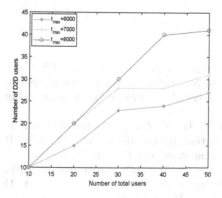

Fig. 5. The number of D2D users varies with the number of total users

Optimal Algorithm (SOA) switch the associated nodes between time slots and bring much switching loss. In Non-cooperative Algorithm (NA) with the worst performance, all users connect with BS and can't perform the distance advantage between users. We further assess the performance of the proposed algorithm, in Fig. 4, by increasing the number of RBs when $|\mathcal{U}| = 20$ and the value of L is 10 Mb. Figure 4 shows that, at all values of the number of RBs, the LOA yields a significant advantage over LOA and NA. When the number of RBs is small, the LOA has better performance. The reason for this result is the less spectrum resources there is, the more important the excellent relay is. In Fig. 5 we set different value for $t_{\max}$ when the value of L is 10 Mb and the number of RBs is 10. The results of Fig. 5 shows how the number of D2D users varies with the number of total users that require the content. From the Fig. 5, we can see that by adjusting the value of $t_{\max}$ the proportion of D2D users can be controlled.

5   Conclusions

In this paper, we integrated long-term D2D communication into cognitive WVN and studied the joint RUEs selection and DUEs association problem from a long-term perspective. Specifically, we have proposed long-term RUEs selection algorithm to choose users act as RUEs and long-term coalition formation game to determine the association relationship of D2D users. Comparing with the existing related works, in this paper we determine the RBs allocation scheme and D2D users' association relationship from a long-term perspective. Numerical results have shown that the proposed scheme yields notable gains relative to both short-term optimal scheme and non-cooperative scheme. The future work will continue to study the issues raised in this work when there are multiple base stations and different users reuse spectrum resources.

Acknowledgment. This work is partially supported by National Natural Science Foundation of China under Grant 61501041, and the Open Foundation of State Key Laboratory under Grant ISN19-19.

References

1. Vu, T.X., Lei, L., Vuppala, S., Kalantari, A., Chatzinotas, S., Ottersten, B.: Latency minimization for content delivery networks with wireless edge caching. In: 2018 IEEE International Conference on Communications (ICC), Kansas City, MO, pp. 1–6 (2018)
2. Ren, J., Yu, G., Cai, Y., He, Y., Qu, F.: Partial offloading for latency minimization in mobile-edge computing. In: GLOBECOM 2017 - 2017 IEEE Global Communications Conference, Singapore, pp. 1–6 (2017)
3. Jeon, S., Jung, K., Chang, H.: Fully distributed algorithms for minimum delay routing under heavy traffic. IEEE Trans. Mob. Comput. **13**(5), 1048–1060 (2014)
4. Ren, J., Yu, G., Cai, Y., He, Y.: Latency optimization for resource allocation in mobile-edge computation offloading. IEEE Trans. Wireless Commun. **17**(8), 5506–5519 (2018)

5. Mokhtarian, K., Jacobsen, H.: Minimum-delay multicast algorithms for mesh over-lays. IEEE/ACM Trans. Netw. **23**(3), 973–986 (2015)
6. Qu, J., Cai, Y., Xu, S.: Power allocation in a secure-aware device-to-device com-munication underlaying cellular network. In: 2016 8th International Conference on Wireless Communications & Signal Processing (WCSP), Yangzhou, pp. 1–5 (2016)
7. Zhao, Y., Song, W.: Truthful mechanisms for message dissemination via device-to-device communications. IEEE Trans. Veh. Technol. **66**(11), 10307–10321 (2017)
8. Xu, C., Gao, C., Zhou, Z., Chang, Z., Jia, Y.: Social network-based content delivery in device-to-device underlay cellular networks using matching theory. IEEE Access **5**, 924–937 (2016)
9. Giatsoglou, N., Ntontin, K., Kartsakli, E., Antonopoulos, A., Verikoukis, C.: D2D-aware device caching in mmWave-cellular networks. IEEE J. Sel. Areas Commun. **35**(9), 2025–2037 (2017)
10. Lin, X., Andrews, J.G., Ghosh, A.: Spectrum sharing for device-to-device commu-nication in cellular networks. IEEE Trans. Wireless Commun. **13**(12), 6727–6740 (2014)
11. Sheng, M., et al.: On transmission capacity region of D2D integrated cellular net-works with interference management. IEEE Trans. Commun. **63**(4), 1383–1399 (2015)
12. Mitra, K., Zaslavsky, A., Åhlund, C.: Context-aware QoE modelling, measurement, and prediction in mobile computing systems. IEEE Trans. Mob. Comput. **14**(5), 920–936 (2015)
13. Zhou, Z., Yu, H., Xu, C., Zhang, Y., Mumtaz, S., Rodriguez, J.: Dependable content distribution in D2D-based cooperative vehicular networks: a big data-integrated coalition game approach. IEEE Trans. Intell. Transp. Syst. **19**(3), 953–964 (2018)

Assessing the Feasibility of the Citizens Broadband Radio Service Concept for the Private Industrial Internet of Things Networks

Seppo Yrjölä[1]([⊠]) and Al Jette[2]

[1] Nokia, Oulu, Finland
seppo.yrjola@nokia.com
[2] Nokia, Naperville, IL, USA

Abstract. 5G emerges with ultra-dense deployments of small cell networks to serve various vertical sectors' location specific service requirements. While the development of technical solutions for network densification is progressing, less attention is paid to the spectrum models for the new ultra-dense networks and location specific service offerings. This paper examines the problems currently faced by industry in acquiring spectrum to support the Industrial Internet of Things (IIOT). Industrial applications where such spectrum is needed were assessed and their requirements identified. The US Citizens Broadband Radio Service (CBRS) spectrum sharing model to support the IIOT needs is introduced and how it addresses the IIOT requirements were evaluated based on four different real-life use cases. This study developed a view of options for the spectrum supply side, how this could interface with demand from private networks. Results showed that the CBRS model is well suited for several IIOT use cases based on having smaller licensed areas for PALs allowing a low-cost path for acquiring exclusive use spectrum along with a no cost option of using GAA spectrum. The leasing rules defined for CBRS PALs also provides an excellent minimal overhead option for enterprises to lease spectrum to other neighboring enterprises. Furthermore, the CBRS concept was found to leverage all the three forces of the long tail framework: Democratizing the tools of production through access to affordable spectrum, cutting the costs of consumption by democratizing distribution with web-scale automatization and connecting supply and demand via marketplace via SAS.

Keywords: Citizens Broadband Radio Service · Industrial Internet of Things · Spectrum sharing · Use case · 5G

1 Introduction

Digitalization has been transforming and disrupting industries at an unprecedented pace [1] and the diffusion of information technology into the physical industries is poised to revive the economy, create jobs, and boost incomes [2]. New 5th generation wireless network technologies (5G) are foreseen to enable this through wireless services provided at gigabit speeds, millisecond latency, support of wide range of novel applications

© ICST Institute for Computer Sciences, Social Informatics and Telecommunications Engineering 2019
Published by Springer Nature Switzerland AG 2019. All Rights Reserved
A. Kliks et al. (Eds.): CrownCom 2019, LNICST 291, pp. 344–357, 2019.
https://doi.org/10.1007/978-3-030-25748-4_26

connecting devices and objects, and versatility by virtualization enabling innovative business models across multiple sectors [3]. Present connectivity market has been characterized by incumbent network operators whose business is structured around service mass provisioning with high advance investments in infrastructure and exclusive long-term spectrum licenses [4]. At the same time, the responsibility of delivering resources is being transformed from centralized mobile network operator (MNO) centric system into a more dynamic mode of operation due to the deployment of software defined networks (SDN), network function virtualization (NFV), cloudification, spectrum sharing concepts, and the development of vertical service and application ecosystems [5].

A wide variety of users, machines, industries, public services and organizations will each have their special demands, and the 5G network is expected to fulfill these needs. Furthermore, 5G could be the enabler for new innovative business opportunities and lower the barrier to collaborate across domains. For example, for the industrial control and factory automation 5G can enable fully automated and flexible production and manufacturing systems consisting of sub-processes and subassemblies from several stakeholders. Consequently, this shift to more on-demand and decentralized local network services will require changes in the network's architecture especially in the management and orchestration levels [6] across resources from service integration to spectrum.

The industrial internet of things (IIOT) is a major component for next generation wireless systems and is being studied by many organizations globally. The International Telecommunication Union (ITU) [7] identifies industry automation and smart home/building as key usage scenarios of international mobile telecommunications (IMT) for 2020 and beyond. The European Commission focused on this critical need under the banner of *Industry4.0* [8]: "*Industry 4.0 refers to the intelligent networking of machines and processes for industry with the help of information and communication technology.*" Furthermore, the Industrial Internet Consortium (IIC) was formed to accelerate the development, adoption and widespread use of interconnected machines and devices and intelligent analytics [9], and the 5G Alliance for Connected Industries and Automation (5G-ACIA) was established for addressing, discussing, and evaluating relevant technical, regulatory, and business aspects with respect to 5G for the industrial domain. [10]. Today most of the IIOT equipment is short range devices using unlicensed spectrum. In the future, these short range IoT devices will likely remain a majority of the need, however, there is a significant enterprise need for larger wide-area coverage, supporting mobility, increased security and privacy, and assured certainty and Quality of Service (QoS).

While auctions have resulted in significant income for the governments, their impact on society goes beyond revenues. For example, competition, which will ultimately lead to greater innovation with better and cheaper services, will likely contribute to greater future governmental revenues compared to the sole auction revenues [11]. Future networks are expected to be increasingly locally deployed by new entrant stakeholders, e.g., facility owners or service providers [12]. Furthermore, local high-quality 5G wireless networks are gaining increasing attention as the solution to deliver guaranteed quality of service, particularly concerning the low latency requirements, in various vertical sectors' and enterprises' use cases [13]. Private mobile networks as stand-alone solutions or for collaboratively serving MNOs' customers [14] are particularly envisaged to operate in shared spectrum bands [15].

These trends are expected to result in defining spectrum access rights increasingly over appropriate geographical areas, e.g., national, regional, city or hyper-local, like for use in a factory [16]. The regulators foresee the need for more flexibility in 5G spectrum authorization approaches including the commons approach (general authorization, unlicensed), licensed shared use between different users, geographical sharing, or more dynamic approaches to spectrum sharing in time and space, with the help of geolocation databases [16]. Sharing-based spectrum management approaches facilitate more efficient spectrum use by allowing two or more radio systems to operate in the same frequency band. Prominent sharing concepts under standardization and pre-commercial trials are the US based Citizens Broadband Radio Service (CBRS) [17] and the European Licensed Shared Access (LSA) [18].

While the spectrum sharing models and CBRS concept have been widely studied in the technology, trial validation, regulation and business contexts, e.g., [19–22], to the best knowledge of the author, IIoT real life use case assessment and the options for related business model antecedents that could potentially develop has not been proposed in the literature. This paper will examine four selected IIoT use cases where such spectrum is needed, identify their requirements in the context of spectrum allocation, and assess the applicability of the CBRS spectrum sharing concept [17].

The rest of the paper is organized as follows. In Sect. 2, the economic drivers behind IIOT, its application characteristics and selected use cases are shortly reviewed. Next, an overview of the spectrum options for IIoT applications is given and the CBRS spectrum sharing systems presented. Section 4 presents and discusses the results of the use case assessment. Finally, the conclusions are drawn in Sect. 5.

2   Industrial Internet of Things (IIoT)

Productivity growth in the digital industries covering technology, content, finance & insurance, professional & technical services over the last 15 years has been strong, e.g., 2.7% in U.S. At the same time, productivity in the physical industries consisting of manufacturing, construction, mining, utilities, healthcare, hotels, restaurants, transportation, wholesale and retail trade grew just 0.7% annually, leading to weak overall economic growth over the last decade [23]. Companies and countries that provide spectrum and resources to support digital automation needs of industry verticals have been predicted to gain a significant financial benefit. For example, [24, 25] estimated that in 2025 the value creation potential of the IIoT can be between 1.2 B$ and 3.7 B$ in the factory segment only. The European 5G-PPP organization [3] is also focused on the economic benefits of IIOT in the manufacturing sector as a pivotal driver for economic growth: *The manufacturing sector is a pivotal driver for growth of the economy. In fact, it accounted in the period 2010–2012 for about 60% of productivity growth and 67% of exports in Europe.* At the same time, China has put in place a *Made in China 2025* action plan to future-proof their manufacturing industry to handle information technology highlighting *Manufacturing is the main body of the national economy* [26].

2.1 IIOT Application and Use Case Characterization

To analyze which use cases are best supported by the spectrum sharing solution and CBRS concept, it is necessary to first characterize the applications by the properties of the network that are required to deliver them [27]. The key dimensions considered in this study are depicted in Fig. 1.

Fig. 1. Characterization of IIOT applications.

Data transmitted for industrial applications may need to be protected due the *sensitivity* of the data for commercial or *security* reasons whereas the other dimensions by which the applications are characterized relate to technical requirements. Indoor and outdoor *coverage* relates to the setting in which the application is deployed as well as the requirements for mobility on the devices used for the application. Industrial applications can include the use of wireless connectivity in wide outdoor area such as on mines or in indoor spaces such as logistic hubs. *Bandwidth* requirements refer to the throughput level necessary to enable the application. Remote, tandem operation of machinery in a factory through virtual/augmented reality not only requires high QoS but also very high data transmission rate. This is particularly the case where the processing of a vast amount of information is done in the cloud. *QoS* refers to the application's requirement in terms of the network speed and consistency as well as requirements on latency and resilience. High QoS is needed for applications that are mission-critical such as process automation control or safety measures in industrial setting, where human and robots may come into close collaboration. QoS "dependability" is a key dimension that distinguishes a private network from a public or license-exempt network solution and can be broken down into five properties: reliability, availability, maintainability, safety, and integrity [28]. In some applications, there can be a significant *transmission asymmetry* between bandwidth required for uplink and downlink transmissions, e.g., high-definition monitoring and analytics will require a very high bandwidth in the uplink direction.

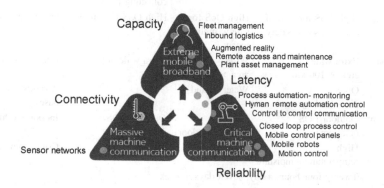

Fig. 2. Key network characteristics and IIoT applications.

Based on the 3GPP [28] and 5G-ACIA [10] use case categorization and the introduced characterization, the following IIoT application types were found optimal for *private network solution,* as depicted in Fig. 2: mobile robotics that require cloud computing access; remote operation of mobile machinery including augmented reality; sensors on machinery for personnel safety; remote operation of stationary machinery and high definition monitoring and tracking. The current use of private networks and IIoT is characterized by narrowband sensor networks and driven by simple functional requirements. The trend is to greater use of broadband sensors, while narrowband sensors keep a role, with enhanced downlink and uplink transmission capacity and latency shown in Fig. 2. Network performance needs for industrial automation are evolving and going towards [10]: higher reliability and availability of communication service; lower latency with higher synchronicity between devices; higher data rates with extended coverage; enhanced mobility and device density with high localization accuracy; and improved security. These evolutions are likely to be slow, because of implementation constraints and related costs. Potential for integration into systems and processes is increasing, while it is still limited, as well as network adaptability to configuration changes.

2.2 Use Cases

The characteristics of the selected real-life use cases, a mine, a harbor, a windmill park and a logistic hub, of this study are summarized in Table 1. While there are many common characteristics across the cases there are differences that may accentuate over time as the requirement for increased broadband capability develops. As previously noted, this is likely to place a significant data requirement in the uplink, which in turn creates more demanding transmission conditions and will increase bandwidth requirements and decrease the size of cells.

Table 1. Use case characteristics.

	Mine	Harbor	Windmill park	Hub
Activity	Extraction of ore, On-site processing, loadout facilities	Ship docks for goods and passengers	Turbines and substations generating and collecting power	Logistics facilities: collection, storage and sorting
Specificities	High QoS for critical operations and safety	High QoS for safety, multi-tenant	High QoS for safety	Fast pace process
Dimension	Extra wide area × 100 km^2	Wide area × 10 km^2	Extra wide area	Local × 1 km^2
Area spec.	Open pit + Underground	Water + land		Airports, Military bases
Environment	Outdoor LOS (+Indoor)	Outdoor NLOS (+Indoor)	Outdoor LOS (+Indoor)	Indoor + (Outdoor)
Specificities	High temperature + moisture	Marine env.		Cold chain center
Term	Twenty-four hours a day, seven days a week			

The purpose of the study was not to undertake a detailed network planning on each use case but to give a sense of what would be the requirements and differences in terms of spectrum between the use cases. For this, a set of common assumptions were made for the deployment of indoor and outdoor environments as depicted in Table 2. Device parameters would give a traffic density per km^2 of 200 Mbps for broadband devices and 7.5 Mbps for narrowband devices. Using 12 Mbps per sector, this would require 16 sectors to be deployed per km^2, assuming a relatively uniform geographic distribution of devices.

Table 2. Assumptions for line-of-sight (LOS) and non-LOS LTE sectors and device density in outdoor site scenarios.

Parameter	LOS	NLOS
Equivalent radius (km)	1.0	0.4
Coverage (omni antenna) (km^2)	3.1	0.5
Spectrum carrier (MHz)	10	10
Spectral efficiency (b/sec/Hz)	1.5	1.5
Maximum capacity loading (%)	80%	80%
Plan sector throughput (Mbps)	12	12
Narrowband devices (10 Kbps) per km^2	750	
Broadband devices (5 Mbps) per km^2	40	

3 Spectrum Models for IIoT

Industry seeks standardized solutions for IIOT to avoid expensive proprietary equipment. Standards have worked well to provide low cost devices for wireless local area networks (Wi-Fi) and LTE equipment where global low-cost devices have allowed for the technology to become universally prevalent. On the other hand, spectrum regulation has created a gap for industry to acquire spectrum which may require power and QoS levels above those of unlicensed spectrum for usage within an enterprise at a reasonable cost. There is an immense global effort to provide additional spectrum to meet the growing IIOT demand, and the following allocation approaches has been considered as depicted in Fig. 3: sub-licensing, leasing or trading of MNO spectrum; auction spectrum in smaller areas via regional micro-licensing, e.g., in Germany [29]; and shared spectrum approaches in particular CBRS and LSA.

Spectrum allocation decisions have major impact on the connectivity service model, as depicted in Fig. 3. Historically, self-provision has been the norm for private networks involving entities having access to licensed spectrum and buying, installing and operating the infrastructure themselves or using it through a lease/rental agreement with a third-party supplier that also facilitates access to the necessary frequencies. Being unable acquire appropriate spectrum would be a barrier to such an approach. On the other hand, technologies like LTE or 5G form a key part of public mobile networks and these platforms could provide private services utilizing an existing MNO's network and its spectrum portfolio and operating the private network as a virtual instance.

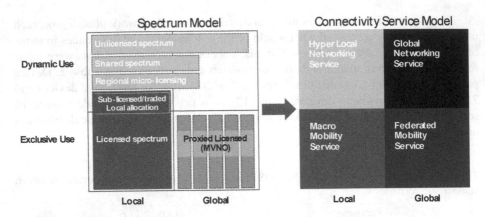

Fig. 3. Spectrum allocation model options determines the solution and the opportunity.

Alternatively, the private LTE network could be supplied by a third-party federated mobility managed service provider, wholesaler not generally providing public mobile services. Under this model, a wholesale mobile network could offer the private LTE service as a virtual instance on dedicated network infrastructure using its own spectrum. The spectrum would be dedicated for the private LTE service. The wholesaler would manage coverage and capacity related spectrum issues to ensure that its wholesale network meets the QoS requirements of the industrial sites it serves.

3.1 Citizens Broadband Radio Service

The US regulator FCC proposed a novel CBRS approach in the 3.5 GHz band to allocate up to 150 MHz low-cost shared spectrum in the 3550–3700 MHz band [17]. The spectrum currently has existing tier 1 incumbents who are given priority and protection from interference within the band from lower tiers, as depicted in Fig. 4. Besides incumbents, the FCC has setup two additional tiers of users: tier 2 priority access license (PAL) users and tier 3 general authorized access (GAA) users.

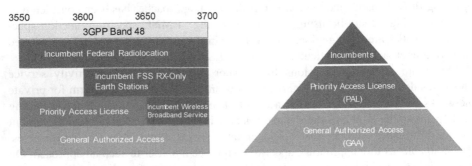

Fig. 4. CBRS spectrum sharing model.

The rules provide two paths for IIOT spectrum, PAL and GAA. According to the recent final FCC rules [30], the PAL spectrum will be auctioned off in smaller regional areas, in roughly 3200 counties, for ten years licenses with possibility of renewal. In comparison, to date many spectrum auctions in the US are done over 416 Partial Economic Areas that has made it expensive and not viable for an enterprise to acquire. The cost to acquire a single 10 MHz PAL will be determined by a spectrum auction but should be reasonably priced compared to other cellular spectrum auctions. Furthermore, the rules allow for a single PAL licensee to hold up to four channels in any licensed area at any given time, providing up to 40 MHz of spectrum protected from interference [30].

The final rules also allow the PAL holder to lease their PAL spectrum beyond their deployment coverage but within their PAL area. Moreover, PAL may be partitioned and disaggregated. Thus, in an industrial area a PAL holder may lease to other neighboring enterprises within their PAL area use of their PAL spectrum. A second path for IIOT spectrum in the CBRS band is to use 80 GHz of opportunistic, licensed by the rule GAA spectrum. Additionally, spectrum not currently used by an incumbent or by a PAL holder is available for GAA users on a shared basis. For both GAA and PAL, the base stations (CBSDs) must register with a SAS (Spectrum Access System) and request a spectrum grant. The SAS will attempt to find suitable spectrum for the CBSD and ensures higher tier users are protected by lower tier users, thought the unprotected GAA spectrum may have significant interference from other GAA users.

4 CBRS Use Case Analysis

CBRS suitability to selected use cases is assessed below based on interference risks, PAL/sub-leasing/GAA use of spectrum, initial network planning simulations (bandwidth, CBSDs), and use case specific limitations and is summarized in Table 3.

4.1 Mine

Private LTE networks in mines are almost unique in that they are completely separated from any potential public LTE network, not just by geographic distance but also by natural barriers, and as such there is unlikely to be any interference caused by using spectrum which is otherwise used by these networks. There would be very limited scope for any interference or alternative use of the spectrum, meaning that GAA licensing would be sufficient to ensure that the private LTE network could operate with sufficient reliability and consistency. Furthermore, in case needed a PAL sublicense could be obtained cheaply since it would present a very low opportunity cost for the PAL holder.

Some remote location of private LTE equipment may suffer from difficulties in receiving the heartbeat that appears to be a limitation of the CBRS framework. Under these conditions private network operator will require a sublease of alternative LTE spectrum from mobile network operators, or operate using unlicensed spectrum, again because the location will restrict the potential interference. Utilization of alternative spectrum layer will additionally offer resiliency needed in operating critical application.

4.2 Harbor

The networks to be deployed will be dense and require large bandwidths, particularly given the streaming video applications, future automated vehicles and high reliability. As they are also in areas open to the public, particularly surrounding the cruise ship terminals and marinas, it is likely that MNOs have deployed their own networks to cover at least part of the port area. This mix of requirements from private LTE operators and MNOs will mean that the only spectrum that could be used for the private LTE network would be that covered by CBRS: MNOs will not be willing to sublease their spectrum since they will either be using it or have a reasonable belief they will use it in the near future; while unlicensed spectrum would not have sufficient reliability and security to support the applications needed.

Given the need for greater reliability, private LTE operators would need access to PAL, but given the county-wide licensing regime, it would likely be unprofitable for them to buy a PAL themselves. Therefore, we would expect a PAL leasing arrangement would be most suitable. The case where the port elects to acquire a PAL itself at a low price may be limited by the bandwidth available as it may be necessary for multiple licenses to be used to meet the demands of various networks. This is a fairly significant drawback to the use of CBRS for this licensing.

4.3 Windmill Park

Unlike ports, most large windmill parks are situated away from built-up and populated areas. This means that higher frequency spectrum is unlikely to either cause interference or suffer from interference from other sources, given the short propagation distance. The lack of interference will in turn lead to a low opportunity cost associated with the use of CBRS spectrum; PAL holders may be unlikely to be inconvenienced if they allow the windmill parks to sublease their spectrum. This may not be true for other potential LTE bands; the wider propagation of these could lead to surrounding villages suffering interference.

The location of a windmill park is an enabler and a framer. To be at their most efficient, wind turbines tend to be tall buildings located high on hills, which means that any signals sent by transmitters placed on the turbines themselves are ideally located to achieve maximum range. This will reduce the willingness of mobile operators to allow for subleasing of their spectrum and may also reduce the possibilities for PAL leasing as well, depending on how PAL holders plan to use the spectrum in surrounding areas. Windmill parks can be very large and spread over a large geographic area, particularly where the geography requires straight lines of turbines. Where this lies in two different counties, it will require multiple agreements to sublease PAL for CBRS spectrum, and it may not be possible to obtain compatible licenses in the two different counties. Given this, while windmill parks appear to be an ideal case for CBRS spectrum sharing, given the lack of interference and likely low opportunity cost, there are several logistical issues to overcome before the benefits can be realized.

4.4 Hub

The use of CBRS spectrum in the logistics hub considers indoor and outdoor use cases. Within the fulfilment centers and sorting buildings, CBRS spectrum can be used indoors with minor risk of interference to the surrounding region. However, the outdoor usage has the same issues as described for harbor use case above; logistics hubs are likely to be in built-up areas, with many other potential uses for the spectrum, and as a result the opportunity cost of PAL leasing may be very high. Unlike the case of harbor, however, the use of spectrum outdoors at the logistics hub is likely to be restricted to fixed point-to-point links, and this further reduced the likelihood of CBRS spectrum being used, since higher frequencies, such as 26 GHz can be used instead with little loss in quality.

The spectrum needs of a logistics hub are therefore most likely to be met with a mix of CBRS and other bands. The relevant mix will depend on the usage of CBRS in surrounding regions, derived from the demand for fixed wireless access or other services. Indeed, this demand will affect CBRS in two ways: will interference clash with high demand (preventing PAL leasing or GAA at all) or will low demand mean there is no PAL sublicensing at all?

Table 3. CBRS suitability to use cases.

	Mine	Harbor	Windmill park	Hub
CBRS suitability	++	−	++	Indoor: +++ Outdoor: −
Interference risks	Low	High	Low	Indoor: low Outdoor: high
Use of spectrum	GAA sufficient PAL license in some cases	PAL necessary depends on other services in the county	GAA PAL sub-license	GAA PAL outdoor
Sub-leasing	Very low opportunity cost for PAL holder	High opportunity cost	Low opportunity cost for PAL holder	High opportunity cost
Bandwidth (10 MHz carrier)	2 or 3	2 or 3	2 or 3	3
CBSD sites (3 sectors)	18/12	Land 45/32 Water 8/5	19/13	14
Limitations	Reception of heartbeat in some cases	County-level PAL awards unfeasible	Wide coverage from masts	Other spectrum usage on site

4.5 Discussions

The heterogeneity of industrial use cases, applications and requirements leads to a flexibility requirement in spectrum award and use, which makes CBRS relevant and suitable for small-cell deployments and private LTE applications in industry. In its latest regulatory updates, CBRS appears as a favorable license scheme to industry as it responds to a growing need to bridge the gap between very large projects with direct mobile operator involvements and large numbers of smaller projects that are too small for mobile operators to consider, but too complex for enterprises to handle on their own. Flexibility of spectrum use and license periods of 10 years with renewal possibility are an incentive to investment. Furthermore, the CBRS concept was found to leverage all the three forces of the long tail framework [31]: *Democratizing the tools of production* through access to "free" spectrum, *cutting the costs of consumption* by democratizing distribution with web-scale automatization and *connecting supply and demand* via marketplace via SAS.

On the other hand, CBRS displays certain characteristics that appear to be unfavorable to some IIoT use cases. Using CBRS spectrum incurs incremental costs, while providing a low incremental reliability. In this way, the value of CBRS spectrum is based on the operator' willingness to pay and thus largely based on the value put in the incremental reliability and flexibility compared to unlicensed spectrum options. Another CBRS constraint is the mandatory heartbeat mechanism that makes its application unreliable or not feasible in some use cases. Furthermore, FCC policy decisions may make PAL-leasing unrealistic for the some IIoT use cases. Counties are appropriate for rural ISP type operators but not for micro-operators serving distinct local facility.

The cost of using CBRS consists of the *licensing* cost of spectrum, where PAL holders must consider the opportunity cost of not having access to the subleased spectrum to assess this cost, and so this is not driven directly by the private LTE network; and the cost of additional *spectrum control and monitoring equipment*. Environmental Sensing Capability (ESC) [17] is mandatory to detect military radar operations before operating CBRS. This may be provided by the SAS operator, or by a third-party ESC operator, and these costs will be passed on to end users through higher service fees. Although these costs may be small, they will still be non-zero, meaning that private network operators must receive a clear benefit from using CBRS spectrum. Users should consider how to quantify the benefits by thinking of willingness to pay in terms of QoS and reliability through business continuity insurance, existing investment in failsafe and security and the cost of loss of productivity These benefits are going to differ for every specific case, and in almost every example deriving a robust estimate of benefits will be impossible for a third party given the need for confidential operating information and business decisions. Required spectrum monitoring and controlling equipment may present a barrier to entry and know-how cost is also high, even if supplied by the SAS: If costs and related SAS fees become too high, alternative technologies will be used instead.

5 Conclusions

Current spectrum regulation forces vertical business to mostly rely on the MNOs. Direct services to verticals are possible only through shared spectrum or unlicensed bands, and spectrum brokering emerging as an alternative. Main stream spectrum regulation for 5G promotes 100% spectrum assignment to MNOs. The growing need for regional quality spectrum by vertical IIOT application is creating a conflict between verticals and traditional operators.

Verticals' IIOT use case specific applications differs from mobile broadband use case in several ways. Local private network deployments are typically in distinct limited coverage areas. Dedicated spectrum resources of 10–30 MHz bandwidth are needed. Leased or shared use of bands may be possible as access to spectrum through auctions is not considered feasible. Timely access to free/low cost spectrum with variable license period are requirements driven by business needs. IIOT applications have distinct application specific technical performance requirements (e.g., throughput, latency, transmission symmetry), and several IIOT applications set also high requirements for local data security and privacy.

The heterogeneity of industrial use cases, applications and requirements leads to a flexibility requirement in spectrum award and use, which makes CBRS relevant and suitable for small-cell deployments and private LTE applications in industry. In its latest regulatory updates, CBRS appears as a favorable license scheme to industry. Flexibility of spectrum use and license periods of 10 years with renewal possibility are an incentive to investment. This paper studied four industry Internet of things use cases: a mine, a harbor, a windmill park and a logistics hub. The results show that use case that appear to make the most of CBRS spectrum use is the logistics hub in indoor environment. There may be a good availability of spectrum in mines and windmill park areas with low interference risks, but these use cases may face an issue by not being able to access the heartbeat signal. Both the harbor use case and the outdoor environment of the logistics hub appear as being less likely to make the most of CBRS spectrum.

In addition to access to affordable quality spectrum, there could be other regulatory issues to address when operating private networks, including: other licensing requirements for operation of telecommunications systems and any non-spectrum license or authorization fees payable. Policy makers should try to keep the overall complexity and costs as low as possible, and PAL holders should be incentivized to price PALs just above opportunity cost to encourage as much use as possible.

Acknowledgment. This research has been financially supported by Business Finland in micro-operator project. The author would like to acknowledge the contributions of uO5G project consortium and colleagues A. Schoentgen, T. Lavender, T. Miller from PLUM consulting who provided insight and expertise during our collaborative private LTE project that greatly assisted the research.

References

1. Downes, L., Nunes, P.: Big Bang Disruption: Strategy in the Age of Devastating Innovation Big-bang disruption, vol. 91. Penguin Group, New York (2014)
2. Mandel, M., Swanson, B.: The coming productivity boom. The Technology CEO Council (2017)
3. 5GPPP: 5G empowering vertical industries: Roadmap paper. The 5G Infrastructure Public Private Partnership, Brussels (2016)

4. Ahokangas, P., Matinmikko, M., Yrjölä, S., Okkonen, H., Casey, T.: "Simple rules" for mobile network operators' strategic choices in future cognitive spectrum sharing networks. IEEE Wirel. Commun. **20**(2), 20–26 (2013)
5. Yrjölä, S., Ahokangas, P., Matinmikko-Blue, M.: Novel context and platform driven business models via 5G networks. In: The 2018 IEEE PIMRC Proceedings, Genova, Italy (2018)
6. NGMN Alliance: NGMN 5G White Paper (2015)
7. ITU-R: M.2083, IMT Vision - Framework and overall objectives of the future development of IMT for 2020 and beyond. https://www.itu.int/rec/R-REC-M.2083-0-201509-I/en. Accessed 02 Jan 2019
8. Federal Ministry for Economic Affairs and Energy: Platform Industrie 4.0. https://www.plattform-i40.de/I40/Navigation/EN/Industrie40/WhatIsIndustrie40/what-is-industrie40.html. Accessed 02 Jan 2019
9. Industrial Internet Consortium (IIC). https://www.iiconsortium.org/members.htm. Accessed 02 Jan 2019
10. 5G Alliance for Connected Industries and Automation (5G-ACIA): White Paper, 5G for Connected Industries and Automation (2018)
11. Cramton, P.: Spectrum auction design. Rev. Ind. Organ. **42**(2), 161–190 (2013)
12. Zander, J.: Beyond the ultra-dense barrier: Paradigm shifts on the road beyond 1000x wireless capacity. IEEE Wirel. Commun. **24**(3), 96–102 (2017)
13. Guirao, M.D.P., Wilzeck, A., Schmidt, A., Septinus, K., Thein, C.: Locally and temporary shared spectrum as opportunity for vertical sectors in 5G. IEEE Netw. **31**(6), 24–31 (2017)
14. Ahmed, A.A.W., Markendahl, J., Ghanbari, A.: Evaluation of spectrum access options for indoor mobile network deployment. In: IEEE 24th International Symposium on Personal, Indoor and Mobile Radio Communications Proceedings, London, pp. 138–142 (2013)
15. ETSI: TR 103 588 Feasibility study on temporary spectrum access for local high-quality wireless networks (2018)
16. European Commission: RSPG18-005 Strategic spectrum roadmap towards 5G for Europe: RSPG second opinion on 5G networks. Radio Spectrum Policy Group (2018)
17. FCC: Part 96 - Citizens Broadband Radio Service (2015)
18. ECC: Report 205 Licensed Shared Access (LSA). European Conference of Postal and Telecommunications Administrations, Electronic Communications Committee (2014)
19. Sohul, M., Yao, M., Yang, T., Reed, J.: Spectrum access system for the citizen broadband radio service. IEEE Commun. Mag. Year **53**(7), 18–25 (2015)
20. Palola, M., et al.: Field trial of the 3.5 GHz citizens broadband radio service governed by a spectrum access system (SAS). In: IEEE International Symposium on Dynamic Spectrum Access Networks Proceedings, pp. 1–9 (2017)
21. Mustonen, M., Matinmikko, M., Palola, M., Rautio, T., Yrjölä, S.: Analysis of requirements from standardization for Licensed Shared Access (LSA) system implementation. In: IEEE International Symposium on Dynamic Spectrum Access Networks Proceedings, pp. 71–81 (2015)
22. Yrjölä, S., Ahokangas, P., Matinmikko, M.: Evaluation of recent spectrum sharing concepts from business model scalability point of view. In: IEEE International Symposium on Dynamic Spectrum Access Networks Proceedings, pp. 241–250 (2015)
23. Mandel, M.: Long-Term U.S. Productivity Growth and Mobile Broadband: The Road Ahead. Progressive Policy Institute (2016)
24. McKinsey Global Institute: The internet of things: mapping the value beyond the hype (2015)

25. McKinsey: Manufacturing the future: The next era of global growth and innovation. https://www.mckinsey.com/~/media/McKinsey/Business%20Functions/Operations/Our%20Insights/The%20future%20of%20manufacturing/MGI_Manufacturing%20the%20future_Executive%20summary_Nov%202012.ashxa. Accessed 02 Jan 2019

26. The State Council People's Republic of China: Made in China 2025. http://www.gov.cn/zhengce/content/2015-05/19/content_9784.htm. Accessed 02 Jan 2019

27. Lavender, T., et al.: Use cases, spectrum requirements and valuation of spectrum for private LTE. PLUM Consulting (2018)

28. 3GPP: TR 22.804 V1.0.0 Study on Communication for Automation in Vertical Domains (2017)

29. Bundesnetzagentur (BNetzA), German National Regulatory Authority: Electronic Communications Services, Spectrum for wireless access for the provision of telecommunications services. https://www.bundesnetzagentur.de/EN/Areas/Telecommunications/Companies/FrequencyManagement/ElectronicCommunicationsServices/ElectronicCommunicationServices_node.html. Accessed 02 Jan 2019

30. FCC: Increasing Incentives for Investment and Innovation in 3.5 GHz Band. https://www.fcc.gov/document/increasing-incentives-investment-and-innovation-35-ghz-band. Accessed 02 Jan 2019

31. Anderson, C: The Long Tail: Why the Future of Business Is Selling Less of More. Hyperion (2006)

Workshop on Open Radio Platforms for 5G Research and Beyond

Enhanced Resource Management for Web Based Thin Clients Using Cross-Platform Progressive Offline Capabilities

George Alex Stelea[1]([✉]), Maurizio Murroni[2] [ID], Vlad Popescu[1] [ID],
Titus Balan[1], and Vlad Fernoaga[1]

[1] Transilvania University, bd Eroilor 29A, Brasov, Romania
george.stelea@unitbv.ro
[2] University of Cagliari, Via San Giorgio 12/2, 09124 Cagliari, Italy

Abstract. Web based thin clients are applications delivering content from the Internet or Intranet and accessed via the browser on the running end device. These clients are portable and cross-device compatible and have a large spectrum of applications, can perform from tele-measurement tasks to management and information centralization. The capability of web-based thin clients to function offline is a requirement that is indispensable even today for many companies because offline-enabled thin clients allow the users to continue working without workflow disturbance, preventing the loss of data, even when the connection to the Internet is missing or malfunctioning. This paper is dedicated to a "barrier-free" cross-platform responsive and progressive web based thin client, presenting its architecture and development, as well as the offline capabilities using caching techniques and its advantages in resource management and information back-up and security.

Keywords: Web based thin client ·
Cross-platform progressive offline capabilities · Enhanced resource management

1 Introduction

Thin clients are necessary when fat clients are too expensive and upscale or because they need more computing power or energy than available from the low-end terminals (e.g. tablets or smartphones). A web based thin client is an application program functioning according to the client-server model [1], where all the software is running on the server, and only the presentation is delivered on the device (e.g. GIS or tele-measurement applications where the thin client is basically the web browser) [2]. In the modern Web, oriented towards portables, Cloud computing and virtualization the trend is towards thin clients. Unlike traditional desktop software, web based thin clients do not need important installation and execution processes on the user's machine [3]. Instead, the data processing and evaluation mainly takes place on a remote web server [4] and only the result of the data processing is transmitted to the user's local client computer for display or output [5], usually via a web browser which handles the

© ICST Institute for Computer Sciences, Social Informatics and Telecommunications Engineering 2019
Published by Springer Nature Switzerland AG 2019. All Rights Reserved
A. Kliks et al. (Eds.): CrownCom 2019, LNICST 291, pp. 361–372, 2019.
https://doi.org/10.1007/978-3-030-25748-4_27

communication with the web server (via the HTTP protocol) as well as the representation of the user interface [6].

On the server or virtualized desktop, the inputs are processed and the output is sent back to the client, who only has to display them. The current generation of terminal servers or virtualization solutions also allows the use of hardware beyond a printer and works with optimized methods for playback of audio or video data. Although the trend in the age of IoT networks and mobile/smart devices today is that everyone is online 24 h a day [7], there are still many situations in which an Internet connection is malfunctioning or temporary missing (e.g. on the go, in cellular radio communications). An offline-enabled thin client allows the user to maintain the session – continuing to work even in this case without being disturbed in his workflow and without the loss of data.

In this paper, we present a "barrier-free" [8] cross-platform responsive and progressive web based thin client using the new JavaScript [9] and HTML5 [10] specifications with the Online Application Caching API, "Service Workers" and Bootstrap Framework, presenting its architecture, development environment and its advantages in resource management, rich context information administration and enhanced security. The paper is structured as follows: Sect. 2 presents the concept and the technology to achieve the cross-platform content to terminal adaption and the aim of this practice, Sect. 3 outlines the progressive web methodology, describing the combined possibilities offered by most modern browsers with the benefits of mobile use, Sect. 4 details the offline capabilities of the proposed solution using a caching system, retrieving and intercepting network requests from the cache and delivering push messages, indicating the increased maintainability and testability, while Sect. 5 describes the conclusions and the future work to be done.

2 Content to Terminal Adaption

The size and resolution of displays on laptops, desktops, tablets, smartphones and TV sets can vary considerably [11]. For this reason, the appearance and operation of a web based thin client are very dependent on the different requirements of the devices [12]. The aim of the content to terminal adaption practice is that thin clients adapt their presentation so that they present themselves as clear and user-friendly as possible to each viewer [13]. The criteria for the customized appearance are, in addition to the size of the display device, also the available input methods (touch screen, mouse) or the bandwidth of the Internet connection [14]. In order to achieve the responsive and adaptive design of the solution we have chosen the Bootstrap Framework and CSS3 Media Queries - as shown in Figs. 1 and 2 - that allow different designs depending on certain characteristics of the output environment.

```
1  <!DOCTYPE html>
2  <html lang="en">
3    <head>
4      <meta charset="utf-8">
5      <meta http-equiv="X-UA-Compatible" content="IE=edge">
6      <meta name="viewport" content="width=device-width, initial-scale=1">
7      <!-- The above 3 meta tags *must* come first in the head; -->
8      <title>Cross-platform responsive</title>
9      <!-- Bootstrap -->
10     <link href="css/bootstrap.min.css" rel="stylesheet">
11     <link href="css/style.css" rel="stylesheet">
12     <link href="css/lightbox.css" rel="stylesheet">
13     <!-- HTML5 shim and Respond.js for IE8 support of HTML5 elements and media queries -->
14     <!-- WARNING: Respond.js doesn't work if you view the page via file:// -->
15     <!--[if lt IE 9]>
16       <script src="https://oss.maxcdn.com/html5shiv/3.7.3/html5shiv.min.js"></script>
17       <script src="https://oss.maxcdn.com/respond/1.4.2/respond.min.js"></script>
18     <![endif]-->
19     <!-- jQuery (necessary for Bootstrap's JavaScript plugins) -->
20     <script src="https://ajax.googleapis.com/ajax/libs/jquery/1.12.4/jquery.min.js"></script>
21     <!-- Include all compiled plugins (below), or include individual files as needed -->
22     <script src="js/bootstrap.min.js"></script>
23     <script src="js/lightbox.js"></script>
```

Fig. 1. Bootstrap Framework installation and call in the web applications <head> section.

```
@media (max-width: 767px) {
  .hidden-xs {
    display: none !important;
  }
}
@media (min-width: 768px) and (max-width: 991px) {
  .hidden-sm {
    display: none !important;
  }
}
@media (min-width: 992px) and (max-width: 1199px) {
  .hidden-md {
    display: none !important;
  }
}
@media (min-width: 1200px) {
  .hidden-lg {
    display: none !important;
  }
}
```

Fig. 2. CSS3 media queries.

In order to achieve optimal recognition using CSS for all display formats, the media information with media queries was requested before loading the application. It was not necessary to record the appropriate screen size for each individual device, but rather the types of devices, media features and breakpoints. The terminal then automatically loads the correct part of the CSS file and displays the content as the size of the screen allows. The thin client's user interface (UI) uses custom web services for graphical indicators, meters, text boxes, inputs and buttons, which are also adaptable to various devices, as shown in Fig. 3.

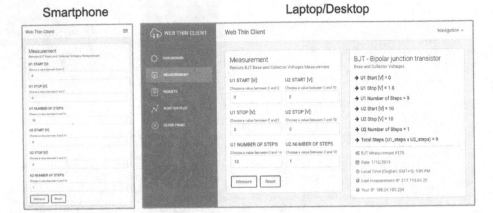

Fig. 3. "Thin client" solution display of a tele-measurement on a smartphone and laptop/desktop device resolution.

This is how cross-platform content to terminal adaption was implemented, and, because there are used only standardized technologies, the software will also be compatible with subsequent devices that would be built using this standards.

3   Management Trough Progressive Web Methodology

Using this application model we aimed to combine the possibilities offered by most modern browsers with the benefits of mobile use [15]. The term "progressive" refers to the fact that, from the point of view of the user experience, the thin clients progressively adapts itself to the device and has many features, including speed and device optimization, that were previously reserved only for native software. We have combined, this way, the advantages of a classic thin client and of a custom desktop/mobile application. To implement the previously described feature, standardized HTML5, CSS3 and JavaScript was used, as shown in Fig. 4.

```
async function registerSW() {
  if ('serviceWorker' in navigator) {
    try {
      await navigator.serviceWorker.register('./sw.js');
    } catch (e) {
      alert('ServiceWorker registration failed.');
    }
  } else {
    document.querySelector('.alert').removeAttribute('hidden');
  }
}
```

Fig. 4. JavaScript asynchronous registration function.

In addition, we have used "Service Workers" [16] to serve through optimal caching of the online functionalities, as presented in Fig. 5. A service worker is a programmable network proxy, allowing the network requests control from the application. It is terminated when not in use, and restarted/executed on access and custom events. The HTTPS protocol was used for secure communication between the web client and the web server.

For example, the user starts the application in a browser, enters the URL of the web server and sends the first request. The web server accepts this request and passes it to the thin client, which initially acts like a web application. The thin client then generates or loads the HTML source code of the requested resources, which are sent back to the user's browser by the web server (HTTPS response).

Due to the responsive design, the users see a software product layout adapted to its terminal. Although the web based thin client is accessed through an URL, the users can drag an icon to their mobile device home screen or receive push notifications and use the application offline. Progressive enhancement technology allows for each user the best possible user experience based on their technical capabilities.

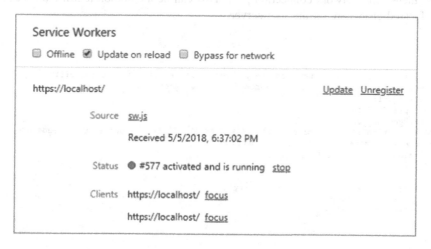

Fig. 5. Browser console - service worker verification.

The main advantages of using progressive enhancement on the thin client are:

- Custom Management and Updates – data and information is easy to manage and the resources are always up-to-date thanks to the data update process offered by Service Workers;
- Decentralized backups - a user can work in a different workplace every day without any restrictions using local storage and local databases;
- Endpoint Management - is optimized because on the thin client is running only the software necessary for server communication;
- Progressive - it works for every user, regardless of the browser chosen because it is built at the base with progressive improvement principles;

- Responsive - it adapts to the various screen sizes: desktop, mobile, tablet, or even dimensions that can later become available;
- Independent of connection availability - Service Workers allow the application to run online, in intermittent connections (with longer interrupts) or with low quality connections.
- Secure - exposed over HTTPS protocol to prevent the connection from displaying unwanted information or altering the contents;
- Discoverable - it is identified as an "application" thanks to the W3C manifesto and the Service Worker registration scope that allows search engines to find them;
- Linkable - easily shared via the URL, not requiring complex installations.

4 Offline Capabilities

The engine of the offline capabilities is based on a cache manifest file as presented in Fig. 6. The manifest file is a list of all the resources that the thin client needs to access when there is no network connection [17]. This can be a common text file (or a JSON file) located elsewhere on the web server.

```
{
    "name": "Cross-platform responsive and progressive with offline capabilities",
    "short_name": "Cross-platform responsive and progressive",
    "start_url": "./",
    "display": "standalone",
    "background_color": "#2b2284",
    "description": " The manifest file is a list of all the resources that needs access
to use when there is no network connection",
    "theme_color": "#2b2284",
    "icons": [
        {
            "src": "./img/icons/icon-72x72.png",
            "sizes": "72x72",
            "type": "image/png"
        }
    ]
}
```

Fig. 6. JSON cache manifest file.

In online mode the thin client reads the list of URLs from the manifest file, downloads the resources, caches them locally, and automatically keeps the local copies up to date as they change. If the resources are accessed without a network connection, the thin client automatically uses the local copies.

The Caching API is made of:

- the Web Storage Specification: includes an API for client-side storage of session-specific data and an API for storing session-spanning data;
- the Web SQL Database: a client-side JavaScript database;
- Web Workers: to execute parallel "background" processes in the client.

When the resource is recalled, the thin client checks to see if the manifest has changed. If so, all the necessary resources will be downloaded again. The cache will be updated only if the files that are registered with it have changed. To trigger an update, the contents of the manifest file must be changed. In Fig. 7 is presented the updated manifest in the browser console:

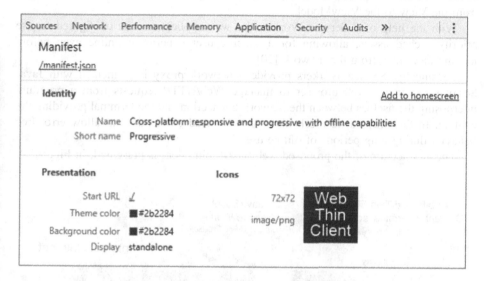

Fig. 7. Browser console – thin client manifest file.

To increase maintainability and testability, of the resources the thin client under consideration was built to separate the markup and JavaScript code using the Model-View-ViewModel (MVVM), also known as Model View Presenter. The MVVM pattern splits the application into three parts Model, View and ViewModel [18] as shown below in Fig. 8.

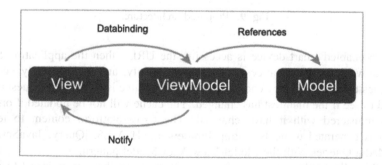

Fig. 8. Model-View-ViewModel pattern.

An important feature of a ViewModel is that it does not know the View. The bridge between the two concepts is realized via data binding, fields in the View are bounded to properties of the ViewModel. The same applies to events that occur in the View, such as clicks on buttons and operations in the ViewModel. The data binding mechanism handles both the updating of GUI (Graphical User Interface) [19] elements in the View, as the associated properties in the ViewModel change, and the transfer of user changes from the View to the ViewModel.

With the help of a Service Worker, the web application was configured to use priority cached assets, allowing for a specific user experience online even before loading more data from the network [20].

Technically, Service Workers provide a network proxy implemented with Java-Script script in the web browser to manage Web/HTTP requests from a program, interposing themselves between the network connection and the terminal providing the content. In this way cache mechanisms can be used efficiently and allow error-free behavior during long periods of offline use.

The architecture of the proposed web based thin client is presented, in Fig. 9.

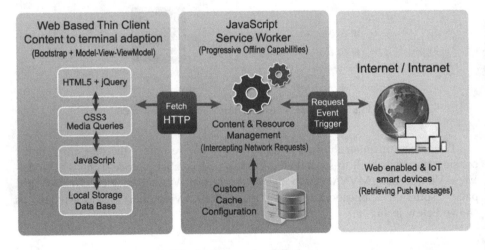

Fig. 9. Proposed architecture.

A web-enabled smart device is accessing the URLs, then the application Service Worker downloads the resources, caches them locally, and automatically keeps the local copies up to date as they change. When the resource is recalled, a request event is triggered to see if the manifest has changed. The cache will not be updated if only files that are registered with it have changed. The Cross-platform content to terminal adaption is generated using Bootstrap Framework, HTML5, jQuery, JavaScript and CSS3 Media Queries with the Model-View-ViewModel pattern.

The thin client subsequently carries out all the tasks that are assigned to it independently. Offline accessibility is managed by the Service Worker, which once installed in the navigation browser intercepts network requests and performs appropriate actions depending on whether the network is available or not.

This feature allows the user to access the resources even without connection and also to improve the Quality of Experience (QoE) and Quality of Service (QoS). Even if there is a connection, some files will not need to be uploaded from the server since they have already been stored locally.

In order that JavaScript can be executed, it has always been a prerequisite that the resource was opened in a browser - using this method, on the other hand, allows a JavaScript file to be executed even if the associated resource is not open at all.

The script also has the option of loading new content in the background - for example, a previous executed measurement result. If the data are accessed at a later time, the content will be already available, as a back-up.

In addition, the thin client only works over secure HTTPS connections [21]. This has security reasons in the first place, but it was also build in a farsighted mode, because it's likely that new standards will impose that resources to sensitive features or hardware will only run over HTTPS connections in the future [22].

As previously mentioned, the JavaScript file is detached from the actual web application, and runs in the background - invisible to the visitor - and it is registered by the thin client in the first online access of the resources. This is done through the Service Worker API's "register()" method, as shown in Fig. 10.

```
if ('serviceWorker' in navigator) {
    // Register a service worker hosted at the root of the
    // site using the default scope.
    navigator.serviceWorker.register('/sw.js').then(function(registration) {
        console.log('Service worker registration succeeded:', registration);
    }).catch(function(error) {
        console.log('Service worker registration failed:', error);
    });
} else {
    console.log('Service workers are not supported.');
}
```

Fig. 10. Service worker API's - JavaScript "register()" method.

The Service Worker also runs only in its own thread, does not allow direct manipulation of the parent's DOM (Document Object Model) [23], and has the message-based interface [24]. It acts as a controller, proxy or interceptor: it has its own cache and can switch between every outgoing network request [25]. The offline capability is realized by an automatic preconfigured decision if an answer can be requested from the cache or forward the request to the network.

5 Conclusions and Future Work

The novelty and the biggest advantage of the presented web based thin client solution over classic clients is the easier operation, with effectively reduced overhead, only running the software needed to access centrally operated applications. It can operate consistently regardless of the applications that are actually being used, in accordance with the balanced Edge Computing/Cloud Computing paradigm. This also allows a very simple resource management of centralized or decentralized control systems and has a large spectrum of potential applications, being able to perform from tele-measurement tasks to management and rich context information centralization. The web based cross-platform thin client enhanced with content to terminal adaption does not need to be installed on the device - this has enormous advantages, as most of the equipment is quickly reaching the limits of storage space. Since the presented solution does not need to be installed, the operators are also independent of commercial software stores that would take shares for marketing. Because it is developed with standardized technologies, also a big asset it that it has an increased security and reduced computing power needs because the thin client does not imply third party plugins or additional dependencies which increase the risks of security breaches and often require additional resource allocation.

The architecture of our solution was presented together with the development environment and its advantages, being oriented towards achieving a decoupling of developments in the still relatively young market of mobile devices - thus, it is also likely to run on future devices. In addition, concurrent access is achieved, as an almost unlimited number of thin clients can be managed by simply assigning configurations. Quickly turning resource events and handlers on and off results in a significant service advantage for the end user, especially for remote clients and lengthy installations.

The capability of the web-based thin client solution to function offline is a main advantage, especially for companies with users who occasionally have a bad Internet connection - this is an elementary aspect because in this scenario the software consistently adopts an offline first approach. Push notifications are also available, allowing users the same access to their personal interface, configuration, directories, and installed programs, regardless of which physical thin client workstation they log on to. The endpoint management is optimized because on the thin client is running only the software necessary for server communication.

Future work will be devoted to enhancing the proposed solution, helping the user to work in a different workplace every day without any restrictions, using local storage and local databases, creating also the back-up systems in a distributed environment.

Acknowledgement. This paper has been supported by the Autonomous Region of Sardinia-Italy (Piano Sulcis Research Grant SULCIS-821101).

References

1. Oluwatosin, H.S.: Client-server model. IOSR J. Comput. Eng. 16(1), 67–71 (2014). p-ISSN 2278-8727
2. Kim, J., Baratto, R.A., Nieh, J.: pTHINC: a thin-client architecture for mobile wireless web. In: Proceedings of the 15th International Conference on World Wide Web (WWW 2006), pp. 143–152. ACM, New York (2006)

3. Tian, Y., Song, B., Huh, E.: Towards the development of personal cloud computing for mobile thin-clients. In: 2011 International Conference on Information Science and Applications, Jeju Island, pp. 1–5 (2011)
4. Al-Hammouri, A., Al-Ali, Z., Al-Duwairi, B.: ReCAP: a distributed CAPTCHA service at the edge of the network to handle server overload. Trans. Emerg. Telecommun. Technol. **29** (4), e3187 (2017)
5. Stelea, G.A., Fernoaga, V., Gavrila, C., Robu, D.: Web-service based thin client for tele-measurement. Int. Sci. Conf. eLearn. Softw. Educ. **2**, 128–134 (2018)
6. Pohja, M.: Comparison of common XML-based web user interface languages. J. Web Eng. **9** (2), 95–115 (2010)
7. Charland, A., Leroux, B.: Mobile application development: web vs native. Commun. ACM **54**(5), 49–53 (2011)
8. Mahmood, A., Casetti, C., Chiasserini, C.F., Giaccone, P., Harri, J.: Efficient caching through stateful SDN in named data networking. Trans. Emerg. Telecommun. Technol. **29** (1), e3271 (2010)
9. Richards, G., Lebresne, S., Burg, B., Vitek, J.: An analysis of the dynamic behavior of JavaScript programs. SIGPLAN Not. **45**(6), 1–12 (2010)
10. Vaughan-Nichols, S.J.: Will HTML 5 restandardize the web? J. Comput. **43**(4), 13–15 (2010). https://doi.org/10.1109/MC.20
11. Orsini, G., Bade, D., Lamersdorf, W.: CloudAware: empowering context-aware self-adaptation for mobile applications. Trans. Emerg. Telecommun. Technol. **29**(4), e3210 (2017)
12. Rajesh, N.A.: Responsive web design. Int. J. Eng. Comput. Sci. **4**(3)
13. Robu, D., Fernoaga, V., Stelea, G.A., Sandu, F.: Tele-measurement with virtual instrumentation using web-services. Des. Technol. Electron. Packag. (SIITME) **23**, 387–394 (2017)
14. Baturay, M.H., Birtane, M.: Responsive web design: a new type of design for web-based instructional content. Procedia – Soc. Behav. Sci **106**, 2275–2279 (2013). ISSN 1877-0428
15. Joorabchi, M.E., Mesbah, A., Kruchten, P.: Real challenges in mobile app development. In: ACM/IEEE International Symposium on Empirical Software Engineering and Measurement, pp. 2851–2864 (2013)
16. Smutny, P.: Mobile development tools and cross-platform solutions. In: Proceedings of the 13th International Carpathian Control Conference (ICCC), pp. 653–656 (2012). https://doi.org/10.1109/carpathiancc.2012.6228727
17. Maddah-Alim, M.A., Niesen, U.: Fundamental limits of caching. IEEE Trans. Inf. Theory **60**, 2856–2867 (2014). https://doi.org/10.1109/TIT.2014.2306938
18. Leff, A., Rayfield, J.T.: Web-application development using the Model/View/Controller design pattern. In: Proceedings Fifth IEEE International Enterprise Distributed Object Computing Conference, pp. 118–127 (2001). https://doi.org/10.1109/edoc.2001.950428
19. Banerjee, I., Nguyen, B., Garousi, V., Memon, A.: Graphical user interface (GUI) testing: systematic mapping and repository. J. Inf. Softw. Technol. **55**(10), 1679–1694 (2013). ISSN 0950-5849
20. Bhamare, D., Samaka, M., Erbad, A., Jain, R., Gupta, L.: Exploring microservices for enhancing internet QoS. Trans. Emerg. Telecommun. Technol. **29**, e3445 (2018)
21. Clark, J., Oorschot, P.C.: SoK: SSL and HTTPS: revisiting past challenges and evaluating certificate trust model enhancements, pp. 511–525 (2013)
22. Georgiev, M., Iyengar, S., Jana, S., Anubhai, R., Boneh, D., Shmatikov, V.: The most dangerous code in the world: validating SSL certificates in non-browser software. In: CCS 2012, pp. 38–49. ACM, New York (2012)

23. Jensen, S.H., Madsen, M., Moller, A.: Modeling the HTML DOM and browser API in static analysis of JavaScript web applications. In: Proceedings of the 19th ACM SIGSOFT Symposium and the 13th European Conference on Foundations of Software Engineering, pp. 59–69 (2013). https://doi.org/10.1145/2025113.2025125

24. Arnbak, A., Asghari, H., Van Eeten, M., Van Eijk, N.: Graphical user interface (GUI) testing: systematic mapping and repository. Commun. ACM **57**(10), 47–55 (2013). https://doi.org/10.1145/2660574

25. Leu, J., Chen, C., Hsu, K.: Improving heterogeneous SOA-based IoT message stability by shortest processing time scheduling. IEEE Trans. Serv. Comput. **7**(4), 575–585 (2014)

Stabilized Distributed Layered Grant-Free Narrow-Band NOMA for mMTC

Hui Jiang[1]([⊠]), Qimei Cui[1], and Rongting Cai[2]

[1] National Engineering Laboratory for Mobile Network Technologies,
Beijing University of Posts and Telecommunications, Beijing 100876, China
{jianghui,cuiqimei}@bupt.edu.cn
[2] State Key Laboratory of Networking and Switching Technology,
Beijing University of Posts and Telecommunications, Beijing 100876, China
cairongting@bupt.edu.cn

Abstract. The main challenge of supporting Internet of Things (IoT) in 5G network is to provide massive connectivity to machine-type communication devices (MTCDs), with sporadic small-size data transmission. Narrowband technology is energyefficient with extended coverage, on a narrow bandwidth, for low-rate and low-cost MTCDs. Grant-free transmission is expected to support random uplink communication, however, this distributed manner leads to high collision probability. Nonorthogonal multiple access (NOMA) can be used in grantfree transmission, which multiplies connection opportunities by exploiting power domain. However, coordinated NOMA schemes where base station performs coordination is not suitable for grant-free transmissions. In this paper, based on a detailed analysis of the novel distributed grant-free NOMA scheme proposed in our previous work, a stabilized distributed narrow-band NOMA scheme is proposed to reduce collision probability, which derives the optimal (re)transmission probability for each MTCD. With the stabilized scheme, the system can be always stable and its throughput can be guaranteed whatever the new arrival rate is. Simulation results reveal that, when the system is overloaded, for uplink throughput, our proposed scheme outperforms by 45.2% and 87.5%, respectively, compared with the distributed NOMA scheme without transmission probability control and the coordinate OMA scheme considering transmission control.

Keywords: Stability · Grant-free · Distributed NOMA · Narrowband · Massive MTC · IoT

1 Introduction

Narrowband IoT (NB-IoT) can enable low-rate energyefficient communication with extended-coverage, which is standardized by Third Generation Partnership Project (3GPP) [1,2]. More specifically, it can support throughput of 20–50 kbps and extend the network coverage by up to 20 dB [3], which means a 100 times

© ICST Institute for Computer Sciences, Social Informatics and Telecommunications Engineering 2019
Published by Springer Nature Switzerland AG 2019. All Rights Reserved
A. Kliks et al. (Eds.): CrownCom 2019, LNICST 291, pp. 373–385, 2019.
https://doi.org/10.1007/978-3-030-25748-4_28

bigger coverage. The coverage enhancement margin makes NB-IoT communications immune against propagation and indoor penetration losses. In addition, devices can reduce their uplink transmission power to prolong the battery lifetime, which fits massive MTC (mMTC) well.

Unlike human-type communications, which involve a small amount of high-rate devices with large-sized data [4], mMTC is generally characterized with massive low-computational-capability devices and sporadic transmissions. Thus uplink access is a serious challenge for mMTC [5]. Devices access the network via a four-step random access (RA) procedure in conventional grant-based schemes, however, it is inefficient to establish dedicated bearers for small data transmission in the scenario of mMTC, since the consequential signaling overheads are proportional to the number of devices. Therefore, grant-free schemes are more promising to mMTC.

Grant-free is gaining attention recently, which allows devices transmitting without base stations (BSs)' radio resources granting [6, 7], which is perfect for mMTC due to their low signaling overhead. Conventionally, slotted ALOHA [8] based on orthogonal multiple access (OMA) is used for uplink communications. However, it seriously suffers from the nuisance of collision resulting from contention based access by multiple devices. Fortunately, by exploiting power domain, nonorthogonal multiple access (NOMA) enables multiple devices to share one time-frequency resource. Therefore, NOMA based grant-free can support significantly increased connections [9, 10]. However, most of existing studies on NOMA focus on coordination with known channel state information (CSI) at both transmitter and receiver sides, to optimize subchannel and power allocation [11, 12], which is not suitable for grant-free transmissions.

To address these challenges, we adopt a distributed NOMA, power division multiple access [13], in narrowband system, and propose a low-complexity distributed layered grant-free narrowband NOMA scheme to realize a hybrid transmission. With this scheme, the inherent drawback of grant-free random access, high collision probability due to its distributed manner, can be greatly alleviated. The key of the proposed scheme is, based on predetermined inter-layer received power difference, firstly dividing the extended-coverage cell into several regions. Secondly, power domain NOMA can be used to drastically reduce the number of MTCDs that compete for grant-free transmission in each region. However, grantfree transmission probably lead to unstable system without effective control scheme. To further guarantee the stability of the system, no matter what the new arrival rate is, we apply an optimal (re)transmission probability self-control scheme. With the stabilized scheme, the system can be always stable and its throughput can be guaranteed whatever the new arrival rate is. Simulation results reveal that, when the system is overloaded, for uplink throughput, our proposed scheme outperforms by 45.2% and 87.5%, respectively, compared with the distributed NOMA scheme without transmission control and the coordinate OMA scheme considering transmission control.

The main contributions of this paper are as follows.

- A detailed analysis of the novel distributed grant-free NOMA scheme proposed in our previous work is made, which reveals that grant-free transmission probably lead to unstable system without effective control scheme.
- We propose a distributed layered grant-free NOMA based hybrid transmission scheme for narrowband system. Moreover, considering the stability of the system, we derive the optimal self-control (re)transmission probability for each MTCD.
- The system is modeled as a Markov chain, from which both the average throughput and MTCD delay can be effectively calculated. Simulation results demonstrate that the analysis matches well with the simulation, and the proposed self-control (re)transmission probability scheme works well whatever the new arrival rate of the system is.

The rest of the paper is organized as follow. In Sect. 2, we introduce the system model. In Sect. 3, stability of the proposed grant-free NOMA scheme in our previous work is analyzed. In Sect. 4, we propose a stabilized scheme with (re)transmission probability control, and the performance evaluation is listed. In Sect. 5, we present simulation results and the paper is concluded in Sect. 6.

2   System Model

Consider a time slotted narrowband cellular network as shown in Fig. 1, with a single BS located in the origin, serving Q MTCDs in the area. We assume that the MTCDs are uniformly distributed in a circle of radius D, which is much longer than that of other systems. All MTCDs share a narrow bandwidth of B_T for uplink data transmissions. The available system bandwidth is divided into frequency resource blocks (subchannels), each of bandwidth B. Thus, the total number of frequency resource blocks is given as $M = B_T/B$. In such a system, each MTCD always starts its transmission at the beginning of a time slot. At the end of the time slot, BS boradcasts the feedback message for all MTCDs.

In this paper, we use *Connection Opportunity* (CO) to represent a connection resource. In OMA systems, the number of COs in a time slot is determined only by the number of available subchannels. Due to the limited frequency spectrum, the number of COs is inadequate for massive grantfree transmission. As seen in Fig. 1(a), if two users in a cell simultaneously access the BS with the same subchannel, collision happens.

In our previous work [14], a distributed NOMA concept is applied. Suppose that there are L predetermined aiming received power levels that are denoted as $v_1 > v_2 > ... > v_L > 0$, where $v_l = \Gamma(\Gamma+1)^{L-l}$ $(l = 1, 2, ..., L)$, and Γ is the target signal to interference-plus-noise ratio (SINR) at the BS for all MTCDs, which guarantees the throughput performance of each MTCD. Therefore, according to the predetermined received power levels, the single-BS cell can be divided into L concentric layers, which is reasonable to the narrowband system for its broad converage. MTCDs in different layers have different aiming received power, as shown in Fig. 1(b). The outsider layer denotes the smaller received power. MTCDs decide their transmission power according to locations and CSI. For example, for an MTCD k, d_k is the distance to the BS, if it belongs to set $K_l = \{k|D_{l-1} < d_k \leq D_l\}$, where $D_0 = 0$, $D_l = D\sqrt{\frac{l}{L}}$, then its aiming received power level is v_l. Based on its knowledge of channel gain of different sunchannel i, refered as $g_{i,k}$ $(i = 1, ..., M)$, its transmission power is decided as $P_k = \frac{v_l}{\max\limits_i g_{i,k}}$.

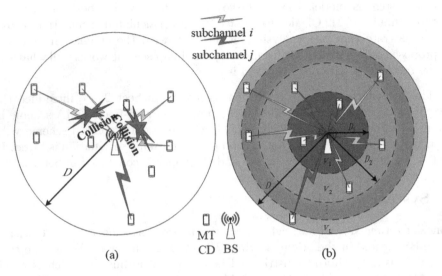

Fig. 1. (a) Grant-free transmission; (b) principle diagram of a distributed layered NOMA, where different colored circle and rings indicate the layers of different aiming received powers, and darker color represents bigger received power [14]. (Color figure online)

The key of the proposed scheme is dividing the extended-coverage cell into different regions (layers) based on predetermined inter-layer received power difference, thus, power domain NOMA can be used to drastically reduce the number of MTCDs that compete for grant-free transmission in each region. To be specific, given M subchannels and Q_T MTCDs, under the distributed layered grant-free NOMA scheme, it is equal to the situation that there are L layers, each layer has M subchannels. Since the MTCDs are uniformly distributed, to simplify mathematics, we assume that the number of active MTCDs in each layer is the same (i.e. $Q = \frac{Q_T}{L}$). The assumption can be easily extended to the general scenario with different number of MTCDs for each layer. For a subchannel, if there are multiple MTCDs who choose the same power level, the signals cannot be decoded, which is called power collision. Otherwise, due to the clear power gap between the predetermined received power levels, the BS can decode any signal which is the only one who chose a subchannel with some predetermined received power level. If the transmitted packet experiences collision, it will retransmit through a subchannel in the next time slot immediately, which is called fast retrial [15], that is, fast retrial and infinite retransmissions are assumed.

In this paper, an MTCD with a stored packet to transmit is called *activer*, while one without packet to transmit is called *inactiver*. *Activer* and *inactiver* are convertible, that is, an *inactiver* becomes *activer* once is has the arrival of a new packet, and an *activer* becomes an *inactiver* after a successful transmission. To simplify the system, there are some assumptions as follows:

(1) Each MTCD has only one buffer to store a packet;
(2) When a new packet arrives at an *inactiver*, it changes to an *activer* in the next time slot, and is treated equally as any other *inactivers*;
(3) Each *activer* can decide whether to transmit in the next time slot or not with transmission probability p_t;

3 Stability Analysis of the Proposed Grant-Free NOMA Scheme

In this section, using the Foster-Lyapunov criterion [17], the stability of the proposed grant-free NOMA scheme with fast retrial and infinite retransmissions is analyzed. According to the model description and analysis in the previous section, we can analyze each layer region, respectively and identically. Thus, in the following sections, we take one layer region as an example.

3.1 Drift of the Proposed Grant-Free NOMA

For a grant-free system with M subchannels, since an MTCD in each layer select a subchannel at random, each subchannel is selected with equal probability $\frac{1}{M}$. Let N_k denote the number of *activers* in one layer at time slot k, and A_{k+1} denote the number of inactivers having new arrival at time slot k. According to the principle of fast retrial, N_{k+1} can be given by

$$N_{k+1} = [N_k]_M + A_{k+1}, \tag{1}$$

which is actually a Markov chain, where $[n]_M$ denotes the number of the backlogged (due to stopping transmitting) and collided (due to transmitting collsion) packets among n transmitted packets, when the number of subchannels is M. If fast retrial is applied, these $[n]_M$ packets are all to be retransmitted in the next slot.

Since N_k is non-negative, we can analyze it with Lyapunov function. Given $N_k = n$, the drift of N_k can be expressed as [18]

$$d_n = \mathrm{E}\,[N_{k+1}|N_k] - N_k = \mathrm{E}\,[N_{k+1}|N_k = n] - n. \tag{2}$$

From Eq. (1), we can find that

$$\mathrm{E}\,[N_{k+1}|N_k = n] = \mathrm{E}\,[[n]_M] + \mathrm{E}\,[A_{k+1}|N_k = n] = \mathrm{E}\,[[n]_M] + \bar{A}_n, \tag{3}$$

where $\bar{A}_n = \mathrm{E}[A_{k+1}|N_k = n]$, which is an important function impacting the stability of the system with fast-retrial ans infinite retransmissions.

With M subchannels, let X_m denote the number of *activers* that choose some subchannel m ($m \in [1, ..., M]$) at the same time slot in the layer region. Since a subchannel is chosen by each *activer* uniformly at random, provided that there are $N_k = n$ *activers* to transmit their packets, the probability that any subchannel m is perfectly chosen by one MTCD is

$$\Pr(X_m = 1) = \binom{n}{1} \tfrac{1}{M} \left(1 - \tfrac{1}{M}\right)^{n-1} = bin(1, n, \tfrac{1}{M}) = \tfrac{n}{M} q^{n-1}, \qquad (4)$$

where $q = 1 - \tfrac{1}{M}$ and $bin(*)$ denotes a binomial probability. Thus, the average number of packets that are collided among n *activers* is given by

$$\mathrm{E}[[n]_M] = n - \sum_{m=1}^{M} \Pr(X_m = 1) = n(1 - q^{n-1}). \qquad (5)$$

According to Eqs. (2), (3) and (5), the drift of N_k can be expressed as

$$d_n = \bar{A}_n - nq^{n-1}. \qquad (6)$$

From Eq. (6) we can find that, since $nq^{n-1} \to 0$ as $n \to \infty$, the drift of N_k cannot be negative if the average number of *inactivers* having new arrival packets is a constant value, i.e., $\bar{A}_n = \lambda$. In other words, the proposed grant-free NOMA transmission with fast retrial cannot be stable with a constant arrival rate $\bar{A}_n = \lambda > 0$. In other words, the system needs to control the new arrival rate in order to stay stable.

3.2   New Arrival Rate Control Scheme for Stability

Control criteria of the new arrival rate can relate to the number of activers at the present time slot, Nk, which is usually not directly obtained but can be estimated in many ways. In our previous work about stochastic online learning [19], a new kind of maximum likelihood technique, the number of active users, N_k, can be reliably estimated, based on themaximum likelihood estimation at the BS, by taking advantage of stochastic online learning technique.

Suppose there exists the activer number threshold, which denotes the sign of a overloaded system with fast-retrial. That is, if $N_k = n > \bar{N}$, the new arrival rate control should be implemented. For simplicity and rationality, we assume that $\bar{N} = M$ in a grant-free system with M subchannels. Then, according to Eq. (6), we consider the following average new arrival rate control function

$$\bar{A}_n = \begin{cases} \lambda, & if \ n < M \\ \lambda q^{n-M}, & if \ n \geq M \end{cases}. \qquad (7)$$

Let set $S = \{0, ..., M - 1\}$, if the number of transmitting packets n is bigger than the threshold, i.e., $n \in Z - S = \{M, M+1, ...\}$, with Eq. (7), it can be found that $d_n = \lambda q^{n-M} - n q^{n-1} = (\lambda q^{-M+1} - n) q^{n-1}$. So, if $\lambda q^{-M+1} \leq M < n \in Z-S$, that is, $\lambda \leq M q^{M-1}$, is a sufficient condition for

$$d_n < 0, \ n \in Z - S. \tag{8}$$

For $n \in S$, we can find that

$$d_n = \lambda - n q^{n-1} \leq \lambda, \ n \in S. \tag{9}$$

Since set S is finite, we can see that Eqs. (8) and (9) satisfy the Foster-Lyapunov criterion [17], which denotes a stable system. When M increases, which is a usual situation with narrow-band mMTC system, we can have $\lim_{M \to \infty} q^{M-1} = \lim_{M \to \infty} (1 - \frac{1}{M})^{M-1} = \frac{1}{e}$, so $M q^{M-1}$ is approaching $\frac{M}{e}$, which means that the average new packet arrival rate should be less than $\frac{M}{e}$ when only new arrival rate control scheme is applied.

However, average new arrival rate control scheme is not easy to implement in practice, since the probability to generate a new packet cannot be controlled. Thus, we take another feasible control strategy, based on (re)transmission probability control scheme.

4 Stabilized Scheme with Transmission Probability Control

4.1 Transmission Probability Control

In this scheme, when an activer has a packet to (re)transmit, it transmits with a controlled transmission probability p_g instead of 100% fast retrial. For a grant-free system, since an MTCD in each layer select a subchannel from all M subchannels at random, once the MTCD decides to transmit, each subchannel is selected with equal probability $\frac{1}{M}$. We assume the number of packet transmissions in one time slot is a Poisson random variable. The average packet attempting rate is N. Since the probability for an attempting MTCD to choose a subchannel is equal, the number of packet arrivals at a subchannel is also Poisson with rate $\frac{N}{M}$. So the system throughput for a subchannel m can be expressed as

$$T_m = \frac{N}{M} e^{-\frac{N}{M}}, \tag{10}$$

and the overall throughput at a time slot is $T = \sum_{m=1}^{M} T_m = N e^{-\frac{N}{M}}$.

From Eq. (10), it is easy to find that the maximal throughput for any subchannel m is achieved when $\frac{N}{M} = 1$, i.e., $N = M$, and the maximal value is $e^{-1} \approx 0.368$. Thus the maximal throughput of all M subchannels is $M e^{-1}$, which can be regarded as the system capacity upper limit. Obviously, for a stable system, average new packet arrival rate λ should satisfy $\lambda \leq M e^{-1}$, which is also consistent with the conclusion of the previous section.

Thus, for each layer, if the overall MTCD transmission rate is M, the maximal throughput is achieved theoretically. Therefore, if the number of *activers* in the layer is knowed as N, with new packet arrival rate $\lambda \leq Me^{-1}$, the maximum throughput of the system can be achieved stably, if the transmission probability for each *activer* in the layer is controlled adaptively as $p_t = \min\{1, \frac{M}{N}\}$. Thus, the controlled transmission probability for the MTCDs in the layer at time slot k is

$$p_t(k) = \min\{1, \frac{M}{N_k}\}. \tag{11}$$

4.2   State Transition Probability

As indicated in the previous section, an *inactiver* having a new packet becomes an *activer* at the start of the next time slot. N_k denotes the number of *activer* at the start of time slot k. Assuming that there are Q MTCDs, and an *inactiver* has a probability of p_g to generate a new packet, which meets the stability requirement, that is, $p_g \leq \frac{M}{(Q-N_k)e} \leq \frac{M}{Qe}$. In the following, for simplity purpose, we assume that $p_g = \frac{M}{Qe}$, which denotes a maximal new packet generating probability. With a given controlled (re)transmission probability $p_t(k)$, N_{k+1} is only depends on N_k. Thus, the number of *activers* in each time slot k, denotes as N_k $(k = 1, 2, ...)$, is a Markov chain with $p_t(k)$, which is given according to Eq. (11), with state space $\{0, 1, 2, ..., Q\}$.

Let $p_{i,j}$ denote the state transition probability of N_k from state i to state j, i.e., $p_{i,j} = \Pr(N_{k+1} = j | N_k = i), 0 \leq i, j \leq Q$. Let D_k denote the number of successful departure packets (or the number of *activers* transmitted and decoded successfully) at time slot k, and satisfy $D_k \in [0, \min\{M, N_k\}]$. Let $A_{k+1}, 0 \leq A_{k+1} \leq Q - N_k$ denote the number of *inactivers* having new packet arrivals in time slot k, which means these A_{k+1} MTCDs will become *activers* at time slot $k+1$. The state transition of N_k $(k = 1, 2, ...)$ satisfies

$$N_{k+1} = N_k - D_k + A_{k+1}. \tag{12}$$

For simplicity, with $D_k = d$, $N_k = i$, $N_{k+1} = j$, A_{k+1} can be written as

$$A_{k+1} = j - i + d. \tag{13}$$

Next we will discuss how to obtain the state transition matrix P, which consists of state transition probability $p_{i,j}$. For $i = 0$, that is when there is no *activers* at this time slot, there is no departure packet, that is, $d = 0$. So according to Eq. (13), $N_{k+1} = A_{k+1}$. For $i = 1$, that is, when there is only one *activer* at this time slot, we have $p_t = 1$ according to our controlled (re)transmission probability scheme denoted as Eq. (11). Thus in this situation, the only *activer* will always transmit and be successful, that is, $d = 1$. So we have $N_{k+1} = A_{k+1}$. Therefore, it is easy to get the transition probabilities when $i = 0$ and $i = 1$ as

$$p_{0,j} = p_{1,j} = \binom{Q}{j} p_g{}^j (1 - p_g)^{Q-j} = bin(j, Q, p_g). \tag{14}$$

With new arrival rate control scheme, $p_g = \frac{M}{Qe}$.

Apart from $i = 0$ and $i = 1$, when $i \geq 2$, the state transition probability $p_{i,j}$ can be expressed as

$$p_{i,j} = \sum_{d=0}^{\min\{M,i\}} \Pr\{N_{k+1} = j | N_k = i, D_k = d\} \cdot \Pr\{D_k = d | N_k = i\}, \quad (15)$$

which is the conditional probability of the number of successfully transmitted *activers* at time slot k, D_k.

For the first multiplier factor of Eq. (15), $\Pr\{N_{k+1} = j | N_k = i, D_k = d\} = \Pr\{A_{k+1} = j - i + d | N_k = i, D_k = d\}$ when $i - d \leq j \leq Q - d$, otherwise $\Pr\{N_{k+1} = j | N_k = i, D_k = d\} = 0$. Since *inactivers* have the new packet arrival probability as p_g, then A_{k+1} is binomial as

$$\Pr\{A_{k+1} = a | N_k = i, D_k = d\} = \Pr\{A_{k+1} = a | N_k = i\} = bin\,(a, Q - i, p_g) \tag{16}$$

For the second multiplier factor of Eq. (15), we assume that there are T_k transmitted MTCDs when there are N_k *activers* at time slot k, so $\Pr\{D_k = d | N_k = i\}$ can be expressed as the conditional probability of T_k, that is,

$$\begin{aligned}
&\Pr\{D_k = d | N_k = i\} \\
&= \sum_{t=d}^{i} \Pr\{D_k = d | N_k = i, T_k = t\} \cdot \Pr\{T_k = t | N_k = i\}. \\
&= \sum_{t=d}^{i} \Pr\{D_k = d | T_k = t\} \cdot \Pr\{T_k = t | N_k = i\}
\end{aligned} \tag{17}$$

For the first multiplier factor of Eq. (17), $\Pr\{D_k = d | T_k = t\}$ can be derived from combinatorial problem where T_k balls is randomly distributed to M boxes, resulting in exactly D_k boxes with perfect one ball, the probability can be easily obtained as [16]

$$\Pr\{D_k = d | T_k = t\} = \frac{(-1)^d M! t!}{M^t d!} \cdot \sum_{x=d}^{\min\{M,t\}} \frac{(-1)^x (M-t)^{t-x}}{(x-d)!(M-x)!(t-x)!}. \tag{18}$$

For the second multiplier factor of Eq. (17), since with given (re)transmission probability $p_t(k)$, the number of transmitted MTCDs T_k is binomial with given number of *activers* N_k, that is,

$$\Pr\{T_k = t | N_k = i\} = bin\,(t, i, p_t(k)), \tag{19}$$

where $p_t(k) = \min\{1, \frac{M}{i}\}$ according to (re)transmission probability control scheme denoted by Eq. (11).

With Eqs. (18) and (19), $\Pr\{D_k = d | N_k = i\}$ can be calculated according to (17). So the whole transition probability can finally be obtained from (14) and (15), which can be obtained from Eqs. (16), (17), (18) and (19).

4.3   Performance Evaluation Criteria

After the derivation and analysis above, with known state transition matrix P, for a Markov chain, the steady state probability $\pi = [\pi_0, \pi_1, \pi_2, ..., \pi_Q]$ can be obtained through solving the following equations

$$\begin{cases} \pi = \pi \cdot P \\ \sum_{q=0}^{Q} \pi_q = 1 \end{cases}. \tag{20}$$

The elements of P are calculated from Eqs. (14) and (15).

Since the average number of *activers* can be given as $\bar{N} = \sum_{q=0}^{Q} q \cdot \pi_q$, we can obtain the first system performance evaluation criteria, the average throughput, which can be calculated as

$$\bar{D} = \sum_{d=0}^{M} d \cdot \Pr(D_k = d), \tag{21}$$

where $\Pr(D_k = d)$ can be easily calculated by the formula of full probability, which is

$$\begin{aligned}
&\Pr(D_k = d) \\
&= \sum_{i=d}^{Q} \Pr(D_k = d | N_k = i) \cdot \Pr(N_k = i) \\
&= \sum_{i=d}^{Q} \Pr(D_k = d | N_k = i) \cdot \pi_i.
\end{aligned} \tag{22}$$

For the second system performance evaluation criteria, the average number of backlogged packets, since at the end of a time slot, there are D_k MTCDs who successfully transmitted their packets, which can be obtained with Eq. (22), we have $B_k = N_k - D_k$ MTCDs become backlogged in the time slot k, who will continue to be a part of *activers* in the next slot. So the average number of backlogged MTCDs can be calculated as

$$\bar{B} = \bar{N} - \bar{D}. \tag{23}$$

For a time-tolerate mMTC system, the average number of backlogged MTCDs is not at the highest priority. But the system should maintain stability, which requires that the number of backlogged packets plus new-generated ones not to be out-of-control, that is, the sum should be less than the system capacity upper limit.

5   Simulation Results

In this section, we present simulation results to evaluate the performance of proposed stabilized scheme. The list of key mathematical symbols used in this paper are summarized in Table 1. The new packet arrival rate λ is normalized by Me^{-1}, which is the system capacity limit of one layer. The performance of the algorithm is characterized by the normallized average throughput and average backlog. The normallized average throughput of one layer is the average number of successful packets in a time slot normalized by M.

From Fig. 2, we notice the normalized system throughput is in proportion to the normalized new arrival rate $\bar{\lambda}$ when $0 \le \bar{\lambda} \le 1$, which is expected. The system throughput can be viewed as the packet departure rate. It is easy to know, for a stabilized system, the departure rate almost equals to the arrival rate. Thus Fig. 2 shows the system is stable for all new packet arrival rates of $0 \le \bar{\lambda} \le 1$, which has achieved the equal effect with new packet arrival rate control scheme. However, in Fig. 2, we also plot the normal average throughput for $\bar{\lambda} > 1$, which is the advancing effect of the proposed stabilized scheme. It is obvious when $\bar{\lambda} > 1$, the system exists increasing backlogs phenomenon,

Table 1. Notation summary

Notation	Description	Value
B_T	Total bandwidth of the system in Hz	180
B	The bandwidth of subcarrier in Hz	3.75
M	The number of subchannels	48
D	The radius of the cell in km	10
L	The number of NOMA power level	5
λ	The mean new packets arrival rate	5
Q_T	The number of MTCDs in one slot	500
Γ	Target SINR in dB	6

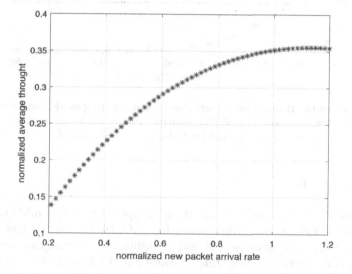

Fig. 2. Normalized average throughput of the stablized distributed grant-free NOMA scheme.

according to the theoretical analysis above. However, the normallized average throughput can be guaranteed by our stabilized algorithm. And we can notice the throughput is stabilized around the maximal possible rate of e^{-1}.

In Fig. 3, when the system is overloaded, our proposed transmission scheme with transmission probability control outperforms pure grant-free NOMA without transmission probability control, non-NOMA schemes with/without transmission probability control, high-complexity&overhead coordinated OMA scheme, for different values of Q in terms of conditional throughput. For example, when the system is overloaded, uplink throughput is expected to increase by 45.2% and 87.5%, respectively, compared with the distributed NOMA scheme without transmission control and the coordinate OMA scheme considering transmission control. System throughput can be guaranteed by our stabilized algorithm.

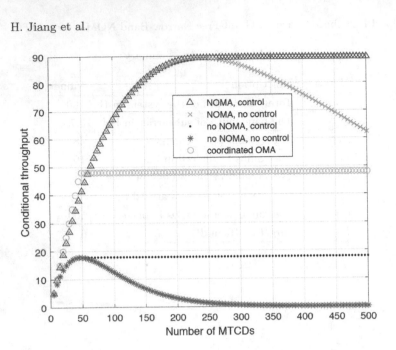

Fig. 3. Conditional throughput comparison between proposed grant-free NOMA with/without transmission probability control, non-NOMA schemes with/without transmission probability control, coordinated OMA scheme, for different values of Q.

6   Conclusion

In this work, to support more connectivity in uplink grantfree mMTC, we proposed a novel distributed layered grant-free NOMA framework based on distributed NOMA. The proposed hybrid transmission scheme can significantly reduce signaling overhead comparing to coordinated schemes. We prove that the scheme with fast-retrial and without transmission probability control is unstable. For the stability analysis, the Foster Lyapunov criterion is considered. For the proposed grant-free NOMA system, a stabilization algorithm was proposed. We give a theoretical analysis on the stablized algorithm performance. The simulation results show that the performance of the stabilized algorithm is much better than the non-stabilized algorithm. With the stabilized algorithm, the system is always stable when the new packet arrival rate is less than system capacity. Even when the arrival rate is higher than capacity, system throughput can still be guaranteed.

Acknowledgment. The work was supported in part by National Nature Science Foundation of China Project under Grant 61631005, Beijing Natural Science Foundation (No. L182038) and National Youth Top-notch Talent Support Program and the 111 Project of China (B16006).

References

1. 3GPP RP 151621: NarrowBand IOT (NB-IOT), September 2015
2. 3GPP TS 36.213 V14.0.0: Evolved universal terrestrial radio access (E-UTRA): Physical layer procedures (Release 13), September 2016
3. Ratasuk, R., Mangalvedhe, N., Ghosh, A., Vejlgaard, B.: Narrowband LTE-M system for M2M communication In: Proceedings of IEEE Vehicular Technology Conference (VTC Fall), Vancouver, Canada, September 2014
4. Cui, Q., et al.: Preserving reliability of heterogeneous ultradense distributed networks in unlicensed spectrum. IEEE Commun. Mag. **56**(6), 72–78 (2018)
5. Gu, Y., Cui, Q., Ye, Q., Zhuang, W.: Game-theoretic optimization for machine-type communications under QoS guarantee. IEEE Internet of Things J. **6**(1), 790–800 (2019)
6. Zhang, J., et al.: PoC of SCMA-based uplink grant-free transmission in UCNC for 5G. IEEE J. Sel. Areas Commun. **35**(6), 1353–1362 (2017)
7. Dhillon, H.S., Huang, H., Viswanathan, H., Valenzucla, R.A.: Fundamentals of throughput maximization with random arrivals for M2M communications. IEEE Trans. Commun. **62**(11), 4094–4109 (2014)
8. Roberts, L.G.: ALOHA packet system with and without slots and capture. ACM SIGCOMM Comput. Commun. Rev. **5**(2), 28–42 (1975)
9. Dai, L.L., Wang, B.C., Yuan, Y.F., Han, S.F., Chih-Lin, I., Wang, Z.C.: Non-orthogonal multiple access for 5G: solutions, challenges, opportunities, and future research trends. IEEE Commun. Mag. **53**(9), 74–81 (2015)
10. Yuan, Y., et al.: Non-orthogonal transmission technology in LTE evolution. IEEE Commun. Mag. **54**(7), 68–74 (2016)
11. Zhu, J., Wang, J., Huang, Y., He, S., You, X., Yang, L.: On optimal power allocation for downlink non-orthogonal multiple access systems. IEEE J. Sel. Areas Commun. **35**(12), 2744–2757 (2017)
12. Lei, L., Yuan, D., Ho, C.K., Sun, S.: Power and channel allocation for non-orthogonal multiple access in 5G systems: tractability and computation. IEEE Trans. Wirel. Commun. **15**(12), 8580–8594 (2016)
13. Choi, J.: NOMA-based random access with multichannel ALOHA. IEEE J. Sel. Areas Commun. **35**(12), 2736–2743 (2017)
14. Jiang, H., Cui, Q., Gu, Y., Qin, X., Zhang, X., Tao, X.: Distributed layered grant-free non-orthogonal multiple access for massive MTC. In: 2018 IEEE 29th Annual International Symposium on Personal, Indoor and Mobile Radio Communications (PIMRC), Bologna, pp. 1–7 (2018)
15. Choi, Y.J., Park, S., Bahk, S.: Multichannel random access in OFDMA wireless networks. IEEE J. Sel. Areas Commun. **24**(3), 603–613 (2006)
16. Szpankowski, W.: Packet switching in multiple radio channels: analysis and stability of a random access system. Comput. Netw. **7**(1), 17–26 (1983)
17. Hajek, B.: Random Processes for Engineers. Cambridge University Press, Cambridge (2015)
18. Kelly, F., Yudovina, E.: Stochastic Networks. Cambridge University Press, Cambridge (2014)
19. Cui, Q., et al.: Stochastic online learning for mobile edge computing: learning from changes. IEEE Commun. Mag. **57**(3), 63–69 (2019)

5G CrowdCell with mm-Wave SDR Based Backhaul

Milan Savić⭕, Miloš Božić⭕, Branko Bukvić⭕,
Dušan N. Grujić$^{(\boxtimes)}$⭕, Zydrunas Tamosevicius, Karolis Kiela,
and Andrew Back⭕

Lime Microsystems, Surrey Tech Centre, Occam Road,
The Surrey Research Park, Guildford, Surrey GU2 7YG, UK
{m.savic,d.grujic}@limemicro.com

Abstract. This paper presents the 5G CrowdCell platform developed by Lime Microsystems, as well as future mm-Wave SDR platforms which are under development, and which, among other applications, will be used as the CrowdCell backhaul.

Keywords: 5G · CrowdCell · mm-Wave · SDR · Open source

1 CrowdCell Platform

1.1 CrowdCell

CrowdCell concept was introduced by Vodafone [1]. In short, this concept presents the open source cellular relay platform using general purpose processors (GPP) and software-defined radio (SDR) technologies.

In more details, CrowdCell presents a relay concept, whereby an intermediate "Crowd" enabled device relays traffic between a customer UE and the macro network. Its main benefit is to be a rapid and low cost small cell solution thanks to its Plug-and-Play (P&P) concept by means of using the available 4G coverage.

The concept is also promoted through the CrowdCell Project Group within the Telecom Infra Project (TIP) [2]. The CrowdCell Project Group focus is on creating a CrowdCell by leveraging General Purpose Processing (GPP) platforms, Software-Defined Radios (SDR) and Open Source designs for both hardware and software to minimize costs with a "one design" flexible platform.

Even if CrowdCell delivers fraction of capacity compared to standard small cell, this concept is expected to be widely accepted since CrowdCell cost of ownership is dramatically reduced, because it is detached from physical location. Additionally, as a result of openness of the concept and utilization of GPP, the CrowdCell software is expected to be able to run on any platform (such as Raspberry Pi, standard PC, etc.).

1.2 Lime Microsystems CrowdCell Implementation

Lime Microsystems is at the forefront of CrowdCell implementation, and was named as a provider of end-to-end (E2E) technology for the CrowdCell at the TIP Summit 2018 [3].

© ICST Institute for Computer Sciences, Social Informatics and Telecommunications Engineering 2019
Published by Springer Nature Switzerland AG 2019. All Rights Reserved
A. Kliks et al. (Eds.): CrownCom 2019, LNICST 291, pp. 386–393, 2019.
https://doi.org/10.1007/978-3-030-25748-4_29

Lime Microsystems' CrowdCell platform was acknowledged as the most compliant E2E platform.

Fig. 1. Vodafone CrowdCell prototype developed by Lime Microsystems

Lime Microsystems' CrowdCell implementation (Fig. 1) is based on a standard PC, and LimeSDR, which is an open source, app-enabled SDR platform [5]. It was developed based on the LimeNET platform [5].

In terms of its narrow definition, a CrowdCell utilizes existing mobile network as a backhaul. However, acknowledging flexibility as one of the utmost priorities, other types of backhaul may be used for a CrowdCell platform in a broader sense. Depending on the availability, those could be, for example, an Ethernet connection, or even, preferably, high-performance flexible dynamic mm-wave backhaul (Fig. 2).

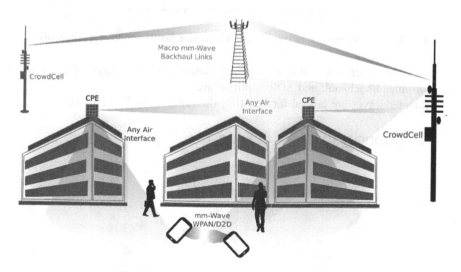

Fig. 2. CrowdCell with mm-Wave Backhaul

Lime Microsystems is developing mm-Wave SDR which will provide the base platform for such backhaul, among other applications. This platform is detailed in the following.

2 mm-Wave SDR Platform

Key enablers of the Lime Microsystem platforms are own integrated transceivers.

LMS7002M transceiver, which is Lime's second-generation field programmable RF (FPRF) transceiver IC, covers all the way from 100 kHz to 3.8 GHz, with 2×2 MIMO and extended functionality [7]. This IC provides the foundations for the LimeSDR platform family.

LMS8001 is a single chip up/down RF frequency shifter with continuous coverage up to 10 GHz, utilizing 4 highly flexible channels [8, 9]. This IC enables the Companion platform [10, 11].

With a goal of increasing continuous frequency coverage, the integrated transceiver roadmap extends to 100 GHz (Fig. 3).

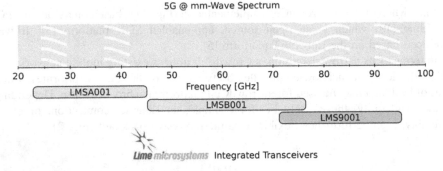

Fig. 3. Lime Microsystems' mm-Wave transceiver family (under development)

Lime Microsystems' mm-Wave integrated transceiver family targets the 5G, multi-gigabit fronthaul/backhaul, and SDR applications.

Millimeter waves are crucial for boosting the capacity in 5G (Fig. 4).

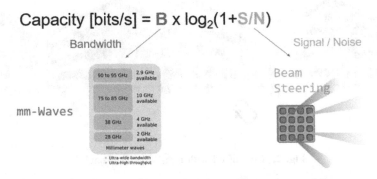

Fig. 4. Millimeter waves enable boosting capacity in 5G

The task of determining optimal architecture, interface, technology, beamforming architecture, is not trivial and is application specific [12].

Zero-IF architecture is chosen, with IQ baseband signals determined as an optimal interface between the mm-wave radio, and readily available baseband processing platforms.

SiGe Bi-CMOS was determined as an optimal technology for mm-wave ICs. Technology break-down of the complete analog-beamformer platform is illustrated in Fig. 5. Millimeter wave radio implemented in Bi-CMOS is connected to the commercially available CMOS baseband. Conceptually, the implementation is illustrated in Fig. 6. This architecture will generally be suitable for UE and CPE applications. In case higher power is required, as with BTS applications, GaN or III-V semiconductor front end can be added.

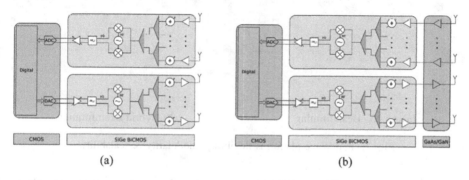

(a)   (b)

Fig. 5. Millimeter wave platforms technology break-down

Separate TX and RX chips are targeted. Primarily, 4 beam-steering channels per chip are targeted. Scaling the number of channels will be straight-forward in future designs, if required.

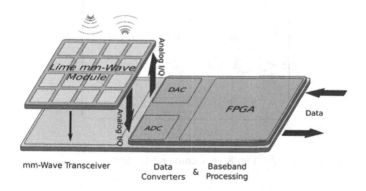

Fig. 6. Lime mm-Wave module & main board concept

The key to flexibility of the millimeter wave systems is modularity. Namely, number of the phased array elements, as well as the beamforming architecture should be flexible in order to meet requirements of various applications and scenarios (Fig. 7).

Each of the Lime Microsystems' mm-Wave chips includes frequency conversion. However, by simple modification of a single metal mask, baseband and mixer can be bypassed. Such a chip could be used to increase the number of channels in purely analog beamforming portions of the systems.

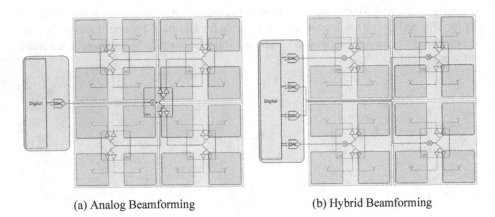

     (a) Analog Beamforming   (b) Hybrid Beamforming

Fig. 7. mm-Wave platforms flexibility, modularity & scalability

Architectures of the RX and TX chips are presented in Figs. 8 and 9, respectively.

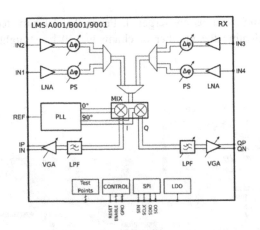

Fig. 8. RX chip architecture

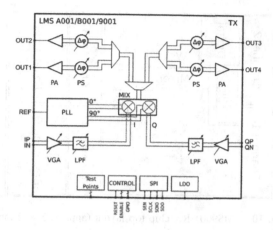

Fig. 9. TX chip architecture

Receiver chain starts with the low noise amplifier (LNA). In order to reduce the number of pins, the LNA input is single-ended, followed by the integrated balun. LNA is implemented as differential, two-stage cascode, providing the trade-off between robustness, bandwidth, gain and noise.

Phase-shifters are implemented as 5-bit vector modulators. Quadrature signal is generated by the Quadrature All-Pass Filter (QAF), and followed by the quadrature Variable Gain Amplifiers (VGA).

Power combiner is implemented as two-stage differential Wilkinson divider.

Signal combined from all channels is fed into a quadrature down-conversion mixer.

Baseband blocks (I and Q) are following the mixer. They consist of 8-steps digitally controllable low-pass filter, with 1 GHz maximum bandwidth, and the variable gain chain. Fast AGC and DC offset cancellation are implemented within the baseband block.

Frequency synthesizer is a fractional-N PLL with multiple fundamental frequency VCO cores. Loop filter is 3rd order with programmable component values integrated on chip. Quadrature is generated by the multistage Poly Phase Filter (PPF).

Power amplifier provides high linearity output signal over the complete band, and provides single-ended output, reducing the pin count.

The TX and RX chips targeting the 71–95 GHz range (LMS9001) are in the top-layout design phase.

In order to avoid grating lobes in a phased array, distance between antenna elements is usually chosen to be equal to the half of the wavelength ($\lambda/2$). To facilitate efficient module design, channel inputs/outputs of the chips are targeted to be equally spaced at the $\lambda/2$ of the central frequency. In addition, chip size must be smaller than module with 2 by 2 integrated antennas, which is generally λ by λ, which imposes harsh area constraints. Hence, careful top level design is necessary.

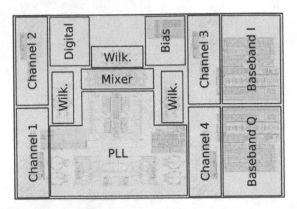

Fig. 10. LMS9001 RX chip top layout (approx. 3 × 2 mm)

Top layout of the LMS9001 RX chip with denoted block is presented in Fig. 10. Top layout of the TX chip is similar.

Depending on the target application, different packaging options are considered (Fig. 11). Module with antennas would enable easy, cheap multi-element modular solutions. BGA package would be appropriate for applications where high-performance antennas and/or high-performance external amplifiers are required. Advanced packaging technologies as eWLB (embedded Wafer Level BGA) will also be considered.

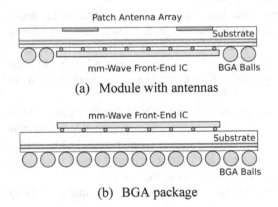

Fig. 11. Lime mm-Wave packaging options

3   MyriadRF

Lime Microsystems is committed to open source and in 2012 founded the MyriadRF open source initiative [13], a multi-stakeholder community and an umbrella for open source hardware (OSHW) and free and open source software (FOSS) wireless, RF and related projects. Over the years MyriadRF has grown to become a vibrant community with hundreds of active members and projects spanning amateur radio, radioastronomy, test & measurement, cellular networks, and satellite communications.

All LimeSDR hardware designs are published via MyriadRF under open source licences, together with associated FPGA gateware and microcontroller firmware, plus supporting host driver software and utilities. This includes hardware designs that range from the low cost SISO LimeSDR Mini USB peripheral [14], to the high performance 4×4 MIMO LimeSDR QPCIe [15]. In addition to which Lime Microsystems have open sourced their adaptive digital pre-distortion implementation for power amplifier linearisation, LimeADPD [16], which is targeted to the LimeSDR QPCIe.

Lime Microsystems believes that the democratisation of innovation is key to the future of wireless communications and in addition to founding MyriadRF, has hosted numerous public workshops, provided free developer hardware, and provided significant support for community events such as Electromagnetic Field (EMF) Camp [17].

References

1. Vodafone: Vodafone taps into mobile network of things to improve coverage. https://www.vodafone.com/content/index/what/technology-blog/crowdcell.html. Accessed 16 Jan 2019
2. Telecom Infra Project (TIP): CrowdCell Project Group. https://telecominfraproject.com/crowd-cell/. Accessed 16 Jan 2019
3. Telecom Infra Project (TIP): Accelerating the Future of Telecom Tech. https://telecominfraproject.com/accelerating-the-future-of-telecom-tech/. Accessed 16 Jan 2019
4. Vodafone: Open source CrowdCell ready for deployment. https://www.vodafone.com/content/index/what/technology-blog/openran-crowdcell-ready-for-deployment.html. Accessed 16 Jan 2019
5. Myriad-RF: LimeSDR. https://myriadrf.org/projects/limesdr/. Accessed 16 Jan 2019
6. Lime Microsystems: LimeNET. https://limemicro.com/initiatives/limenet/. Accessed 16 Jan 2019
7. Lime Microsystems: LMS7002M. https://limemicro.com/technology/lms7002m/. Accessed 16 Jan 2019
8. Lime Microsystems: LMS8001. https://limemicro.com/technology/lms8001/. Accessed 16 Jan 2019
9. Myriad-RF: LMS8001 Documents. https://github.com/myriadrf/LMS8001-docs. Accessed 16 Jan 2019
10. Lime Microsystems: LMS8001 Companion. https://limemicro.com/products/boards/lms8001-companion/. Accessed 16 Jan 2019
11. Myriad-RF: LMS8001-Companion. https://github.com/myriadrf/LMS8001-Companion. Accessed 16 Jan 2019
12. ADI: RF Technology for 5G mmWave Radios. https://www.analog.com/en/education/education-library/webcasts/rf-technology-for-5g-mmwave-radios.html. Accessed 22 Jan 2019
13. MyriadRF initiative main website. https://myriadrf.org/. Accessed 28 Jan 2019
14. LimeSDR Mini. https://wiki.myriadrf.org/LimeSDR-Mini. Accessed 28 Jan 2019
15. LimeSDR QPCIe. https://wiki.myriadrf.org/LimeSDR-QPCIe. Accessed 28 Jan 2019
16. Lime Adaptive Digital Pre-distortion. https://wiki.myriadrf.org/LimeADPD. Accessed 28 Jan 2019
17. Electromagnetic Field 2018 GSM Network. https://limemicro.com/community/electromagnetic-field-gsm-network/. Accessed 28 Jan 2019

Performance Analysis of Full Duplex Wireless Multi-hop Networks

Miguel Sílvio André Francisco$^{(\boxtimes)}$ and José Marcos Câmara Brito

Instituto Nacional de Telecomunicações, Santa Rita do Sapucaí, Brazil
miguelarcanjo03@gmail.com, brito@inatel.br

Abstract. The Fifth Generation (5G) mobile communication standard is expected to come on-line in 2020. Among the performance requirements of 5G, stands out the capacity, that is expected to be 1000-fold of the Fourth Generation (4G). Several technologies have been considered to achieve this goal, among them Full Duplex communications. This paper analyzes the performance of a Full Duplex multi-hop wireless network combined with directional and omnidirectional antennas. Several performance metrics were considered to evaluate the performance of the network: throughput, capacity, block and drop probability. The Markovian model presented here considers buffers in the network nodes and is an extension of a simpler model previously presented in [14], in which no buffer was considered.

Keywords: 5G · Multi-hop network · Full Duplex communications · Performance analysis · Markovian models · Directional antennas · Omni-directional Antennas

1 Introduction

The Fifth Generation (5G) of mobile communication networks is currently under standardization. Several performance requirements are defined for 5G networks. One challenging requirement is the capacity, which is expected to be 1000 times greater than the capacity of the Fourth Generation (4G) networks [1]. Several new technologies have been proposed to 5G networks in order to achieve the performance requirements of that network. Cognitive Radio is an important technology to improve the spectral efficiency and capacity of the network. Another important technology that can be used in conjunction with cognitive radio to increase the capacity of the network is In-Band Full Duplex (IBFD) or simply Full Duplex (FD) communication.

FD communication technology enables a device to transmit and receive simultaneously at the same frequency band; thus, this technology can potentially double the spectral efficiency and, consequently, the network capacity [1]. Some important issues about FD communication are Self-Interference (SI) [2–6], the need of new radios [7,8], the need of new Medium Access Control (MAC) Protocols [8–12] and the transmission modes [13].

© ICST Institute for Computer Sciences, Social Informatics and Telecommunications Engineering 2019
Published by Springer Nature Switzerland AG 2019. All Rights Reserved
A. Kliks et al. (Eds.): CrownCom 2019, LNICST 291, pp. 394–408, 2019.
https://doi.org/10.1007/978-3-030-25748-4_30

The performance analysis of FD wireless networks is important to define the best configuration and transmission mode of the network. In general, the performance analysis of FD networks is based on simulations only [7,8,8–13]. In [14], the authors proposed an analytical Markovian model to analyze the performance of a multi-hop wireless FD network. However, that model considered that the nodes of the network had no buffer. The results presented in [14] showed that the performance of FD networks are very limited if the nodes do not have a buffer. Thus, an analytical model considering buffer in the nodes is important to understand the real benefit of full-duplex networks and the best configuration of these networks.

In this paper, we propose an analytical Markovian model to analyze the performance of a full duplex multi-hop wireless network (and also a half duplex network) combined with directional and omnidirectional antennas considering buffer in the nodes. The size of the buffer is defined by the parameter b. The performance metrics are the same previously discussed in [14]: throughput, capacity, block, and drop probability.

The remainder of this paper is organized as follows: Sect. 2 describes the network scenario considered; Sect. 3 presents the proposed Markovian model; in Sect. 4 we define and compute the performance metrics; Sect. 5 shows the numerical results and, finally, Sect. 6 presents the conclusion and future works.

2   Network Scenario and Assumptions

The network considered in this paper is the same proposed in [8]: a wireless multi-hop network, with data in only one way, composed of 4 nodes (Source, 1, 2, Destination), as illustrated in Fig. 1. Each node can communicate only with its neighbor (for example, node S can communicate only with node 1 and node 1 can communicate only with nodes S and 2).

Following [8], the network nodes can be configured with two parameters: the communication type (HD-Half Duplex or FD-Full Duplex) and the antenna type (DA-Directional Antennas or ODA-Omni-directional Antennas). HD nodes can only transmit or receive at a given time, and FD nodes can transmit and receive simultaneously. When an ODA node transmits, the transmission interferes with the reception of the previous neighbor reception (for example, a transmission of node 2 can interfere with the reception of node 1); DA nodes do not have this problem. Thus, it is possible to have four types of nodes:

* A[Half,Omni]: This type of node is configured with HD communication and one ODA to transmit and receive, just like a conventional node.
* B[Full,Omni]: This type of node is configured with FD communication and two ODAs, one to transmit and one to receive, as proposed in [7].
* D[Full,Direc]: This type of node is configured with FD communication and two DA for Transmission (TX), TX1 to transmit from 0 to π, TX2 from π to 2π and one ODA for Reception (RX) (TX1 and TX2 cannot be used simultaneously), as proposed in [8].

⋆ C[Half,Direc]: This type of node is configured with HD communication and the same antenna configuration used in D[Full,Direc].

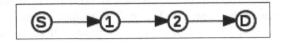

Fig. 1. Network scenario [8].

We suppose that the network is composed only by one type of node (A, B, C or D). Thus, we have four different operation modes, denoted by mode A[Half,Omni], when the nodes are type A; mode B[Full,Omni], when the nodes are type B; mode C[Half,Direc], when the nodes are type C; and mode D[Full,Direc], when the nodes are type D. These modes operate differently, as illustrated in Fig. 2.

For Mode A[Half,Omni], the network needs three steps to complete a transmission: (i) transmission from S to 1; (ii) transmission from 1 to 2; and (iii) transmission from 2 to D. In this mode, nodes cannot transmit and receive at the same time because they are HD and node 2 cannot transmit while node 1 is receiving because they use ODA. So, in this mode, only one node can transmit at a given time, and the packet is transmitted until the destination without stops.

For Mode B[Full,Omni], the network needs two steps to complete a transmission: (i) transmission from S to 1 and from 1 to 2; (ii) transmission from 2 to D. In this mode, nodes are FD, so, in step (i), node 1 can receive and transmit at the same time; however, because the nodes use ODA, node 2 cannot transmit in step (i), because its transmission would interfere with the reception of node 1.

For Mode C[Half,Direc], the network needs two steps to complete a transmission: (i) transmission from S to 1 and from 2 to D; (ii) transmission from 1 to 2. In this mode, DA is used, so, nodes S and 2 can transmit at the same time without interference, but, because all nodes are HD, no node can transmit and receive at the same time.

For Mode D[Full,Direc], the network needs only one step to complete a transmission: (i) transmission from S to 1, 1 to 2 and 2 to D. In this mode, all nodes can transmit simultaneously, because they are FD and use DA.

3   Markovian Model

In this section, we propose a multidimensional Continuous-Time Markovian Chain (CTMCs) to model each operation modes. Transitions in the chain occur due to arrival or transmission of a packet. The arrival processes follow a Poisson distribution with average value λ packets/s, the service time follows an exponential distribution with mean value $1/\mu$ s, resulting in a service rate equal to μ packets/s.

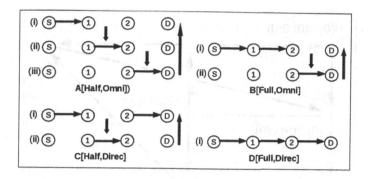

Fig. 2. Transmission process for each operation mode [8].

Each state in the chain is defined by six variables and is represented by dimensions $x = \{i(wi), j(wj), k(wk)\}$, where i represents the existence (1) or not (0) of a transmission from node S to 1; wi represents the number of packets waiting in node S; j represents the existence (1) or not (0) of a transmission from node 1 to 2; wj represents the number of packets waiting in node 1; k represents the existence (1) or not (0) of a transmission from node 2 to D; wk represents the number of packets waiting in node 2.

To simplify the notation the subset $\{hop, node\}$ will be denoted as a server, so $\{i, wi\}$ is defined as server i, $\{j, wj\}$ is defined as server j, and $\{k, wk\}$ is defined as server k. For example, $x = \{0(2), 1(0), 0(0)\}$ represents a state where server i has two packets in the buffer and server j is transmitting. As an example, Fig. 3 illustrates the state transition diagram of mode A[Half,Omni] with buffer size (b) equal to 1. In this mode, the packet is transmitted to the destination without stops; thus, only node S needs a buffer to queue incoming packets. The others transition state diagrams will not be presented due to the complexity, but all the possible state transitions for all modes are explained in Tables 1, 2, 3 and 4, where is also given the set of possible states.

The stationary probabilities, $\pi(x)$ can be calculated from the global balance equations and the normalization equation, which are given by:

$$\pi Q = 0, \sum_{x \in S} \pi(x) = 1. \tag{1}$$

where π is the steady-state probability vector, Q is the transition rate matrix, and S is the set of all possible states. S and Q can be constructed using the transition patterns explained in Tables 1, 2, 3 and 4, where PA means Packet Arrival and TX_a_b means transmission from the server a to b.

When the steady-state probabilities are determined from (1), the performance of the system can be evaluated for different metrics. The derivations of mathematical expressions for these metrics are presented in the following section.

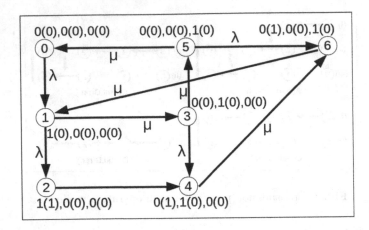

Fig. 3. State diagram of mode A[Half,Omni] with buffer 1.

4   Performance Metrics

In this section the following performance metrics are defined: blocking probability, drop probability, capacity, throughput and the average number of packets in the network.

4.1   Blocking Probability

The blocking probability, P, is the summation of steady-state probabilities for all state where the buffer of the server i is full and, therefore, no packet can enter the network. This parameter is calculated by:

$$P = \sum_{x \in S} \pi(x), \; if \quad wi = b. \tag{2}$$

4.2   Capacity

The capacity C, is the average number of successful transmissions per time unit. This metric is calculated by:

$$C = \sum_{x \in S} \pi(x)\mu, \; if \quad k = 1. \tag{3}$$

4.3   Drop Probability

The drop probability D, is the probability that, once a packet enters the network, it does not complete the transmission with success, meaning it is dropped. This parameter is calculated by:

$$D = 1 - ST. \tag{4}$$

Table 1. Mode A[Half,Omni] $S = \{x|0 \leq i,j,k \leq 1; 0 \leq wi \leq b; wj = 0; wk = 0; i + wi \leq b+1; i+j+k \leq 1\}$

Event	Destination state	Rate	Condition
PA	$i+1$	λ	$i = wi = j = wj = k = wk = 0$
PA	$wi+1$	λ	$i = 1; wi < b; j = wj = k = wk = 0$
PA	$wi+1$	λ	$i = 0; wi < b; j = 1; wj = k = wk = 0$
PA	$wi+1$	λ	$i = 0; wi < b; j = wj = 0; k = 1; wk = 0$
TX_i_j	$i-1, j+1$	μ	$i = 1; wi \leq b; j = wj = k = wk = 0$
TX_j_k	$j-1, k+1$	μ	$i = 0; wi \leq b; j = 1; wj = k = wk = 0$
TX_k_d	$k-1$	μ	$i = wi = j = wj = 0; k = 1; wk = 0$
TX_k_d	$i+1, wi-1, k-1$	μ	$i = 0; wi > 0; j = wj = 0; k = 1; wk = 0$

Table 2. Mode B[Full, Omni] $S = \{x|0 \leq i,j,k \leq 1; 0 \leq wi, wj, wk \leq b; i + wi \leq b+1; j + wj \leq b+1; k + wk \leq b+1; i+j+k \leq 2;\}$

Event	Destination state	Rate	Condition
PA	$i+1$	λ	$i = wi = 0; j \leq 1; wj = k = wk = 0$
PA	$wi+1$	λ	$i = 1; wi < b; j \leq 1; wj \leq b; k = 0; wk \leq b$
PA	$wi+1$	λ	$i = 0; wi < b; j \leq 1; wj \leq b; k = 1; wk \leq b$
TX_i_j	$i-1, j+1$	μ	$i = 1; wi = j = wj = k = wk = 0$
TX_i_j	$wi-1, j+1$	μ	$i - 1; wi > 0; j = wj = k = wk = 0$
TX_i_j	$i-1, wj+1$	μ	$i = 1; wi = 0; j = 1; wj < b; k = wk = 0$
TX_i_j	$wi-1, wj+1$	μ	$i = 1; wi > 0; j = 1; wj < b; k = wk = 0$
TX_i_j	$i-1, j+1, k+1, wk-1$	μ	$i = 1; wi \leq b; j = wj = k = 0; wk > 0$
TX_i_j	$i-1, wj+1, k+1, wk-1$	μ	$i = 1; wi \leq b; j = 1; wj < b; k = 0; wk > 0$
TX_i_j dropped	$i-1$	μ	$i = 1; wi = 0; j = 1; wj = b; k = wk = 0$
TX_i_j dropped	$wi-1$	μ	$i = 1; wi > 0; j = 1; wj = b; k = wk = 0$
TX_i_j dropped	$i-1, k+1, wk-1$	μ	$i = 1; wi > 0; j = 1; wj = b; k = 0; wk > 0$
TX_j_k	$j-1, k+1$	μ	$i = wi = 0; j = 1; wj = k = wk = 0$
TX_j_k	$j-1, wk+1$	μ	$i = 1; wi \leq b; j = 1; wj = k = 0; wk < b$
TX_j_k	$j-1, wk+1$	μ	$i = 0; wi \leq b; j = 1; wj = 0; k = 1; wk < b$
TX_j_k	$wj-1, k+1$	μ	$i = wi = 0; j = 1; wj > 0; k = wk = 0$
TX_j_k	$wj-1, wk+1$	μ	$i = 1; wi \leq b; j = 1; wj > 0; k = 0; wk < b$
TX_j_k	$wj-1, wk+1$	μ	$i = 0; wi \leq b; j = 1; wj > 0; k = 1; wk < b$
TX_j_k dropped	$j-1$	μ	$i = 1; wi \leq b; j = 1; wj = k = 0; wk = b$
TX_j_k dropped	$j-1$	μ	$i = 0; wi \leq b; j = 1; wj = 0; k = 1; wk = b$
TX_j_k dropped	$wj-1$	μ	$i = 1; wi \leq b; j = 1; wj > 0; k = 0; wk = b$
TX_j_k dropped	$wj-1$	μ	$i = 0; wi \leq b; j = 1; wj > 0; k = 1; wk = b$
TX_k_d	$k-1$	μ	$i = wi = 0; j \leq 1; wj \leq b; k = 1; wk = 0$
TX_k_d	$i+1, wi-1, k-1$	μ	$i = 0; wi > 0; j \leq 1; wj \leq b; k = 1; wk \leq b$
TX_k_d	$wk-1$	μ	$i = wi = 0; j \leq 1; wj \leq b; k = 1; wk > 0$

Table 3. Mode C[Half, Direc] $S = \{x|0 \leq i, j, k \leq 1; 0 \leq wi, wj \leq b; wk = 0; i + wi \leq b + 1; j + wj \leq b + 1; i + j + k \leq 2\}$

Event	Destination state	Rate	Condition
PA	$i + 1$	λ	$i = wi = j = wj = k = wk = 0$
PA	$i + 1$	λ	$i = wi = j = 0; wj \leq b; k = 1; wk = 0$
PA	$wi + 1$	λ	$i = 1; wi < b; j = 0; wj \leq b; k \leq 1; wk = 0$
PA	$wi + 1$	λ	$i = 0; wi < b; j = 1; wj \leq b; k = wk = 0$
TX_i_j	$i - 1, j + 1$	μ	$i = 1; wi \leq b; j = 0; wj \leq b; k = wk = 0$
TX_i_j	$i - 1, wj + 1$	μ	$i = 1; wi = j = 0; wj < b; k = 1; wk = 0$
TX_i_j	$wi - 1, wj + 1$	μ	$i = 1; wi > 0; j = 0; wj < b; k = 1; wk = 0$
TX_i_j dropped	$i - 1$	μ	$i = 1; wi = j = 0; wj = b; k = 1; wk = 0$
TX_i_j dropped	$wi - 1$	μ	$i = 1; wi > 0; j = 0; wj = b; k = 1; wk = 0$
TX_j_k	$j - 1, k + 1$	μ	$i = wi = 0; j = 1; wj \leq b; k = wk = 0$
TX_j_k	$i + 1, wi - 1, j - 1, k + 1$	μ	$i = 0; wi > 0; j = 1; wj \leq b; k = wk = 0$
TX_k_d	$k - 1$	μ	$i \leq 1; wi \leq b; j = wj = 0; k = 1; wk = 0$
TX_k_d	$k - 1$	μ	$i = 1; wi \leq b; j = 0; wj > 0; k = 1; wk = 0$
TX_k_d	$j + 1, wj - 1, k - 1$	μ	$i = wi = j = 0; wj > 0; k = 1; wk = 0$

Table 4. Mode D[Full, Direc] $S = \{x|0 \leq i, j, k \leq 1; 0 \leq wi, wj, wk \leq b; i + wi \leq b + 1; j + wj \leq b + 1; k + wk \leq b + 1; i + j + k \leq 3\}$

Event	Destination state	Rate	Condition
PA	$i + 1$	λ	$i = wi = 0; j \leq 1; wj \leq b; k \leq 1; wk \leq b$
PA	$wi + 1$	λ	$i = 1; wi < b; j \leq 1; wj \leq b; k \leq 1; wk \leq b$
TX_i_j	$i - 1, j + 1$	μ	$i = 1; wi = j = wj = 0; k \leq 1; wk \leq b$
TX_i_j	$wi - 1, j + 1$	μ	$i = 1; wi > 0; j = wj = 0; k \leq 1; wk \leq b$
TX_i_j	$i - 1, wj + 1$	μ	$i = 1; wi = 0; j = 1; wj < b; k \leq 1; wk \leq b$
TX_i_j	$wi - 1, wj + 1$	μ	$i = 1; wi > 0; j = 1; wj < b; k \leq 1; wk \leq b$
TX_i_j dropped	$i - 1$	μ	$i = 1; wi = 0; j = 1; wj = b; k \leq 1; wk \leq b$
TX_i_j dropped	$wi - 1$	μ	$i = 1; wi > 0; j = 1; wj = b; k \leq 1; wk \leq b$
TX_j_k	$j - 1, k + 1$	μ	$i \leq 1; wi \leq b; j = 1; wj = k = wk = 0$
TX_j_k	$j - 1, wk + 1$	μ	$i \leq 1; wi \leq b; j = 1; wj = 0; k = 1; wk < b$
TX_j_k	$wj - 1, k + 1$	μ	$i \leq 1; wi \leq b; j = 1; wj > 0; k = wk = 0$
TX_j_k	$wj - 1, wk + 1$	μ	$i \leq 1; wi \leq b; j = 1; wj > 0; k = 1; wk < b$
TX_j_k dropped	$j - 1$	μ	$i \leq 1; wi \leq b; j = 1; wj = 0; k = 1; wk = b$
TX_j_k dropped	$wj - 1$	μ	$i \leq 1; wi \leq b; j = 1; wj > 0; k = 1; wk = b$
TX_k_d	$k - 1$	μ	$i \leq 1; wi \leq b; j \leq 1; wj \leq b; k = 1; wk = 0$
TX_k_d	$wk - 1$	μ	$i \leq 1; wi \leq b; j \leq 1; wj \leq b; k = 1; wk > 0$

Where Successful Transmission (ST) is the probability that, once a packet enters the network, it completes the transmission with success. Calculated by:

$$ST = \frac{C}{\lambda(1-P)}.$$ (5)

Where C is the capacity, and $\lambda(1-P)$ represents the average number of packets that enter the network.

4.4 Throughput

The throughput, denoted by Th, is defined as the ratio between the capacity and the total arrival rate of packets in the network. This metric is computed by:

$$Th = \frac{C}{\lambda}.$$ (6)

With all performance metrics defined, we can now investigate the performance of the network. The results are presented in the next section.

5 Numerical Results

All calculations were done using MatLab, with the following parameters: arrival rate, λ, varying from 1 to 10 packets/s, departure rate, μ, equal to 10 packets/s and two buffer sizes: 5 and 20. The channel is considered error free. The performance metrics depend only on the normalized traffic in the network, or utilization factor, (λ/μ).

Figures 4 and 5 show the blocking probability, it is possible to observe that for modes A[Half,Omni] and B[Full,Omni] the blocking probability remain practically the same for buffer size equal or greater than 5, but for modes C[Half,Direc] and D[Full,Direc] the blocking probability continues to decrease while the buffer size is increased, mode D[Full,Direc] has the lower block probability, followed by mode C[Half,Direc], next is mode B[Full,Omni] and then mode A[Half,Omni]. It is clear that directional antennas have a significant impact on decreasing the blocking probability.

The drop probability is shown in Figs. 6 and 7. It is possible to observe that:

⋆ Mode A[Half,Omni] has no drop because only one packet can be transmitted at a given time in the network.
⋆ Mode B[Full,Omni] has a drop probability about zero even with a buffer size equal to 5.
⋆ Mode D[Full,Omni] get close to zero drop probability only with buffer size equal to 20.

★ For mode C[Half,Direc], the drop probability increase when the buffer size increases; for buffer size equal to 20, the drop probability tends to $1/2$ when the utilization factor tends to 1 ($\lambda = 10$). This occurs because, in this mode, when node 1 finish to send a packet to node 2, it must wait till node 2 send this packet to the destination to be able to send another packet to node 2; meanwhile, node S can transmit to node 1 while node 2 is transmitting, causing, when the utilization factor tends to 1, the node 1 buffer to become full and drop packets. This problem gets worse when the buffer size is increased because node S will always have packets to transmit when node 2 is transmitting, and thus, more packets will be dropped in node 1.

★ It is clear that full duplex communication has a lower drop probability.

Figures 8, 9, 10 and 11 show the System Capacity and Throughput. It is possible to observe that for modes A[half,Omni] and B[Full,Omni] the Throughput and Capacity remain practically the same for buffer size equal or greater than 5, mode C[Half,Direc] have a degradation on the performance while the buffer size is increased due to drop probability explained above, and mode D[Full,Omni] get to a Throughput greater than 0.9 with buffer size equal to 20 when the utilization factor tends to 1 ($\lambda = 10$).

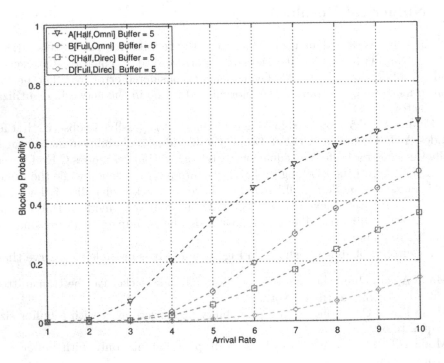

Fig. 4. Blocking probability with buffer 5.

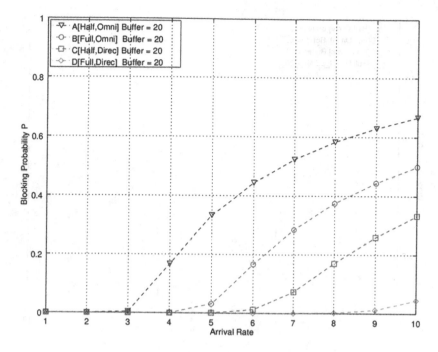

Fig. 5. Blocking probability with buffer 20.

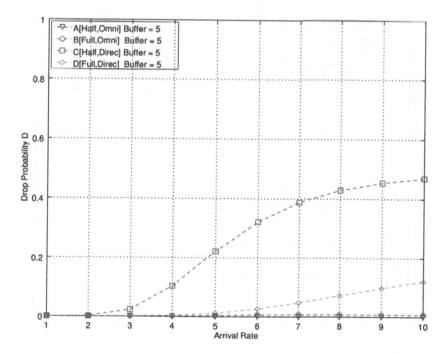

Fig. 6. Drop probability with buffer 5.

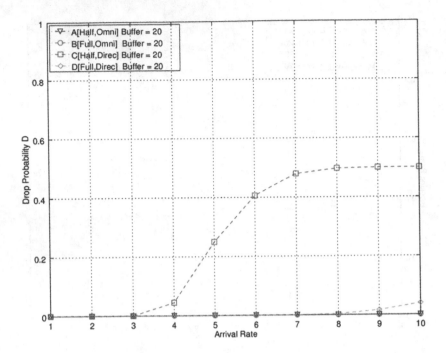

Fig. 7. Drop probability with buffer 20.

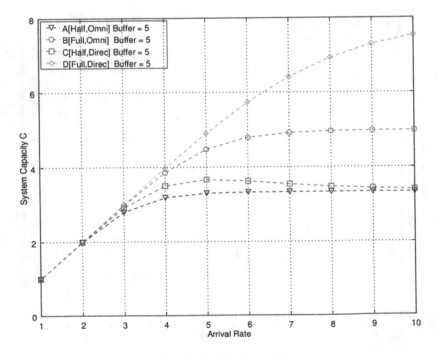

Fig. 8. Capacity with buffer 5.

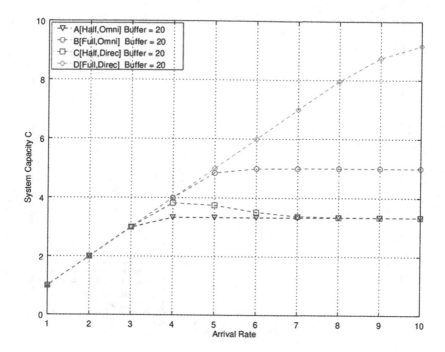

Fig. 9. Capacity with buffer 20.

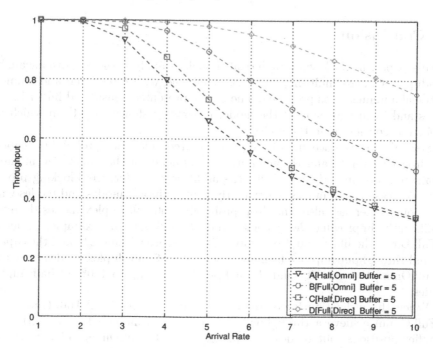

Fig. 10. Throughput with buffer 5.

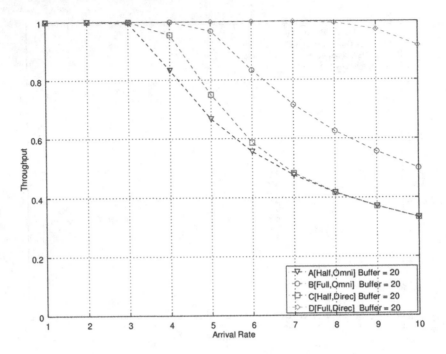

Fig. 11. Throughput with buffer 20.

6   Conclusion

In this paper, a Markovian analytical model to analyze the performance of a full-duplex wireless multi-hop network, combined with directional and omnidirectional antennas, was proposed. The model presented considered buffer in the nodes and is an extension of the model previously shown in [14], in which no buffer was considered in the nodes.

Several performance metrics were computed: blocking probability, capacity, drop probability and throughput. The influence of the buffer size in these parameters was investigated. With the addition of buffer, the blocking probability decrease for all modes, particularly for full duplex modes and modes with directional antennas; also, the drop probability off full duplex modes decrease significantly improving the capacity and throughput, for example, in mode D[Full,Direc] the block and drop probability decrease to almost 0, and the capacity and throughput improve to more than 300% of half duplex modes. In mode B[Full,Omni] the capacity and throughput improve up to 160% of half duplex modes.

When the utilization factor tends to 1 ($\lambda = 10$) modes A[Half,Omni] and B[Full,Omni] achieve a throughput of 1/3 and 1/2 indicated in [8] with only 5 buffer positions, but modes C[Half,Direc] and D[Full,Direc] did not achieve the throughput indicated in [8]. For mode C[half,Direc] the addition off buffer degrades its performance tending to a throughput of 1/3 when the utilization factor tends to 1 ($\lambda = 10$) and mode D[Full,Direc] achieve a throughput higher than 0.9 with 20 buffer positions when the utilization factor tends to 1 ($\lambda = 10$).

Based on the presented analysis it is possible to say that the performance of full duplex communication on the presented network, is at least 166% greater than half duplex communication and can get to more than 300% if used with directional antennas.

Acknowledgment. This work was partially supported by Finep, with resources from Funttel, Grant No. 01.14.0231.00, under the Radiocommunication Reference Center (Centro de Referência em Radiocomunicações - CRR) project of the National Institute of Telecommunications (Instituto Nacional de Telecomunicações - Inatel), Brazil.

References

1. Heino, M., et al.: Recent advances in antenna design and interference cancellation algorithms for in-band full duplex relays. IEEE Commun. Mag. **53**(5), 91–101 (2015)
2. Amjad, M.S., Gurbuz, O.: Linear digital cancellation with reduced computational complexity for full-duplex radios. In: Wireless Communications and Networking Conference (WCNC), pp. 1–6. IEEE (2017)
3. Ahmed, E., Eltawil, A.M.: All-digital self-interference cancellation technique for full-duplex systems. IEEE Trans. Wirel. Commun. **14**(7), 3519–3532 (2015)
4. Huang, X., Guo, Y.J.: Radio frequency self-interference cancellation with analog least mean-square loop. IEEE Trans. Microw. Theory Tech. **65**(9), 3336–3350 (2017)
5. Sim, M.S., et al.: Nonlinear self-interference cancellation for full-duplex radios: from link-level and system-level performance perspectives. IEEE Commun. Mag. **55**(9), 158–167 (2017)
6. Liu, Y., et al.: A full-duplex transceiver with two-stage analog cancellations for multipath self-interference. IEEE Trans. Microw. Theory Tech. **65**(12), 5263–5273 (2017)
7. Jain, M., et al.: Practical, real-time, full duplex wireless. In: Proceedings of the 17th Annual International Conference on Mobile Computing and Networking, pp. 301–312. ACM (2011)
8. Miura, K., Bandai, M.: Node architecture and MAC protocol for full duplex wireless and directional antennas. In: 23rd International Symposium on Personal Indoor and Mobile Radio Communications (PIMRC), pp. 369–374. IEEE (2012)
9. Goyal, S., Liu, P., Gurbuz, O., Erkip, E., Panwar, S.: A distributed MAC protocol for full duplex radio. In: 2013 Asilomar Conference on Signals, Systems and Computers, pp. 788–792. IEEE (2013)
10. Tamaki, K., Raptino, H.A., Sugiyama, Y., Bandai, M., Saruwatari, S., Watanabe, T., et al.: Full duplex media access control for wireless multi-hop networks. In: VTC Spring, pp. 1–5 (2013)
11. Zhou, W., Srinivasan, K., Sinha, P.: RCTC: rapid concurrent transmission coordination in full DuplexWireless networks. In: 2013 21st IEEE International Conference on Network Protocols (ICNP), pp. 1–10. IEEE (2013)
12. Kim, W.-K., Kim, J.-K., Kim, J.-H.: Centralized MAC protocol for wireless full duplex networks considering D2D communications. In: 2016 International Conference on Information and Communication Technology Convergence (ICTC), pp. 730–732. IEEE (2016)

13. Thilina, K.M., Tabassum, H., Hossain, E., Kim, D.I.: Medium access control design for full duplex wireless systems: challenges and approaches. IEEE Commun. Mag. **53**(5), 112–120 (2015)
14. Brito, M.F.J., Couceiro, B.: Modeling and analysis of 5G full duplex wireless radios. In: ICN International Conference on Networks (2018)

Author Index

Printed in the United States
By Bookmasters

Printed in the United States
By Bookmasters